Group 7.

Leaves fan-shaped, thin, notched; veins parallel, no midvein; seeds plum-like

1

Group 8.

Leaves in opposite pairs (or subopposite or whorled); blade simple or compound; edges lobed, toothed, or smooth

2

3

4

5

Group 9.

Leaves alternate, compound (divided into 3 or more leaflets)

6

Group 10.

Leaves alternate, simple; edges lobed

7

8

9

Group 11.

Leaves alternate, simple; edges toothed

10

11

Group 12.

Leaves alternate, simple; edges smooth; deciduous (or evergreen)

12

TREES
OF THE
NORTHERN
UNITED STATES
AND
CANADA

Team Trees in Canada — Production

Project Manager: JOCELYN TOMLINSON

Scientific Coordinator: J. PETER HALL

Dendrologist: KEN FARR

Managing Editor/Text Editor: CATHERINE CARMODY

French Editor: DENIS ROCHON

Illustrations: STEPHEN AITKEN

Photographs*: ANTHONY SCULLION

* Except for the following: bitter cherry flowers (p. 379) by **T.C. Brayshaw**; blue-berry elder fruit (p. 175) by **A.R. Buckley**; water birch fruit (p. 291) and narrowleaf cottonwood tree base (p. 343) by **Bruce Dancik**; red spruce young bark and base (p. 105), maple foliage (p.133), and poison-sumac flowers (p. 229) by **Ken Farr**; whitebark pine cones (p. 49) by **Leo Unger**.

Director of Photography: KEN FARR

Design and Electronic Prepress: STEVEN BLAKENEY

Graphic Assistance: JULIE CHAMPOUX, FRANCINE LANGEVIN

Maps: STEVEN BLAKENEY, DIANE SAMULSKI

Print Production: STEVEN BLAKENEY, MARK KENNEDY, ARNOLD DIENER

TREES

OF THE
NORTHERN
UNITED STATES
AND
CANADA

JOHN LAIRD FARRAR

IOWA STATE UNIVERSITY PRESS / AMES

Published in Canada as *Trees in Canada* by Fitzhenry & Whiteside Limited, 195 Allstate Parkway, Markham, Ontario L3R 4T8, and the Canadian Forest Service, Natural Resources Canada, Ottawa, in cooperation with the Canada Communication Group—Publishing, Supply and Services Canada.

Written, edited, designed and electronically assembled in Canada
Color separations and film production by HBTechnoLith, Ottawa, Canada
Printed and bound in the United States of America

∞ Printed on acid-free paper

Published in the United States by Iowa State University Press, 2121 South State Avenue, Ames, Iowa 50014

Library of Congress Cataloging-in-Publication Data

Farrar, John Laird
    Trees of the northern United States and Canada / John Laird Farrar.
       p.   cm.
    Includes bibliographical references (p.   ) and indexes.
    ISBN 0-8138-2740-X
    1. Trees—United States—Identification. 2. Trees—Canada—Identification. 3. Trees—United States—Pictorial works. 4. Trees—Canada—Pictorial works. I. Title.
QK110.F3   1995
582.160973—dc20   95-22678

Photographs by Anthony Scullion except for the following: bitter cherry flowers (p. 379) by T. C. Brayshaw; blue-berry elder fruit (p. 175) by A. R. Buckley; water birch fruit (p. 291) and narrowleaf cottonwood tree base (p. 343) by Bruce Danelk; red spruce young bark and base (p. 105), maple foliage (p. 133), and poison-sumac flowers and poison-ivy fruit and leaves (p. 229) by Ken Farr; whitebark pine cones (p. 49) by Leo Unger.

Cover: *Front, left,* western hemlock trunk; *right, top to bottom,* sassafras fruit, grand fir cone and axis, striped maple flowers. *Back,* larch branch. Designed by Steven Blakeney; photos by Anthony Scullion; drawing by Stephen Aitken.

Last digit is the print number: 9 8 7 6 5 4 3

Contents

Preface

Trees of the Northern United States and Canada was written to familiarize naturalists, forestry professionals, students, and other tree lovers with the trees of the northern United States—from Virginia to northern California—and Canada. This book is the latest in a natural progression that began with *Native Trees of Canada,* first published in 1917. It was produced to meet a growing demand for information on the identification and characteristics of Canada's trees. Subsequent editions, all produced by the Forestry Branch or its successors in the federal government, included the same type of material, differing mainly in presentation and in information updated in the light of new knowledge.

Changes in the format over the years included the use of photographs to replace drawings of tree parts and the addition of tree silhouettes and large-scale range maps. A valuable companion booklet, *Key to the Native Trees of Canada,* was produced in 1963; it contained dichotomous keys, a glossary of botanical terms, comparative drawings of leaves and needles, fruits, and twigs of the species and genera and silhouettes of crown forms.

In the 1917 edition, about 100 trees and large shrubs were described. During the past 75 years, more species were added; this present publication contains over 300 species, both native and introduced. The book now includes many tree species from outside the region that are widely planted in urban areas and commercial forests.

In this book, a tree is defined as any woody perennial plant that reaches a height of 4.5 m. Color photographs and numerous drawings are provided, and features useful for identification of the tree species are emphasized in the text. Prominent characteristics of the species' range, leaves, twigs, cones, flowers, fruits, bark, and wood are described. Other details are included as additional information or because of their general interest.

Trees of the Northern United States and Canada is intended for the professional forester as well as the amateur naturalist. It will be of interest to students, botanists, schoolteachers, and all those who are fascinated by trees and the forests. We have tried to keep the book to a size that can be taken into the field and used on the spot to identify trees.

To aid in identification, the trees in this book have been organized into 12 groups based primarily on leaf shape and arrangement. An identification key for the groups appears on the endleaves inside the front and back covers; it comprises for each group an icon (a stylized drawing), a description of the group's essential features, and a numbered thumb tab. Dichotomous keys for the groups and large genera are also provided, as well as keys for broadleaf trees and deciduous conifers in winter.

Most native trees are described on two facing pages. Those that only occasionally reach tree height or are rarely encountered have shorter descriptions. Many introduced species are given one or two pages of coverage, and a few common shrubs have several paragraphs. Some genera have many similar species and the more commonly occurring species are briefly described, for example, the willows (*Salix*) and the hawthorns (*Crataegus*).

Trees of the Northern United States and Canada continues the tradition of making it easier for users of the book to identify the trees in our forests, landscapes, and cities. It is more visually comprehensive and wider in scope than *Native Trees of Canada,* while maintaining the accuracy of description set by its predecessor. We hope that this book will follow *Native Trees of Canada* as the most widely consulted and cited reference on forests of the northern United States and Canada.

Acknowledgments

Over the years as the *Trees in Canada* project evolved from a revision of the eighth edition of *Native Trees of Canada* into a much more ambitious undertaking, it has required the expertise, cooperation, and good humor of many. Whether you have only responded to a one-time call for help or have been with me throughout the course of the project, you have my heartfelt gratitude. **Robert I. Fitzhenry**, Chairman of Fitzhenry & Whiteside, suggested the new edition and encouraged its development over the years. **Ian Gillen** advised on almost all aspects of the new edition in its earlier stages, including general layout, page layout, writing style, and wording; his efforts and stimulation were vital. **Betty Farrar**, my wife, provided support and encouragement throughout the eight years of the project, and when my health failed, took care of files and other office work.

I am especially grateful to **Graham R. Powell**, Department of Forest Resources, University of New Brunswick, for his patient and thorough review of the entire manuscript through several drafts, providing much advice on terminology and wording and additional scientific information, especially on the conifer section. **John D. Ambrose**, Curator of Botany, Metropolitan Toronto Zoo, and **John Worrall**, Faculty of Forestry, University of British Columbia, reviewed several drafts of species and genera descriptions; Dr. Ambrose was particularly helpful with the Carolinian species and Dr. Worrall with the species of British Columbia. **Ken Farr** prepared the keys, collected specimens for illustrations, directed the photography, and provided dendrological research on the text and range maps.

Specialists from across the country and in the United States contributed their time and expertise to this project; I acknowledge the following: **Paul Aird**, Faculty of Forestry, University of Toronto, for review of the keys, layout, and parts of the manuscript; **George W. Argus,** Canadian Museum of Nature, for his generous help with the willows text and drawings and in finding specimens; **P.W. Ball**, Department of Botany, University of Toronto, for his detailed examination of text and keys on the oaks and hickories, and help in locating specimens; **Bruce Dancik**, Department of Forest Science, University of Alberta, for review of the manuscript and help in locating western species for photography and illustrative purposes; **James Eckenwalder**, Department of Botany, University of Toronto, for recent information on classification and identification of poplars; **John Furlow**, Department of Plant Biology, Ohio State University, for his detailed review of the alders text; **William F. Grant**, Genetics Laboratory, Macdonald College of McGill University, for help with various species; **Bernard S. Jackson**, Memorial University of Newfoundland Herbarium, for information on species in Newfoundland; **Paul Maycock**, Department of Biology, University of Toronto, for help with the oaks; **Sheila McKay-Kuja**, for comments on the serviceberries; **William H. Parker**, School of Forestry, Lakehead University; **Louis Parrot** and **Jean Smith**, retired, Faculté de foresterie et de géomatique, Université Laval, for research on the French common names; and **A. Reznicek**, University of Michigan Herbarium, for a broad review of the manuscript.

The Canadian Forest Service (CFS) reviewers, headed by **J. Peter Hall**, Science and Sustainable Development Directorate, Headquarters, were as follows: **Ian Corns**, Northwest Region; **Peter de Groot**, Forest Pest Management Institute; **Slavoj Eis**, retired, Pacific and Yukon Region; **Ole Hendrickson**, Headquarters; **J.C. Lees**, Maritimes Region; **Jean-Louis Lethiecq**, retired, Quebec Region; **A.W. Robertson**, Newfoundland and Labrador Region; **Hugh Schooley**, Petawawa National Forestry Institute; and **Gerrit van Raalte**, Maritimes Region.

Jock A. Carlisle, under contract to the CFS, analyzed the text and design of the 8th edition and provided many worthwhile ideas for this present work. **Edward S. Kondo**, Director General, CFS Forest Pest Management Institute, was responsible for initiation of the project.

Bob Duncan, CFS Pacific and Yukon Region, went beyond the call of duty to find and ship specimens of western species to Ottawa for illustration. **W.M. Stiell**, Petawawa National Forestry Institute, provided a history of the various editions of *Native Trees of Canada*, largely based on his personal knowledge.

The National Herbarium of the Canadian Museum of Nature generously provided specimens for many species. In particular, I wish to thank Collections Manager **Mike Shchepanek** and **Albert Dugal**.

x

Trevor Cole, Curator, Dominion Arboretum (Ottawa), Agriculture and Agri-Food Canada, was most helpful with species identification and the logistics of photographing and collecting at the arboretum. Other locations that permitted the collection of specimens for illustration or photography are gratefully acknowledged: Cypress Provincial Park, BC; Devonian Botanic Garden, Devon, AB; Mount Pleasant Cemetery, Toronto; Rondeau Provincial Park, Kent County, ON; Royal Botanical Gardens, Hamilton, ON; University of British Columbia, Botanical Gardens, Vancouver; VanDusen Botanical Garden, Vancouver; Morgan Arboretum, Macdonald College of McGill University, Ste-Anne-de-Bellevue, PQ.

Unfortunately space does not permit detailed acknowledgment of the work of many others involved in the *Trees in Canada* project over the years. Their contributions were often instrumental in ensuring the creation of this complex and multifaceted work. On behalf of the Canadian Forest Service, I thank the following:

M.L. (Moe) Anderson, retired, CFS Petawawa National Forestry Institute; **T.C. Brayshaw**, Curator Emeritus, Royal British Columbia Museum; **Alain Brunet**, St-Jovite, Quebec; **Hubert Bunce**, Reid, Collins and Associates; **D. Burger**, Forest Ecologist; **Henry M. Cathey**, Director, United States National Arboretum, USDA; **J.H. Cayford**, retired, CFS; **Suzanne Chartrand**, translator, Secretary of State; **W. Cheliak**, Program Director, CFS Forest Pest Management Institute; **Bill Cody**, Former Curator of the Vascular Plant Herbarium, Agriculture and Agri-Food Canada; **Misha Dubbeld**, School of Architecture and Landscape Architecture, University of Toronto; **André Fortin**, Faculté de foresterie et de géomatique, Université Laval; **Harry Foster**, Chief Photographer, Canadian Museum of Civilization; **D. Fowler**, CFS Maritimes Region; **Christian Godbout**, Faculté de foresterie et géomatique, Université Laval; **Lizanne Gosselin**, Hull, Quebec; **Kerry Guglielmin**, Ottawa; **Alton S. Harestad**, Department of Biological Sciences, Simon Fraser University; **Monte Hummel**, World Wildlife Fund Canada; **P.E. Irving**, CFS Headquarters; **R. Jaakson**, Department of Geography, University of Toronto; **Brenda Laishley**, CFS Northwest Region; **Janet Lalonde**, formerly CFS Headquarters; **H. Cedric Larsson**, retired, Ontario Ministry of Natural Resources; **Dave Laverie**, Canadian Museum of Civilization; **Brian MacInnes**, manager, Mount Pleasant Cemetery, Toronto; **John McCarron**, Faculty of Forestry, University of Toronto; **E.J. Mullins**, retired, CFS Headquarters; **Garry Poke,** CFS Pacific and Yukon Region; **Chantale Rodrigues**, Aylmer, Quebec; **Steve Scheers**, Ontario Ministry of Natural Resources; **Guido Smit**, School of Architecture and Landscape Architecture, University of Toronto; **Sandy Smith**, Faculty of Forestry, University of Toronto; **Agnes Spanik**, School of Architecture and Landscape Architecture, University of Toronto; **Gerald B. Straley**, Department of Botany, University of British Columbia; **Eric Thompson**, Manager, Morgan Arboretum, Ste-Anne-de-Bellevue, PQ; **Allan Van Sickle**, CFS Pacific and Yukon Region; **Brenda Vanstone**, Faculty of Forestry, University of Toronto; **Roger Vick**, Devonian Botanic Garden, Devon, AB; **Kristian Vitols**, Faculty of Forestry, University of Toronto; **Harold Walther**, Centre for Land and Biological Resources Research, Agriculture and Agri-Food Canada; **Allen Woodliffe**, Ontario Ministry of Natural Resources; **Ben Wang**, CFS Petawawa National Forestry Institute; **Carolyn N. Wild**, Ottawa.

TREES
OF THE
NORTHERN
UNITED STATES
AND
CANADA

Families

In scientific classification the family is the level above the genus. Within the conifers, the identification groups in this book fit the scientific classification into families quite well; but within the broadleaf trees the groups are not related to families. In each broadleaf group, genera that share a common family have been placed together.

Genera and Species

Within the groups, all species belonging to the same genus are placed together, including those that do not fit the group characteristics. These special cases can be identified by checking the group introductions and group keys.

Many genera have their own descriptions. Genera with few or only one species are described in the first species description.

Pines
Genus *Pinus*
Pinaceae: Pine Family

Les pins

Norway maple*
(introduced)

Scots pine **
(naturalized)

Douglas-fir

Tree Names

On the genus page, the common English name of the genus is given first, followed by the scientific name, the family name in scientific and common forms, and the French name.

On the species page, the common English name appears in large bold letters. It may be followed by other common names in smaller letters. Next comes the full scientific name, including the name of the person(s) who named the tree; occasionally synonyms are included. The family names may be added here if there is no genus page. The French name again appears in bold.

An asterisk (*) beside the common name indicates that a tree is **introduced**, not native to Canada. A double asterisk (**) shows that an introduced tree has become **naturalized**; this means it is successfully reproducing itself in this country.

Scientific names (binomials) consist of two terms, the genus and species names. Some species are described at a more detailed level: subgenus, subspecies, variety, cultivar, or hybrid.

A good example of the conventions for naming trees is Douglas-fir, *Pseudotsuga menziesii* (Mirb.) Franco var. *menziesii*.

Preferred common name: Douglas-fir. The hyphen is added because the tree is not a true fir.

Other common names: Coast Douglas-fir, common Douglas-fir, green Douglas, Douglas. These are regional or local names.

Genus name: *Pseudotsuga*. The name is from the Greek "false" (*pseudos*) and the Japanese "hemlock" (*tsuga*). Botanists placed this tree in the hemlock genus (and others) before recognizing its genus as distinct from any other.

Species name: *menziesii*. The species was named for Archibald Menzies, the naturalist who first described it.

Authors' names: (Mirb.) Franco. Mirbel (abbreviated to Mirb.) named the species, but Franco more recently revised its classification into varieties.

Variety: var. *menziesii*. The Rocky Mountain Douglas-fir, *P. menziesii* var. *glauca* (Biessn.) Franco, was recognized as a variety distinct from the species. The original species name therefore became a variety name as well.

Sometimes a species may be divided into **subspecies** (ssp.) rather than varieties, if the differences are great enough. Varieties are naturally occurring subtypes of a species. **Cultivars** (cv.) are varieties bred or selected for human purposes. **Hybrids** can also arise naturally or artificially by crossing between species, or more rarely between genera. Hybrid species are indicated by a multiplication sign in front of the species part of the name, for instance, *Aesculus ×carnea*.

Synonyms: [syn. *P. taxifolia* (Lamb.) Britt.; *P. douglasii* (Lamb.) Carrière]. This tree has been reclassified and renamed several times; different versions of the scientific name may still be found in other reference books. As scientists gain new understanding of complex species, tree classifications continue to be revised, so some names in this book may change again.

Family name: Pinaceae: Pine Family. The Douglas-fir is part of the extensive family collectively called the pines.

French name: Douglas vert. The French common names used in this book have been determined through consultation with authorities at Laval University in Quebec.

var. *menziesii*

var. *glauca*

Range Maps

Most "native" trees have migrated into Canada in the 10 000 to 20 000 years since the melting of the Pleistocene ice sheets. Range maps in the book show where a species grows naturally; the exact boundary of its range is often uncertain. Within the range, its abundance varies from place to place, and a few individuals may grow in places beyond the boundary depending on local soil and climate conditions. These isolated areas are said to be **disjunct** or **outlier** sites.

Introduced species have no natural ranges in Canada. Instead of a range map, a hardiness rating is given (for example, Zones CA2, NA3). Hardiness zones may also be indicated for native species if they are frequently planted outside their natural range. Hardiness is the ability of a tree to survive climatic stress in a given area. It is rated here according to the northernmost zone where the tree is likely to succeed in cultivation. The CA ratings are based on hardiness zones established for Canada by researchers at Agriculture and Agri-Food Canada. They refer to **The Map of Plant Hardiness Zones in Canada** found on page **483**. The NA rating refers to the hardiness zone map of the United States Department of Agriculture, Miscellaneous Publication No. 1475, which is available from the Government Printing Office, Washington, D.C. 20402-9325.

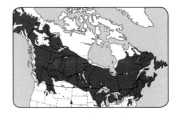

natural range of white birch

Drawings

Genus drawings illustrate features that are common to all, or most, species of the genus, such as the development of seedlings.

Species drawings show small details that are useful for identification. The same features are usually illustrated for all species of a genus, to aid in comparison. Some features are

× 2

lateral bud shown at twice natural size

10 m

icon of Group 6

drawn life size; others are magnified or reduced by amounts that are indicated: for example, × 2 means the drawing is twice natural size, and × ½ means half size. Cross sections of needles and twigs show features that can be seen with the naked eye or with a hand lens.

Silhouettes indicate the height the trees may reach, and the form of the trunk, crown, and branches. The size matches information given in the text, and a representation of a 10-metre-high pole is included beside each silhouette so that the heights of species can be compared.

Icons on the endleaves of this book provide rapid entry into the identification groups. These icons also appear with the botanical keys to the groups. A separate set of icons accompany the winter keys to the genera.

Photographs

The color photographs are an important feature of the book, showing at a glance the structures that can be compared among tree species: cones, flowers, fruit, and young and mature bark. Photos are also used to illustrate common features of a genus.

Descriptions

Each genus description starts with some general information: the size of the genus, the form of the trees, and where they grow in the world. Because species and genus are human concepts, individual plants may not fit conveniently into these categories and the numbers of genera and species are only approximations.

In the introduction to a species, facts about its range are usually noted and other details sometimes given.

Leaves The leaf is the part of the tree that most readers will look at first when identifying a tree. Certain features, however, may require special attention for identification.

Conifer leaves are evergreen on most species. They vary from small scales that completely cover the shoot as in juniper, to pine needles that may be 20 cm long. In spruces, firs, hemlocks, Douglas-fir, larches, and yew, needles are arranged in a spiral along the stem. However, on a horizontal stem the spiral is not obvious because the needle bases are bent and twisted and the needles appear two-ranked, or crowded on the upper side with or without a parting down the middle.

Species with scales may bear short needles on some shoots. Pines have a second type of leaf, a brown scale which usually falls off soon after the buds burst. Deciduous conifers can often be recognized in winter by the leaf scars on their twigs.

For some conifers, needles have been drawn in cross section to show their shape — whether flat, round, four-sided — as well as the midvein and resin ducts when present.

needle leaves

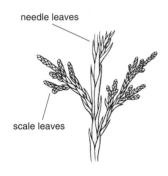

scale leaves

resin duct   midvein

needle cross section

Leaves of most broadleaf species may vary considerably in shape on the same tree. Those preformed in the bud, which grow near the base of the shoot, are usually the most consistent; leaves that form and develop during the summer (neoformed) may vary according to their position in the crown and their place on the shoot.

simple leaf

compound leaf

For a **simple** (undivided) leaf, this book gives the length of the leaf blade only; for a **compound** leaf (one divided into **leaflets**), the measurement combines the lengths of the central stalk and the terminal leaflet. If the central stalk of a compound leaf resembles a stem, leaflets may look like simple leaves. However, there is always a bud in the **axil** of a leaf (the angle between the leaf and the stem), never in the axil of a leaflet.

Many broadleaf species have a pair of **stipules**, ear-shaped structures, attached to the stem beside the base of each leaf. Some stipules are green, leaflike, and persistent (remaining attached to the stem); others are straw-colored and dry, and fall off soon after the leaves appear in spring. On black locust and a few other species, the stipules are persistent **spines**.

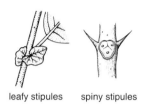

leafy stipules

spiny stipules

The leaf stalk may include leaflike **wings** or warty **glands**. Glands may also occur on the teeth of leaves or stipules.

Buds Buds are a good clue to identification of trees in winter. Buds are here defined as scaly buds, dormant in winter and containing an embryonic shoot for next year's growth; some books refer to any growing tip as a bud.

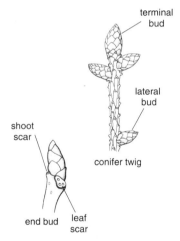

terminal bud

lateral bud

shoot scar

conifer twig

end bud

leaf scar

Many conifers have buds in winter: a **terminal bud** at the end of the twig, and many **lateral buds** along the twig. The outer part of the bud usually consists of **bud scales**, which are modified leaf stalks; they are shed soon after growth resumes in spring. Juniper, thuja (cedar), false cypress and other members of the cypress family do not have leaf buds, although their immature cones are bud-like.

All our broadleaf trees have lateral buds in the leaf axils. In some species, a terminal bud forms at the end of the shoot, halting shoot growth for that year; the terminal bud differs from the lateral buds in size and shape as well as position. In other species, shoot growth stops with abortion of the growing tip, which leaves a lateral bud in place of a terminal bud at the end of the shoot. This **end bud** can often be recognized by the withered shoot beside it, or by a **shoot scar** on the side opposite the leaf scar, where the shoot tip aborted and fell off.

Broadleaf bud scales may be dry like those of conifers and fall off when the bud bursts, or rather fleshy, expanding as the bud bursts. In a few species the outermost "scales" of the bud are really immature leaves, as revealed by the vein pattern, and they develop into green leaves as the bud opens. Such buds are called **naked buds**. Some overwintering flower catkins are also called naked because they are not covered by bud scales.

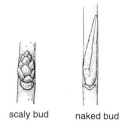

scaly bud

naked bud

Twigs When a new **shoot** completes its growth for the season, it becomes a **twig**. Most conifer twigs remain leafy; twigs of deciduous broadleaf trees lose their leaves but have obvious buds and leaf scars. As soon as growth resumes in spring, the twig becomes the **previous year's twig**, which is sometimes more useful for identification than the new shoot.

twig and buds

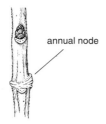

annual node

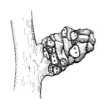

dwarf shoot (ginkgo)

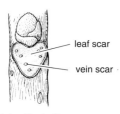

leaf scar

vein scar

lateral bud with
leaf scar

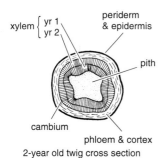

xylem { yr 1
        yr 2

periderm
& epidermis

pith

cambium

phloem & cortex

2-year old twig cross section

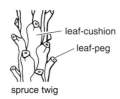

leaf-cushion

leaf-peg

spruce twig

pollen cones
(pine)

In conifers except the cypresses and in all broadleaf trees, the flush of new growth in spring has been **preformed** in the bud; that is, new leaves and shoots start developing inside the buds before they open. As well, in many species, the tips of some vigorous shoots are **neoformed**, or newly created, during the growing season. In a few conifers, for example most of our pines, the whole new shoot is preformed in the bud. However, in juniper, false cypress, and thuja, all new growth is neoformed at the growing point, which is why they have no buds.

At the point between one year's growth and the next, **annual nodes** are formed. These nodes occur in most trees except the cypresses. They are marked by a ring of bud scale scars or dead bud scales and a change in bark features. Often one or more side branches spring from this point.

In some species, some buds may sprout as **seasonal shoots** during the growing season instead of remaining dormant until the following spring. In this way two or more flushes of growth may occur during one season. The resulting **seasonal nodes** can be distinguished from annual nodes by the presence of leaves on the stem below the seasonal nodes.

Long, or extension, shoots are the normal form of growth. **Dwarf shoots** are special small branches that appear in some species. They bear most of the leaves and flowers, but they remain short and slender and rarely produce branches themselves.

When a leaf falls, a **leaf scar** marks this place on the twig. On the leaf scar, **vein scars** mark where veins passed from the leaf stalk into the stem. For winter identification of deciduous trees, the size and shape of the leaf scar and the number and arrangement of its vein scars are especially useful.

Among deciduous conifers, the larches can be recognized in winter by the many small round leaf scars on the long shoots and the cones usually present on some dwarf shoots, as well as by the undivided trunk. The ginkgo can be recognized by its coarse twigs, prominent dwarf shoots, and leaf scars with two vein scars.

Some interior features of twigs may help in tree identification. Four tissues can be seen with the naked eye in cross section: **pith** in the center, then a cylinder of wood or **xylem**, a layer of soft tissue (**phloem** and **cortex**), and a tough surface skin (**periderm** and **epidermis**). The **cambium**, a layer of growing cells between the xylem and the phloem, is too thin to see. Together, the tissues outside the cambium form the **bark** of the twig (the bark of a large stem is different). Drawings of twigs in cross section in the book are included when they help with recognition.

Winter twig drawings show features such as bark ridges, **leaf-pegs**, and **leaf-cushions** (swellings on the stem to which a leaf is attached). **Lenticels** (air pores) appear as wart-like spots on the twig surface.

Cones Conifers bear two kinds of cones, called in this book **pollen cones** and **seed cones**; they are sometimes referred to as male and female cones because of their roles in reproduction. Pollen cones and seed cones may occur on separate trees or the same tree.

Pollen cones are small catkin-like structures. They have a central axis bearing a number of **sporophylls**, which produce pollen to fertilize the seed cones. Pollen cones release their mature pollen grains, then wither away soon afterwards.

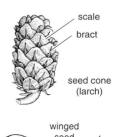
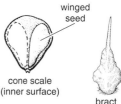

scale

bract

seed cone
(larch)

Seed cones, unlike pollen cones, are a helpful means of identification because they persist on the tree or on the ground around it. Immature seed cones, called **conelets**, are ready for pollination at the time the pollen cones shed their pollen.

Seed cones are composed of **scales** arranged around a central axis; the **ovules**, and the seeds they become when fertilized, are borne on the inner surface of the scales. Just below the scale in many species is a **bract**, which is often obvious in spring but may be covered by the enlarged scales as the cone matures. Seed cones such as those of junipers and false cypresses are not woody but appear globular and fleshy, like berries or capsules.

winged
seed

cone scale
(inner surface)

bract

Other Seed-Bearing Structures The yews and ginkgo are related to the conifers, but they do not have seed cones. Instead, they bear one or two ovules on stalks in the axil of a leaf, and each seed is partly or wholly enclosed in a fleshy **aril**. The pollen, however, is produced in catkin-like cones.

aril
(yew)

Flowers Flowers are the basis of classification for broadleaf trees, as for all other flowering plants. Although they are present only for a short time, identification of a tree may be extremely difficult without them. This book describes flower features that are easily observed, such as size and color, the type of flower cluster, its position on the tree, and the time of flowering.

The typical flower consists of a **calyx** composed of **sepals**, usually green; a **corolla** composed of **petals**, usually white or colored; a number of **stamens** which bear **pollen** for fertilization; and a **pistil** containing one or more **ovules**, which become seeds when fertilized.

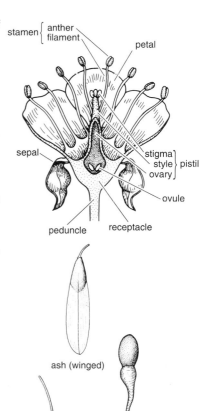

stamen { anther
          filament

petal

sepal

stigma ⎫
style ⎬ pistil
ovary ⎭

ovule

peduncle   receptacle

In some flowers, in this text referred to as **seed flowers** (sometimes known as pistillate or female flowers), only the pistils are functional; in others, the **pollen flowers** (staminate or male flowers), only stamens are functional. More often, both pistil and stamens function in the same flower, and such flowers are said to be **perfect**. Individual trees may bear only perfect flowers, only pollen flowers, or only seed flowers. However, some have two kinds, and some have all three. Seed flowers and pollen flowers may occur on different trees (dioecious), on different parts of the same tree (monoecious), or within one cluster on a tree.

In some species, the flowers are preformed in the bud and appear in spring, often before the leaves. In other species the flowers start to form during the growing season, appearing with or after the leaves. Flower clusters come in a great variety of forms; occasionally the botanical term for the flower cluster is given in parentheses.

ash (winged)

Fruits Fruits are second only to flowers in providing a sure means of identification of broadleaf trees. Mature or immature fruits can be found on or under the tree during much of the growing season, and even into winter. Common terms for fruits are used; the botanical name may be added in parentheses.

sassafras
(berry-like, on a stalk)

red mulberry
(fruit aggregate)

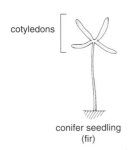

cotyledons

conifer seedling
(fir)

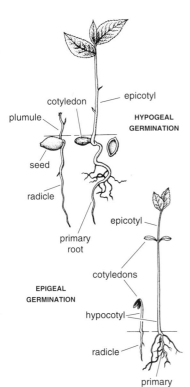

cotyledon   epicotyl

plumule   HYPOGEAL
                           GERMINATION

seed

radicle

                    epicotyl

primary
root

                    cotyledons

EPIGEAL
GERMINATION

          hypocotyl

     radicle

                    primary
                    root

natural layering of
black spruce

The nature of the fruit cluster is determined by the flower cluster. However, a terminal flower cluster may become a lateral fruit cluster if a new shoot develops behind the flower cluster and grows past it. Also, an upright flower cluster may be weighed down by the fruit and become pendulous.

Seeds Seeds are seldom useful in identification, but their germination habits are interesting. Some germinate immediately after being shed. Others need a period of moist cool storage, at 1° to 5°C, before germination. This process, called **stratification**, can be done artificially by filling a pit in the ground (or trays in a cold room or refrigerator) with alternate layers (strata) of peat or other material, and of seeds. Stratification occurs in nature when seeds remain over winter on or in the litter of the forest floor. A few species require light for germination.

Seedlings Conifer seedlings (except those of ginkgo) have 2 to 12 green, needle-shaped **cotyledons** in a whorl around the growing point. Cotyledons are specialized parts of the embryo.
Conifer seeds contain storage tissue that surrounds the embryo. During germination, this albuminous tissue inside the seed coat clings to the cotyledons and feeds the new tree. The cotyledons are raised above the surface by the growing stem (**hypocotyl**), become green, and carry out photosynthesis. This process is known as **epigeal germination**. Eventually the nutritive tissue is used up and the seed coat falls away.

Seedlings of broadleaf trees always have two cotyledons (or **seed leaves**). In some species these are white, fleshy, and full of stored food; they remain in the seed coat, usually underground, as germination proceeds (**hypogeal germination**). In other species, the cotyledons turn green and leaf-like, and carry out photosynthesis for some or all of the growing season (**epigeal germination**, as in the conifers); the seed coat falls away after the cotyledons are raised above the ground.

Vegetative Reproduction Vegetative reproduction is the regeneration of a whole individual plant from part of another. An important result of this process is that each new plant is genetically identical to the parent and is termed a **clone**.

In nature, vegetative reproduction may take place through **sprouting** from stumps or roots, through **rooting** of detached branches, or through **layering**, a process in which roots grow from branches that are buried in organic debris but are still attached to the parent tree.

In horticulture, selected trees and cultivars are usually reproduced by culturing stem and root **cuttings**, and by grafting buds or **scions** (detached stem segments) onto a stock tree called a **rootstock**.

Bark Bark varies as the tree ages, and from tree to tree of the same age, depending on growth rate and growing conditions. The range of this variation can only be suggested by the descriptions and the photographs in this book.

Bark on older stems has two layers: an outer dead layer that is mainly cork, and an inner layer of living tissues in which specialized cells deliver food to all parts of the tree, and where new bark is formed. In some species the freshly exposed part of the outer bark has a distinctive color. Experienced observers can identify many trees by the bark.

Wood Identification of a tree species by its wood is difficult, often requiring a compound microscope. However, in many cases, at least the genus can be determined. Features described in the text refer to the inner **heartwood**; the **sapwood**, the living part of a standing tree trunk, is always lighter in color and not so durable.

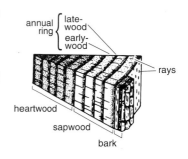

In conifers, resin and **resin ducts** are features of many woods. The relative amount of light **earlywood** and dense, darker **latewood**, together forming the **annual ring**, is characteristic.

For broadleaf trees, the **rays** and **pores** are described. The pores are assemblies of tubular, water-conducting cells. The wood of species with a ring of large pores at the inner part of each annual ring is called **ring-porous**; the wood of those with pores that are more or less equal in size throughout the annual ring are called **diffuse-porous**. Some species have wood with a combination of these features, which is referred to as **semi-diffuse-porous** or **semi-ring-porous**.

Size and Form The size and age a tree species can attain are only approximations. Height varies not just with a tree's age but with the nature of the site and the conditions in the stand of trees. Also, a tree considered to be medium-sized in British Columbia might seem very large in eastern Canada.

The figures given in this book are estimates of the dimensions of most very large trees in an old stand on an average site — they are not the largest on record. The size of the trunk is given as diameter at breast height (**dbh**), 1.3 m (4½ ft) from the ground.

Trees are described in five height classes, and the silhouette drawings have corresponding heights:

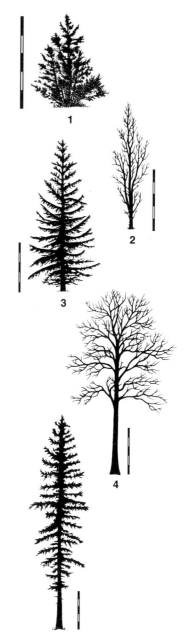

Height class	Tree height m ft		Silhouette height mm
1 Shrubby or very small	up to 10	33	24
2 Small	up to 20	66	32
3 Medium-sized	up to 30	98	40
4 Large	up to 40	131	48
5 Very large	over 40	131	56

Form refers to the shape of the crown, the branches, and the trunk. Although the form of individual trees may vary with age and growing conditions, many species have a characteristic form which can be recognized with practice.

Young rapid-growing trees have cone-shaped crowns with a distinctive leading shoot which outgrows the branches below it. The leading shoot may be upright as in spruce and ash, or oblique and relaxed as in hemlock and elm. As a tree ages, the growth of the leading shoot decreases, both absolutely and in relation to the branches below; the crown becomes rounded or even flat, and eventually there may be no leading shoot.

It is important to note how far the trunk reaches into the crown; in many conifers and a few broadleaf trees, the trunk extends through the crown to the tip of the tree; in others the trunk divides into branches at the base of the crown.

Branches may sweep up or bend down; often they have a double curve, starting at an upward angle, bending down in the middle, then curving up again at the tip. Principal branches grow from the annual nodes. Lesser branches may or may not occur along the internodes (the portion of the stem between nodes), and they are short-lived.

Habitat Habitat describes where the tree is likely to be found, where it will thrive, how it tolerates conditions such as shade, and what other species usually occur with it. The tree may be noted as a **pioneer** species, one that invades newly cleared or burned land. As the habitat is modified over time, **successional** species follow and finally **climax** species form a rather stable and long-lasting association.

As all species grow best in rich, moist, deep, well-drained soils, this feature of habitat is not mentioned in each description.

Notes Notes contain information not mentioned in any of the previous sections, such as a tree's economic and recreational uses, drought and pollution tolerance, susceptibility to insect pests and diseases, and value to wildlife.

Quick Recognition For major species, a few of the most important points for recognition are included in a box so that they stand out clearly. Contrasting features of similar or related species may appear in brackets in this section.

Dichotomous Keys

Keys are a useful tool that can be used to simplify identification of unknown tree species. A set of dichotomous keys follows the species descriptions in this book. Whenever possible, these keys include a combination of leaf, flower, and cone or fruit characteristics to facilitate species identification at any time in the growing season.

In winter, most tree species cannot be assigned to identification groups based on leaf characteristics. To assist winter identification, a separate set of keys, based on twig, bud, and vein scar characteristics, is also included.

How to Identify a Tree

Trees in this book are described in two broad categories, the conifers and the broadleaf trees. Within these two categories, the trees are organized into 12 identification groups, based mainly on the shape and arrangement of their leaves, features that are easily observed.

Identification Based on the Groups

To identify a tree, proceed in three steps:

Step 1. Use the Identification Key to Find the Group.

The identification key inside the front and back covers is made up of icons that illustrate and describe the main features of each group. Work through the 12 groups to find the one to which your tree is most likely to belong; then, check the section of the book where the group is described in detail.

Step 2: Within the Group, Select the Genus.

The simplest way to determine the genus of your tree is to flip through the photos and drawings. If there are several large genera in the group, read the introduction to the group and the genus descriptions. A few species have features that match those of another group but are included with their own genus. These species are noted in the introduction to the group.

Step 3: Within the Genus, Select the Species.

When you have found the genus, look at the photos, drawings, and descriptions of the individual species. The **Quick Recognition** box for major species is also helpful.

Identification Using Botanical Keys

Botanical keys work by the "either/or" principle, where a series of choices is made between two descriptions. These branching pathways are why such keys are called dichotomous. The descriptions go from general to more specific features until the only choice is an individual genus or species.

In this book, numbers are used to lead you through the keys. Each pair of descriptions has the same number. You decide which of the pair most closely resembles your tree, and the one you choose gives you either the name of the tree or a new pair of descriptions to choose from.

Keys to the groups are arranged in numerical order. After each group key, there are keys for each major genus within that group. These keys begin on page 409.

Winter Keys to the Genera (page 451) are for use in identifying leafless trees and broadleaf evergreens in winter. They follow the group keys and are coded with their own set of icons.

Identification in the Field

Experienced observers can sometimes identify a tree from surprising distances, even from a moving car. However, sure identification is only possible at close range. A 10× hand-lens (15× for the oaks) is often useful for examining buds, flowers, and other small parts. A pair of field glasses of the type used by bird watchers is useful in examining leaves, flowers, fruits, or cones that are out of reach on the tree. Also, look on the ground: the leaves, fruits, cones, and other parts found there probably came from the tree you want to identify.

The Conifers

The first part of the book (Groups 1–7) describes the conifers—in botanical terms, trees in the division Pinophyta of the plant kingdom. They are separated from the broadleaf trees (Groups 8–12) on the basis of their reproductive structures. Conifers have no true flowers. Pollen is borne on separate catkin-like structures, sometimes on separate trees. In most genera, the ovules, and the seeds that develop from them, are borne on the scale of a cone (hence, the name conifer from Latin meaning "cone-bearing"); in yew and ginkgo, a single seed is borne on a stalk located in a leaf axil.

The wood (xylem) of conifers differs from that of broadleaf trees in that it is composed mainly of tracheids: long (3–7 mm) narrow dual-purpose cells that provide strength and permit water movement from roots to leaves.

In this book, 19 conifer genera are described. Except for Group 2, all genera within a group are members of the same family: Group 1, Cupressaceae; Group 2, Cupressaceae, Sciadopitaceae, and Araucariaceae; Groups 3, 4, and 5, Pinaceae; Group 6, Taxaceae; and Group 7, Ginkgoaceae.

Other terms commonly used for conifers are gymnosperms, evergreens, needle-bearing trees, and softwoods.

Families and Genera
Groups 1 to 7
The Conifers (Division Pinophyta)

Synopsis of the families and genera in the conifer section of this book, based on Gleason and Cronquist (1991), and Harlow et al. (1991).

Family	Genus
Ginkgoaceae, maidenhair-tree	*Ginkgo,* maidenhair-tree
Taxaceae, yew	*Taxus,* yew
Pinaceae, pine	*Pinus,* pine *Larix,* larch *Cedrus,* cedar *Abies,* fir *Picea,* spruce *Tsuga,* hemlock *Pseudotsuga,* Douglas-fir
Cupressaceae, cypress	*Chamaecyparis,* false-cypress *Juniperus,* juniper *Thuja,* cedar or thuja *Sequoia,* coast redwood *Sequoiadendron,* Sierra redwood *Metasequoia,* dawn redwood *Taxodium,* baldcypress *Cryptomeria,* Japanese-cedar
Sciadopitaceae, umbrella-pine	*Sciadopitys,* umbrella-pine
Araucariaceae, monkey-puzzle	*Araucaria,* monkey-puzzle

Group 1.

Short needles or scales, evergreen; closely spaced in opposite pairs or whorls of 3, often overlapping and obscuring the stem; seeds in cones, some cones berry-like

The trees in Group 1 belong to three genera of the cypress family (Cupressaceae): junipers (*Juniperus*), cedars/aborvitae/thujas (*Thuja*), and false cypress/cedars/cypress (*Chamaecyparis*). Species in Group 1 may have two kinds of leaves: needles and scale leaves. Some species bear only one kind of leaf, others bear both needles and scales on one tree, even on one branch.

Needles are short, divergent, in whorls of 3. Scale leaves are thick, appressed, in alternating pairs, completely covering the shoot; the leaf-covered shoot is square (or rounded) if successive pairs of leaves are equal; the leaf-covered shoot is flattened if one pair is folded and the next pair flat; flat leaves often have a resin gland on the outer surface. On vigorous shoots, the scale leaves are usually elongated, and successive pairs are similar.

During the growing season, neoformed shoot growth continues more or less uniformly; the shoots branch and rebranch; the side shoots may form a characteristic spray or complex. Vegetative shoot tips look much the same summer and winter, being covered with small immature leaves; there are no conspicuous scaly winter buds, no preformed growth, and no distinctive annual nodes.

Immature cones may be seen in winter as solitary naked buds at the tip of a short shoot. Thuja and false-cypress species have mature seed cones that are cone-like, with leathery paired scales. Juniper species have mature seed cones that are berry-like, the "berry" comprising several umbrella-shaped cone scales more or less fused together and forming a fleshy globular structure containing 1 or more seeds.

Junipers

Genus *Juniperus*
Cupressaceae: Cypress Family

Les genévriers

The juniper genus comprises about 60 species of small trees and shrubs distributed throughout the Northern Hemisphere, and on high mountains in the tropics. About 15 species are native to North America; 4 to Canada. Of the Canadian species, only 2 reach tree size, Rocky Mountain juniper and eastern redcedar; the other 2 species are transcontinental shrubs.

Leaves Evergreen, often changing from green to brown or brownish-purple in winter; with 2 leaf forms, needles and scale leaves. All young trees bear needle leaves. Most adult trees have both types, on separate shoots; a few species, e.g. the native shrub, common juniper, have only needle leaves. Needles short, less than 2 cm long, sharp-pointed, spreading, in opposite pairs or whorled; with rows of white resin dots on the surface. Scale leaves small, 2–3 mm long, in alternating similar pairs, partially folded, pressed against and covering the twig; often with a glandular depression on the outer surface; adjacent pairs similar. On upright vigorous shoots scale leaves sometimes twice as long or longer, lance-shaped, with a long tapered point, appressed, widely spaced, exposing a round stem.

Buds The only bud-like structures are immature cones at the tip of some short shoots. Thus shoot tips look the same summer and winter.

Twigs In cross section, shoots covered with scale leaves are 4-sided (or rounded), with sides about equal. Complex of new side shoots form a bundle or plume, often erect, not flattened.

Pollen Cones Catkin-like, ovoid, terminal on short shoots. Pollen cones and seed cones sometimes on separate trees. Pollination takes place in early spring.

Seed Cones Conelets catkin-like, globular, terminal on short shoots, with a very short scaly stalk; scales 3–8, becoming fleshy and uniting during the growing season; mature in 1–3 years.
    Mature cones berry-like, resinous, fragrant, dark blue with a powdery coating, less

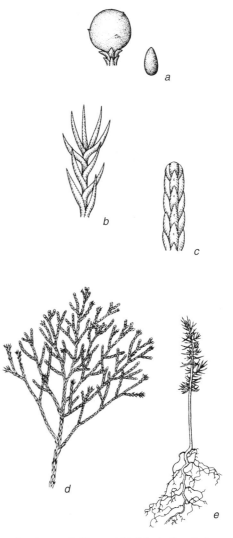

a. Seed cone (left); seed (right). *b.* Needle leaves. *c.* Scale leaves. *d.* Leafy shoot. *e.* Seedling, 1-year-old.

than 10 mm across, with 1 or more seeds. Often borne in large numbers; mature "berries" remain on the tree over winter.

Seeds Ovoid, often grooved, wingless; may be difficult to germinate.

Seedlings Cotyledons usually 2, raised above the surface.

Rocky Mountain juniper
Pollen cones (above) and seed conelets (below), sometimes on separate trees.

eastern redcedar
Mature seed cones berry-like.

Bark Mostly thin and fibrous, separating into long narrow vertical strips, or scales.

Wood Fine-textured, durable, easy to work, reddish-brown, pleasantly fragrant, no resin ducts, latewood not obvious; soft to hard, depending on species.

Size and Form Very small trees or shrubs; crowns vary from strikingly conical to sprawling.

Habitat Usually found on dry sites; intolerant of shade.

Notes Junipers are the most popular of all small trees used for landscape purposes. A great many cultivars have been developed from both native and non-native species; form varies from columnar, to bushy, to sprawling, to pendulous; color from whitish to dark green. Junipers are noted for their ability to survive and grow under a variety of conditions including on dry sites and limestone soils. They may also be propagated by rooted stem cutting from branch tips and by grafts. Junipers can withstand shaping and trimming and thus are suitable for hedges.

Juniper "berries" have a unique, sweetish, resinous taste and are used in flavoring foods and gin. They also supply winter food for birds, which disperse the seeds.

The wood of junipers is favored for lining chests and wardrobe closets because its odor is pleasant and repels moths; the soft kinds are used in the manufacture of lead pencils and for carving. Locally may be used for fence posts.

Susceptible to rusts (*Gymnosporangium*), which form golden gelatinous structures protruding from the branches.

> **Quick Recognition** Leaves of 2 kinds: needle-like, sharp, short, divergent, in whorls of 3; and scale-like, pressed to the twig, in alternating pairs, successive pairs similar. Leaf-covered twig 4-sided in cross section. Seed cones berry-like, pea-size.

Rocky Mountain Juniper

Western juniper

Juniperus scopulorum Sarg.

Genévrier des Rocheuses

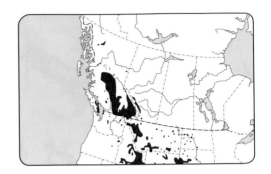

A western species; usually grows on mountains, but also found near sea level in the coastal forest. Useful in preventing erosion on rocky and gravelly slopes; some cultivars suitable for planting on dry sites (Zones C3, NA3).

Leaves Scale leaves with a gland on the outer surface, pale yellowish-green to whitish-green summer and winter; successive pairs barely overlapping. Needles about 12 mm long.

Twigs Leaf-covered twigs rather coarse (for a juniper); pale brown after leaf fall.

Seed Cones Berry-like, up to 8 mm across, fleshy, fragrant, blue with a powdery white coating. Ripen in the 2nd autumn; may persist on the tree for another year or 2; thus 3 or more generations of "berries" may be on the same tree.

Bark Thin, fibrous, reddish- or grayish-brown, divided into flat-topped, interlacing, persistent shreds.

Wood Moderately heavy and hard, weak, strongly aromatic; heartwood reddish-brown, often streaked with white, resistant to decay; sapwood nearly white.

Size and Form Very small trees up to 10 m high and 30 cm in diameter; occasionally 25 m high and 90 cm in diameter; the largest native juniper. A shrub on poor sites. Open-grown trees: crown irregularly conical, coarsely branched; lower branches long, ascending, originating near the base; upper branches short, partly horizontal and partly ascending; trunk often forked. Forest-grown trees: crown slender, branches often drooping. Dead branches often persist on the trunk.

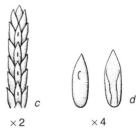

a. Seed cone (left); seeds (right). b. Leafy shoot. c. Scale leaves. d. Outer surface of scale leaf with gland (left), inner surface (right).

Habitat Commonly occurs on dry rocky ridges or sandy soils; in pure but open stands, or as a shrub layer under other species such as Douglas-fir and ponderosa pine.

> **Quick Recognition**   A western species; glands on scale leaves elongated; seeds ripen the 2nd year.

Seed cones (above) mature in 2 years. Young bark (inset right) thin, smooth. Trunk (right) has branches almost to base.

Rocky Mountain Juniper

Common Juniper

Juniperus communis L.

Genévrier commun

A native shrub ranging across Canada; distributed throughout the Northern Hemisphere. Said to have the widest distribution of any tree or shrub. In North America, trunkless with many long branches near the ground, ends upturned. Many cultivars used in landscaping (Zones C2, NA2). Unlike other native junipers has only needle leaves: 12–20 mm long, jointed at the base, with a sharp spine-like tip, arranged in whorls of 3. "Berries" ripen in 3rd year.

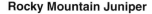

a. Leafy shoot (needle leaves). b. Seed cone.

Creeping Juniper

Juniperus horizontalis Moench

Genévrier horizontal

A native shrub found in open areas across Canada; no trunk, but with very long primary branches that trail over the ground; secondary branches erect, plume-like. Leaves mostly scale-like, prominently whitened, turning purplish over winter. Cultivars suitable for ground cover in landscaping (Zones C2, NA3).

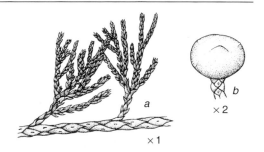

a. Leafy shoot (scale leaves). b. Seed cone.

Eastern Redcedar

Eastern juniper, red juniper

Juniperus virginiana L.

Genévrier de Virginie

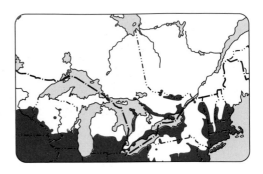

An eastern species; often abundant in pastures and abandoned fields, also in isolated localities along migration routes of seed-eating birds. Cultivars of various forms used for landscape purposes (Zones C3, NA2).

Leaves Scale leaves with or without glandular depressions on the outer surface, 2 mm long, pointed, dark bluish-green (becoming yellowish-brown in winter), successive pairs overlapping. Needles 12 mm long, often on the same branch with the scale leaves.

Twigs Slender, 4-sided.

Seed Cones Berry-like, 3–6 mm in diameter, firm, dark blue with a whitish powder, containing 1 or 2 seeds. Ripen in the 1st autumn.

Bark Thin, fibrous, reddish-brown, separating into long, narrow shreds.

Wood Moderately heavy and hard, weak, strongly aromatic; heartwood bright purplish-red to dull red, resistant to decay; sapwood nearly white.

Size and Form Very small trees, up to 10 m high and 20 cm in diameter. On poor sites, a low shrub. Trunk tapering rapidly, with an irregular cross section. Crown narrowly pyramidal or columnar, becoming open and irregular; branches short, slender, ascending. Root system deep where soil conditions are favorable.

Habitat Found in scattered locations on rocky ridges, and on dry sandy soils.

Notes The name "redcedar" is also used for a western species of *Thuja*.

×2

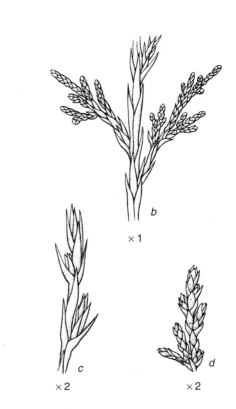

b

×1

c
×2

d
×2

Quick Recognition An eastern species; scale leaves pointed, needle leaves usually present. "Berries" ripen at the end of the 1st growing season.

a. Seed cone. *b.* Leafy shoot. *c.* Needle leaves.
d. Scale leaves.

Seed cones dark blue, coated with a whitish powder.

Young bark reddish-brown.

Eastern Redcedar

Trunk irregular in cross section.

Chinese Juniper*

Juniperus chinensis L.

Genévrier de Chine

Very small trees or shrubs native to China, many cultivars; scale leaves blunt, with pale margins; "berries" ripening in the 2nd year. Hardy as far north as Zones C4, NA4.

a. Leafy shoot with needle and scale leaves.
b. Seed cone.

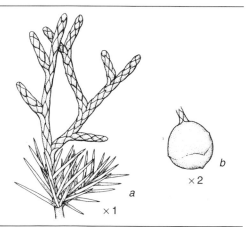

Savin Juniper*

Savin

Juniperus sabina L.

Genévrier sabine

A shrub native to Europe, many cultivars; upper surface of needle whitish with a prominent midvein; green with a resinous depression beneath; "berries" ripening in the 1st year. Characteristic odor. Hardy as far north as Zones C2, NA3.

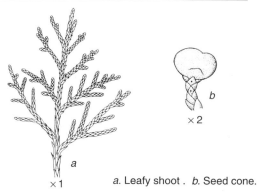

a. Leafy shoot . b. Seed cone.

Cedars/Arborvitae/ Thujas

Genus *Thuja*

Cupressaceae: Cypress Family

Les thuyas

The thuja genus contains 6 species; 2 are native to North America; 4 to eastern Asia.

Leaves Evergreen, remaining green for several years; scale-like, up to 3 mm long; in alternating pairs so as to be in 4 rows along the flat twig; upper and lower leaves flat; lateral leaves folded; each pair overlapping the next; pressed against and covering the twig; falling away with the twig. Leaves on vigorous upright shoots elongated, up to 10 mm long, lance-shaped, with slender pointed tips; successive pairs similar; appressed but widely spaced exposing the round stem; falling away with the outer bark when it is sloughed off.

Buds The only bud-like structures are the immature cones at the tip of some short shoots; form in autumn, evident all winter.

Twigs Tough, flexible, slender, progressively changing from green to brown along the stem. In cross section, shoots with scale leaves are flattened; vigorous shoots with lance-shaped leaves round.

Multibranched shoot complex characteristic in shape; new side shoots arise alternately in 2 ranks from the axils of the lateral leaves, forming a flattened spray; the branching pattern may be repeated once or twice in the course of a growing season. Shoot complex, not individual scale leaves, abscises and falls away after a few years.

Pollen Cones Ovoid to globular, reddish, about 4 mm long; on the same tree as seed cones, usually on different branches. Pollination takes place in early spring.

Seed Cones Conelets green or purplish, about 4 mm long; mature in 1 season.

Mature seed cones ovoid, about 10 mm long; scales leathery, 4–8 pairs; 2 or 3 of the middle pairs bear 2 or 3 seeds; open in autumn; seeds released during the winter.

Seeds Small, flattened; native species winged.

Seedlings Newly germinated seedlings consist of a slender stalk about 10 mm tall, surmounted by 2 flat, linear, blunt cotyledons,

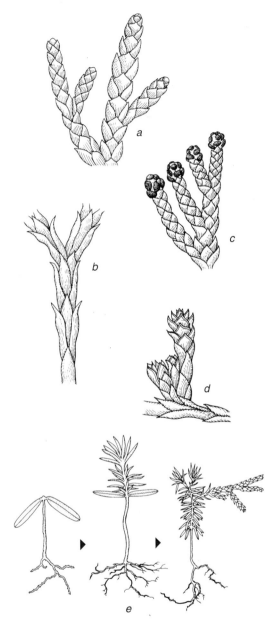

a. Leafy shoot. b. Lance-shaped leaves on vigorous upright shoot. c. Pollen cones. d. Seed conelets. e. Seedling development; cotelydon stage (left), juvenile leaf stage (center), juvenile and first adult leaves (right).

less than 10 mm long. Leafy shoot arises above the cotyledons; juvenile leaves are linear, flat, sharp-pointed, about 1 cm long, often in pairs or whorls of 4, at right angles to the stem, and spaced so that the stem is exposed. The first adult-type scale leaves are borne on side branches.

dcedar

rborvitae

ex D. Don

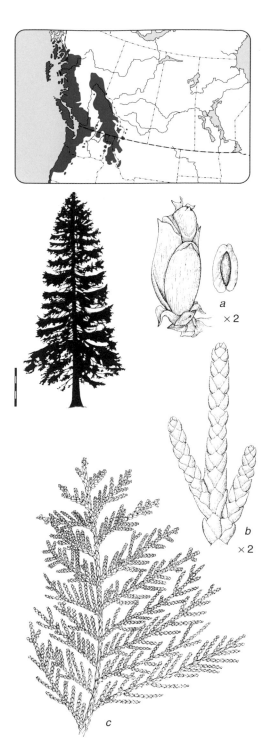

ccurring from sea level
ly planted for landscape
6, NA5).

llowish-green; scale-like
ong, with inconspicuous
outer surface; lance-shaped
long.

overed shoots yellowish-green
surface, often whitened below;
ex elongated, tapering, often
pendulous.

es Ovoid, 12–18 mm long; scales
n with a small, weak, sharp point
ip. Ripen in late summer; shed
e winter.

Thin, reddish-brown, shiny when
; shredded, with age forming narrow
dges.

od Very light, soft, relatively weak,
aight-grained, characteristic odor; heart-
od pinkish or reddish-brown to deep
own, resistant to decay; sapwood
ellowish-white, less resistant to decay.

Size and Form Very large trees, up to 60 m
high, 250 cm in diameter, and 800 years old.
Trunk tapering rapidly, flaring and buttressed
at the base. Crown long, symmetrical, narrowly conical, becoming typically irregular
with age. Principal branches spreading,
drooping, upturned at the ends. An old tree
may have one or more dead spire-like tops
killed by lightning strikes. Open-grown trees
have live foliage extending to the ground,
often so dense that the trunk is obscured.
Root system shallow, wide-spreading, strong.

Habitat A characteristic tree of the forests
of the coast and the wetter parts of interior
British Columbia. At altitudes above 1500 m,
it is a small tree or shrub. Grows best on
moist alluvial sites, but also found on rich dry
soils and in sphagnum bogs. Seldom occurs
in pure stands; usually mixed with other
species such as Douglas-fir, Sitka spruce,
western hemlock, black cottonwood, red
alder, and at higher elevations, Engelmann
spruce and western larch.

a. Seed cone (left); winged seed (right). *b.* Leafy
shoot. *c.* Shoot complex.

Pollen cones on branchlets near base of shoot complex.

Seed conelets on branchlets at tip of shoot complex.

Seed cones mature i

24

1

Western Re
Western thuja, giant a
Thuja plicata Don
Thuya géant

A western species,
to 2000 m. Frequer
purposes (Zones C

Leaves Shiny y
leaves 1–2 mm
resin glands on
leaves 4–5 m

Twigs Leaf-
on the upper
shoot comp
fern-like an

Seed Co
8–10, oft
near the
during t

Bark
young
flat ri

Wo
str
w
b

Vegetative Reproduction Layering, the rooting of attached lower branches where they rest on the surface, is common, especially in bogs or swamps; upper branches on fallen trees and fallen branches may also take root.

Bark Smooth, thin; with age becoming furrowed with long, thin, flat-topped, fibrous strips; inner bark fibrous.

Wood Light, soft, aromatic, no resin ducts; well-marked latewood and heartwood; easily split, resistant to decay.

Size and Form Size varies. Trunks stout, irregular; young crowns symmetrical, conical, becoming irregular with age. Although trunk is distinct to the top of the tree, the leading shoot cannot easily be identified among several upright shoots. Roots shallow, wide-spreading, often rope-like with few branches; lateral rootlets long, thin, reddish-purple.

Habitat Occur on a variety of soils from bogs to dry limestone flats; grow best on moist soil with some drainage. Moderately shade-tolerant.

Notes Valued as ornament
their regular, shapely crowns;
hedge trees because they grow
and produce new shoots after se
ming. Many cultivars have been se
Can be propagated by rooted cutting
An important source of shelter and
for wildlife.
The genus was unknown to the Europeans who colonized North America, which probably accounts for the lack of an appropriate English name. "Cedar" is commonly used, but that name has long been accepted for the genus *Cedrus* and is also often applied to any species with fragrant wood. "Arborvitae" is sometimes applied to the genus; however, the use of the Latin name in English, "thuja", is becoming more common.

Quick Recognition Side shoots flattened, with alternating pairs of flat and folded scale leaves; shoot complex a flattened spray; seed cones small, with 4–8 pairs of leathery scales.

Mature seed cones; scales usually spine-tipped.

Young bark smooth, shiny.

Trunk buttressed at base.

Notes Wood a source of shakes and shingles. Logs have been used to make totem poles and dugout canoes. There are records of canoes 20 m long and 240 cm wide; one on display at the Canadian Museum of Civilization is 16 m long and 180 cm wide. Also valued for poles, posts, boat-building, greenhouse construction, outdoor patios, exterior siding, doors, window sashes, millwork, and interior finishing.

Quick Recognition Compared with eastern white-cedar (which does not occur west of Manitoba): resin glands less conspicuous; leaf-covered shoots silvery on the underside; cone scales often with a small point projecting from the outer surface. Compared with yellow-cedar (*Chamaecyparis*): leaf-covered shoots flat-tened, silvery on the underside; seed cones with leathery paired scales, not berry-like.

Eastern White-Cedar

Northern white-cedar, eastern thuja, eastern arborvitae

Thuja occidentalis L.

Thuya occidental

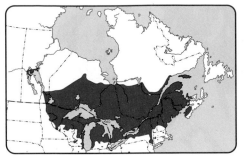

An eastern species, similar to but much smaller than western redcedar. Numerous cultivars are widely used for landscape purposes (Zones C3, NA2); particularly common as a hedge tree.

Leaves   Dull yellowish-green, turning bronze-colored during the winter; scale-like leaves 1–2 mm long, with conspicuous resin glands; lance-shaped leaves 4–5 mm long.

Twigs   Leaf-covered shoots yellowish-green on both sides, flattened; shoot complex broadly fan-shaped, usually horizontal and stiff.

Seed Cones   Ovoid, 7–12 mm long, upright, on a short curved stalk; scales leathery, 5 or 6 pairs. Ripen in late summer; seed dispersal begins at that time; cones are shed over a period of some months.

Bark   Thin, shiny reddish-brown when young; with age separating into long, narrow, flat gray strips.

Wood   Very light, soft, weak, characteristic odor; easily split; heartwood light brown, resistant to decay; sapwood nearly white, less resistant.

Size and Form   Small trees, up to 15 m high, 30 cm in diameter, and several hundred years old; occasionally 25 m high and 90 cm in diameter. Small, stunted trees over 700 years old have been found on limestone cliffs of the Niagara Escarpment in southern Ontario; even older stunted trees have been found in northern Quebec. Open-grown trees: crown long, narrow, dense, conical, almost columnar, neat and trimmed in appearance; branches bending slightly downward before gradually turning upward towards their tips; trunk irregular in cross section, tapering rapidly, often leaning, then curving upward. Forest-grown trees: trunk visible through the open irregular crown; stubs of dead branches on the lower part. Root system shallow, wide-spreading.

Habitat   Occurs mainly in swampy areas where the underlying rock is limestone; also on very shallow dry soils over flat limestone

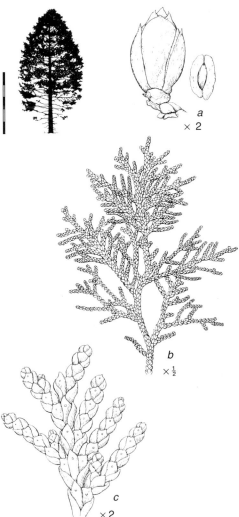

a. Seed cone (left); winged seed (right). *b.* Shoot complex. *c.* Leafy shoot.

rock, and in sphagnum bogs. Grows in small pure stands; more often mixed other species such as eastern white pine, yellow birch, eastern hemlock, silver maple, black ash, and white elm. Slow-growing.

Mature seed cones; scales usually rounded at tips.

Young bark thin, shiny reddish-brown.

Mature bark has flat narrow strips.

Notes Wood is used for small poles, fence posts, cedar-strip canoes. Although the dry wood is resistant to decay, the living trees are subject to heart-rot; consequently, many trees have hollow trunks.

The leaf-covered twigs are a favorite winter food for the white-tailed deer (*Odocoileus virginianus*); winter "yards" for these animals are often in or near groves of eastern white-cedar.

Quick Recognition Leaves scale-like in 4 rows; leaf-covered shoots, flattened, yellowish-green on both surfaces; resin glands usually conspicuous on the outer surface of the flat scale leaves; shoot complex in flat horizontal sprays; seed cones with 5 or 6 pairs of leathery scales.

Oriental-Cedar*

Thuja orientalis L.
[syn. *Biota orientalis* Endl.;
Platycladus orientalis Franco]

Thuya d'Orient

Native to Asia; commonly planted in Canada as an ornamental because of its cylindrical compact form. Hardy as far north as Zones C6, NA6. Distinguishable from the native species by the following features: shoot complex nearly vertical; cone scales thick, fleshy, with conspicuous stout points on their outer surface; and wingless seeds.

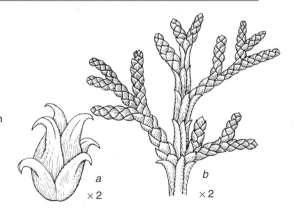

a. Seed cone. *b.* Leafy shoot.

False Cypress/ Cedars/Cypress

Genus *Chamaecyparis*
Cupressaceae: Cypress Family

Les chamaecyparis

The false cypress genus contains 6 species; 3 are found in eastern Asia and 3 in North America; yellow-cedar is the only species native to Canada. Many cultivars are available of native and introduced species of false-cypress; those with pointed spreading juvenile leaves are sometimes called *Retinispora*.

The genus *Chamaecyparis* was unknown to the Europeans who colonized North America; hence it is usually called cedar, probably because the fragrant wood reminded people of the old world cedar, *Cedrus*. False cypress is the name accepted by most authorities.

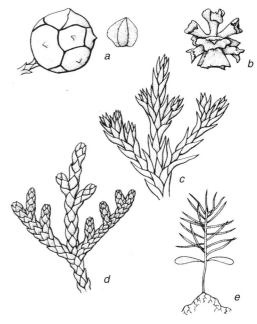

a. Seed cone (left); seed (right). *b*. Open cone. *c*. Juvenile leaves. *d*. Scale-like leaves. *e*. Seedling, 1-year-old.

Yellow-Cedar

Alaska-cedar, Nootka-cypress, Nootka false cypress, yellow-cypress

Chamaecyparis nootkatensis
(D. Don) Spach

Chamaecyparis jaune

A western species, occurring mainly near the Pacific Coast. Frequently planted for land-scape purposes (Zones C6, NA5) because of the attractive crown with long sweeping prin-cipal branches and drooping secondary branches.

Leaves Evergreen; scale-like, small, up to 3 mm long, dull bluish-green, unpleasant res-inous odor when crushed; in alternating pairs so as to be in 4 rows along the twig; upper and lower leaves somewhat flattened, with a resin gland; lateral leaves somewhat folded, blunt-pointed and appressed, or sharp-pointed with tips divergent; pressed against and covering the twig. Leaves on vigorous shoots, elongated, up to 6 mm long, lance-shaped, with slender pointed tips; successive pairs similar; appressed but widely spaced, exposing the round stem.

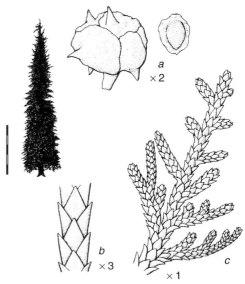

a. Seed cone (left); seed (right). *b*. Lance-shaped leaves on vigorous shoot. *c*. Leafy shoot.

Pollen cones (above); seed conelets (below) at time of pollination.

Crown has spreading branches and loosely hanging branchlets.

Mature seed cones round, each scale with a triangular projection.

Yellow-Cedar

Young bark scaly.

Trunk fluted at base; mature bark grayish.

1

Buds   The only bud-like structures are the immature cones at the tip of some short shoots; form in summer, evident all winter.

Twigs   Tough, very flexible, slender, progressively changing from green to reddish-brown along the stem. In cross section, shoots with scale leaves are 4-sided, slightly flattened, dull green above and below; vigorous shoots with lance-shaped leaves are round.

   Multibranched shoot complex frond-like, flattened, pendulous; new side shoots arise alternately in 2 ranks from the axils of the folded leaves; the branching pattern may be repeated once or twice in the course of a growing season.

Pollen Cones   Ovoid to globular, greenish-yellow, 2−3 mm long; on the same tree as seed cones, often on same branch. Pollen shed in early spring.

Seed Cones   Conelets berry-like, about 4 mm long, green, soft; mature during the 2nd season.

   Mature seed cones globular, stalked, hard, whitish-green with a purple tinge, 8–12 mm in diameter; scales 4–6, thick, with a prominent erect triangular projection; umbrella-shaped, with adhering edges when immature; separating and releasing the seeds when mature. Good seed crops every 4 or more years.

Seeds   Flattened, winged, 2 per fertile scale.

Seedlings   With 2 cotyledons raised above the surface; followed by several whorls of needle-like juvenile leaves; gradually changing to scale leaves.

Bark   Thin, grayish-brown, scaly when young; with age separating into narrow intersecting ridges.

Wood   Light, hard, strong, close-grained, even-textured, mild characteristic odor, no resin ducts; pale yellow; sapwood not clearly distinct from the heartwood.

Size and Form   Medium-sized trees, up to 25 m high and 90 cm in diameter; occasionally much larger. Trunk tapered, often with a broadly buttressed and fluted base. Crown sharply conical; branches spreading, drooping, bearing loosely hanging branchlets. Leading shoot slender, flexible, curved. Above 1500 m, small and shrubby. Root system varies from shallow to deep.

Habitat   Requires plenty of moisture; usually grows singly or in small clumps mixed with other conifers. Shade-tolerant. Slow-growing.

Notes   Wood is used in boatbuilding, greenhouse construction, and for other purposes where its resistance to decay is advantageous; also for carving, pattern-making, hobby work, and canoe paddles; in short supply.

   Can be propagated by seed or by stem cuttings from branch tips.

Quick Recognition   Leaves scale-like, in 4 rows, covering the twig, characteristic unpleasant resinous odor when crushed; alternating pairs quite similar. Shoot complex flattened, drooping. Seed cones berry-like until mature; mature cone scales separate and release the seeds.

Lawson-Cypress*
Lawson false cypress, Port-Orford-cedar

Chamaecyparis lawsoniana
(A. Murr.) Parl.

Chamaecyparis de Lawson

Native to the western United States; restricted to the Pacific Coast. A great variety of cultivars available; frequently planted. Hardy as far north as Zones C7, NA5. Quite similar to yellow-cedar; leaf-covered **twigs** whitened beneath, with resin glands giving an odor of parsley; **seed cones** 5–7 mm in diameter, purplish-green becoming reddish-brown when mature; projection from cone scale reflexed; seeds ripen in 1 season; **bark** thick, separating into long loose strips.

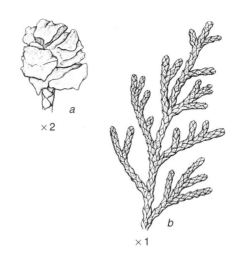

a. Seed cone. b. Leafy shoot.

Sawara-Cypress*
Sawara false cypress

Chamaecyparis pisifera (Siebold & Zucc.) Endl.

Chamaecyparis de Sawara

Native to Japan. Cultivars frequently planted. Hardy as far north as Zones C4, NA3. Leaf-covered **twigs** whitened beneath; scale **leaves** sharp-pointed with a few small glands, sharp resin odor when crushed; **seed cones** about 5 mm in diameter, ripen in 1 season.

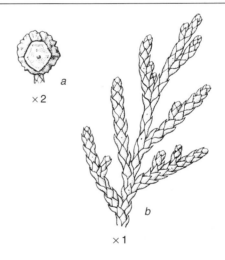

a. Seed cone. b. Leafy shoot.

Hinoki-Cypress*
Hinoki false cypress

Chamaecyparis obtusa (Siebold & Zucc.) Endl.

Chamaecyparis du Japon

Native to Japan. Cultivars frequently planted. Hardy as far north as Zones C5, NA4. Leaf-covered **twigs** whitened beneath; scale **leaves** blunt with very small glands, sweet cedar odor when crushed; **seed cones** about 7 mm in diameter, ripen in 1 season.

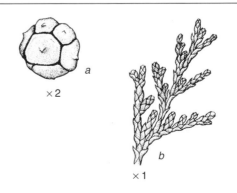

a. Seed cone. b. Leafy shoot.

Group 2.

Introduced species hardy in southwestern British Columbia; leaves various shapes; seeds in cones

Group 2 comprises seven species, all from different genera, with little in common, except that they are non-native, hardy only in the warmer parts of British Columbia (and in sheltered locations in other mild parts of Canada), and have seeds in cones; most lack scaly winter buds. With the exception of umbrella-pine and monkey-puzzle, species in this group belong to genera in the cypress family.

Sierra Redwood*

Giant sequoia, bigtree, redwood, sequoia

Sequoiadendron giganteum
(Lindl.) Buchh.
Cupressaceae: Cypress Family

Séquoia géant

Probably the world's largest tree in terms of stem volume; over 75 m high, 1000 cm in diameter, and 1000 years old; native to the western United States, where its range is limited to a few groves on the western slope of the Sierra Nevada. Frequently planted in Canada for landscape purposes. Hardy as far north as Zones C7, NA6.

Leaves evergreen; those on horizontal shoots of mature trees, scale-like, 3–5 mm long, spirally arranged, appressed, overlapping; those on vigorous upright shoots and young trees, lance-shaped, longer, up to 12 mm, with lines of white dots on both surfaces, rigid, sharp-pointed, spreading from the twig.

Twigs green, becoming brownish-red after leaf fall. No scaly winter **buds**.

Seed cones 5–10 cm long, on the ends of shorter shoots; 25–40 scales, thick, deeply pitted, with reflexed long tip, spreading apart when mature.

Seeds mature in 2 seasons, dispersing during the 3rd season; flattened, winged; 3–5 cotyledons.

Bark smooth, gray; with age, becoming very thick, deeply furrowed with purplish outer scales and brownish-red inner scales.

Crown on young trees conical; on old ones, dense, irregular, rounded.

a. Seed cone (left); seed (right). *b.* Lance-shaped leaves (left); leaf cross section (right). *c.* Shoot complex.

Crown is conical on young trees.

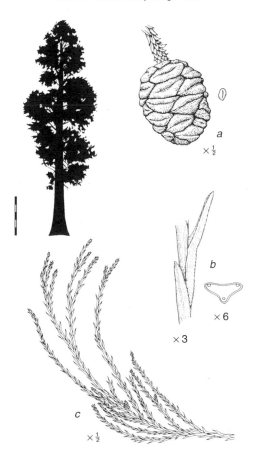

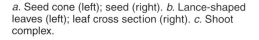

Coast Redwood*

California redwood, redwood, sequoia

Sequoia sempervirens (D. Don) Endl.

Cupressaceae: Cypress Family

Séquoia toujours vert

Native to the United States, where it occurs along the fog belt of the Pacific Coast. The world's tallest conifer; one in California measures 112 m high. Planted in Canada for landscape purposes. Hardy as far north as Zones C8, NA7.

Leaves evergreen; those on horizontal shoots of mature trees, needle-like, 10–20 mm long, lower surface obscurely keeled and marked with 2 lines of white dots, slender-pointed, arranged in 2-ranks; those on young trees and on vigorous upright shoots, wider, 3–5 mm long, spreading in all directions.

New side **shoots** form a flat spray. **Buds** small, scaly.

Seed cones globular, 2–3 cm long, pendulous on the ends of shorter shoots; 14–20 scales, hard, leathery, umbrella-shaped, spreading apart when mature. **Seeds** mature in 1 season, winged; 2 cotyledons. Reproduces vegetatively by stump sprouts; numerous sprouts also arise at the base of dead branches (such as those killed by forest fire), clothing the trunk in green leaves. Burls cut from the trunk and set in water will also sprout.

Bark very thick, reddish-brown, fibrous, deeply furrowed, with rounded scaly ridges.

Crown on young trees conical; on old trees long, narrow, round-topped. Principal **branches** drooping; small secondary branches sometimes abscise and fall away.

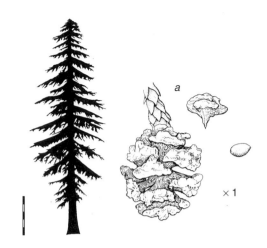

a

×1

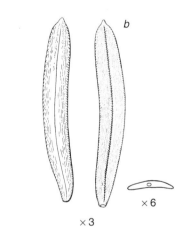

b

×3

×6

c

×½

a. Seed cone (left); cone scale (center); seed (right). *b.* Leaf upper surface (left); lower surface (center); leaf cross section (right). *c.* Leafy shoot

Dawn Redwood*

Metasequoia glyptostroboides
Hu & Cheng
Cupressaceae: Cypress Family

Métaséquoia de Chine

The *Metasequoia* genus existed only in fossil records until a few living trees of *M. glyptostroboides* were found in central China in 1945. Large trees up to 35 m high and 300 cm in diameter. Undoubtedly a conifer, but distinctive because of the 2-ranked opposite leaves and twigs and deciduous habit. Widely planted in North America and Europe. Hardy as far north as Zones C5, NA4.

Leaves deciduous, needle-like, flattened, soft, slightly curved, midvein prominent, upper surface bright green, paler beneath, tapering abruptly to a basal joint attached to a leaf-cushion on the twig; opposite, in 2 ranks. On major shoots, about 3 cm long; falling individually. On deciduous shoots, about 2 cm long, 25–30 pairs; falling attached to the green shoot.

Deciduous **shoots** 2-ranked, in opposite pairs, 6–10 cm long; new shoots each year for several years from the same position on the branch. Major **twigs** reddish-brown, ridged and grooved with the leaf-cushions.

Buds ovoid, small, 3–4 mm long, non-resinous, solitary or in small clusters on opposite sides of the branch.

Seed cones globular, 20–25 mm long, solitary, pendulous on the ends of short side twigs; 20–30 scales, in opposite pairs, woody, outer surface broadly triangular with a horizontal groove. **Seeds** winged; 2 cotyledons.

Bark reddish-brown when young; with age becoming grayish, fissured, peeling off in long strips.

Crown on young trees conical, regular; on old ones rounded.

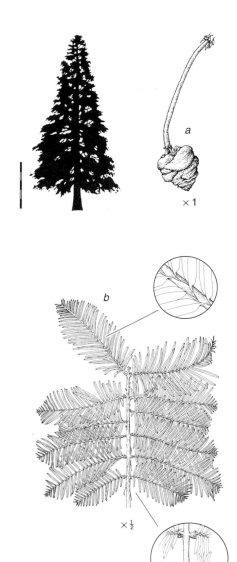

a

×1

b

×½

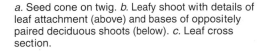

c

×6

a. Seed cone on twig. b. Leafy shoot with details of leaf attachment (above) and bases of oppositely paired deciduous shoots (below). c. Leaf cross section.

Bald-Cypress*
Deciduous-cypress

Taxodium distichum (L.) L. Rich.
Cupressaceae: Cypress Family

Cyprès chauve

A deciduous conifer native to the south-eastern United States. Very large trees up to 45 m high, 300 cm in diameter, and several hundred years old. A stately tree frequently planted far beyond its range for landscape purposes. Hardy as far north as Zones C5, NA4.

Bald-cypress can grow in swamps with its roots flooded most of the year; in such situations it develops large flares at its base and "cypress knees", conical projections up to 1 m high growing from the roots.

Leaves on deciduous shoots, needle-like, 15–20 mm long, flattened, soft, short-pointed, midvein prominent, upper surface yellowish-green, whitened beneath; appearing to be 2-ranked; falling attached to the shoot. On major shoots, shorter, spreading, spirally arranged; falling individually.

Deciduous **shoots** arise from small buds towards the base of previous year's major twigs; alternately arranged; without lateral buds; falling with leaves attached, often persisting on the tree in early winter. Major shoots arise from larger buds toward the tip of previous year's major twigs, green becoming brown the 1st winter, marked by circular scars left by the fallen deciduous shoots. **Buds** small, round, scaly.

Seed cones globular, 15–35 mm in diameter, brownish-purple, solitary or in small clusters, on the ends of previous year's twigs, maturing in 1 season, October to December; 9–15 scales, 4-sided, thick, woody, shield-shaped, breaking away when mature. **Seeds** 3-angled, 3-winged, resin-coated, warty; 4–9 cotyledons. Often reproduces by stump sprouts.

Bark reddish-brown, fibrous, peeling off in long strips. **Wood** from old-growth trees, very resistant to decay; age difficult to determine because of false growth rings.

Crown slender conical; **branches** short. Moderately shade-tolerant. Exceptionally wind-firm.

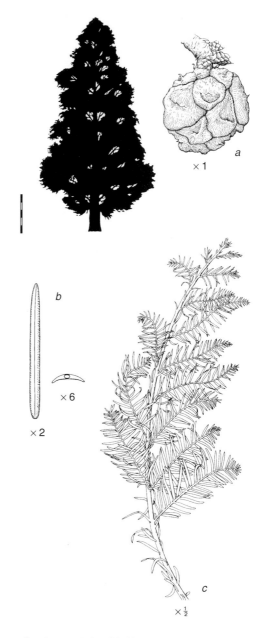

a. Seed cone. *b.* Leaf (left); leaf cross section (right). *c.* Twig with deciduous shoots and individual leaves at base.

Japanese-Cedar*
Cryptomeria, Sugi

Cryptomeria japonica (L. f.) D. Don
Cupressaceae: Cypress Family

Cryptoméria du Japon

Native to eastern Asia; very large trees up to 50 m, frequently planted in Japan for landscape purposes and as a source of wood. More recently various cultivars have been planted in North America and Europe as specimen trees in landscaping. Hardy as far north as Zones C6, NA5.

Leaves evergreen, persisting 4–5 years; narrowly conical, lower part clasping the shoot, upper part diverging from the shoot, pointing forward, curving inward, bluntpointed, 6–12 mm long, slightly twisted, keeled above and below, with white lines parallel to the keels; spirally arranged.

Twigs green, hairless. Shoot tips covered with immature leaves summer and winter.

Seed cones globular, 12–18 mm long, brown, solitary, at the ends of shorter shoots; maturing in 1 season, remaining on the tree for months after the seeds are shed; the tip often proliferating into a new shoot; 20–30 scales, toothed at tip. **Seeds** brown, triangular, with 3 rudimentary wings.

Bark reddish-brown, fibrous, peeling off in long shreds.

Crown regular, conical; principal **branches** in whorls, horizontal or drooping; secondary branches sometimes abscising and falling away; **trunk** tapering, buttressed at base.

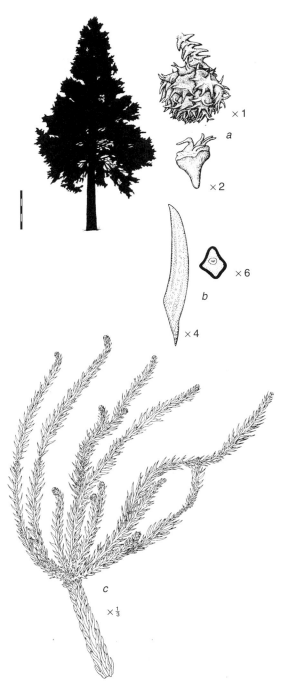

a. Seed cone (above); cone scale (below). *b.* Leaf (left); leaf cross section (right). *c.* Shoot complex.

Umbrella-Pine*

Sciadopitys

Sciadopitys verticillata (Thunb.)
Siebold & Zucc.
Sciadopitaceae: Umbrella-pine Family

Pin parasol du Japon

A medium-sized conifer up to 30 m, with
double leaves in whorls, native to Japan;
limited natural range. Planted as a specimen
tree in landscaping. Hardy as far north as
Zones C7, NA5.

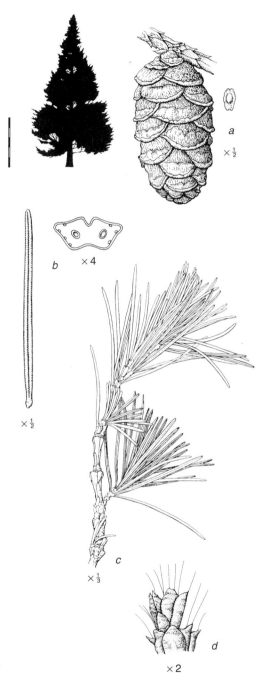

a

$\times \frac{1}{2}$

b $\times 4$

$\times \frac{1}{2}$

c

$\times \frac{1}{3}$

d

$\times 2$

    Leaves evergreen; 2 leaf forms, needles
(principal leaves), and scale leaves. Needles
8–15 cm long, 2–3 mm wide, rigid, margins
thickened, notched at tip; upper surface
glossy dark green, paler beneath with lines
of white dots; grooved on both surfaces; arranged in united pairs, borne on a microshoot (as in pine) in the axil of a scale leaf; in
successive whorls of 10–30 along the major
shoots. Scale leaves triangular, small,
3–7 mm long, set on a leaf-cushion; green
with a reddish-brown membranous tip,
becoming wholly brown toward the end of the
1st season; some in a whorl just below the
principal leaves; others spirally arranged
along major shoots.

    Twigs green, becoming brown, hairless;
ridged and grooved with leaf-cushions.

    Seed cones oblong, 8–12 cm long, short-
stalked, on previous year's twigs; maturing in
the 2nd season. Cone scales fan-shaped,
woody, spirally arranged; surfaces furrowed,
inner surface covered with fine hairs. **Seeds**
9–12 mm long, winged; 2 cotyledons.

    Bark reddish-brown, peeling off in long
strips; newly exposed bark reddish.

    Crown conical; principal **branches**
horizontal, short.

a. Seed cone (left); winged seed (right). *b*. Principal
leaf (left); leaf cross section (right). *c*. Leafy twig.
d. Scale leaves at base of principal leaves.

2

Monkey-Puzzle*
Araucaria, Chilean-pine

Araucaria araucana (Mol.) K. Koch
Araucariaceae: Monkey-puzzle Family

Araucaria du Chili

Large trees up to 35 m. Distinctive with large bract-like leaves and whorled branches; native to South America, unlike most introduced trees in Canada which come from the Northern Hemisphere. Planted in Canada as a landscape tree. Hardy as far north as Zones C8, NA7.

Leaves evergreen; triangular, 2.5–5 cm long, stiff, leathery, margins thickened, tip a sharp prickle; dark glossy green, lines of white dots on both surfaces; spirally arranged, overlapping, clasping the shoot but diverging from it; remaining green for 10–15 years, dead leaves persisting for several years.

Without obvious vegetative **buds**; shoot tips covered with immature leaves in all seasons.

Twigs green, obscured by densely packed scale leaves.

Pollen cones erect, 8–12 cm long, in the leaf axils. **Seed cones** globular, 10–18 cm in diameter, at the ends of side branches; maturing in 2–3 years, then disintegrating; many scales. Pollen cones and seed cones on separate trees. **Seeds** very large, 25–40 mm long, brown, ridged, attached to a scale; 2 cotyledons, remaining within the seed coat during germination. Reproduces vegetatively by sprouts from stumps and from stubs of pruned branches.

Bark thick, resinous.

Crown conical on young trees; irregular, flat-topped on old ones. Principal **branches** generally horizontal; those near the base, drooping; in whorls of 4–8. Secondary branches few, in opposite pairs, deciduous after several years.

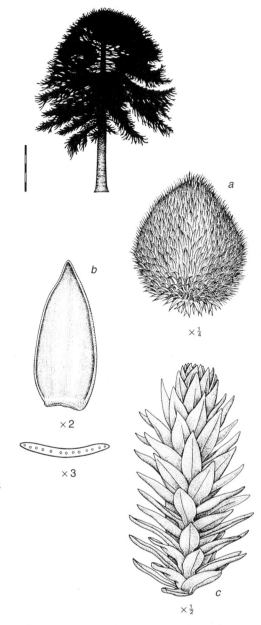

a. Seed cone. *b.* Leaf, inner surface (above); leaf cross section (below). *c.* Leafy twig.

Group 3.

Needles evergreen, in bundles of 2, 3, or 5; seeds in cones

3

Group 3 contains one genus, the pines (*Pinus*). Pines are characterized by evergreen needles in bundles at the tips of microshoots; each twig has many such bundles of needles. Other genera in the pine family are described in Groups 4 and 5.

Pines

Genus *Pinus*
Pinaceae: Pine Family

Les pins

Pines comprise about 95 species of trees and a few shrubs, distributed throughout the Northern Hemisphere; 34 are native to North America; 9 to Canada.

Pines are unusual in that the adult tree has microshoots, 3 types of leaves, budlets within buds, and sometimes 3 kinds of branch nodes on main stems.

Pines are divided into 2 groups — **soft pines** (also called white pines or 5-needle pines) and **hard pines** (also called yellow pines or pitch pines).

Soft pines (eastern white, western white, limber, whitebark, and bristlecone pine) have needles with 1 longitudinal vein; they are borne in bundles of 5, with deciduous basal bundle-sheaths. Their twigs are smooth and the cone scales are usually without prickles.

Hard pines (ponderosa, pitch, red, jack, lodgepole, shore, Austrian, Scots, and mugho pine) have needles with 2 longitudinal veins; they are borne in bundles of 2 or 3, with persistent basal bundle-sheaths. Their twigs are ridged and grooved and the cone scales are often armed with prickles.

Leaves Three types: bud scales or scale leaves; needles in bundles; bundle-sheath scales.

Bud scales or scale leaves: Often deciduous, brown, with a papery texture; inconspicuous; originate as bud scales, later spirally arranged on the major shoots from base to tip. In soft pines, set on an inconspic-uous crescent-shaped leaf-cushion; in hard pines, set on the forward end of a raised elongated leaf-cushion.

Needles in bundles: Evergreen, needle-shaped, in bundles (fascicles) of 2, 3, or 5; in cross-section, with 1 or 2 veins in the core and 2 or more longitudinal resin ducts in the surrounding soft tissue. Each bundle borne on a microshoot set in the axil of a scale leaf; remaining on the tree for 1 year to several decades, depending on species and growing conditions; abscising where the microshoot meets the bark, needles and microshoot falling as a unit.

Bundle-sheath scales: Pale brown, thin papery texture, attached to the microshoot; fully enveloping immature needles, much shorter than mature needles. In soft pines, shed in the 1st season; in hard pines, remaining attached to the microshoot and forming a coherent sheath around the base of the needles, noticeably shorter after the 1st year.

Buds Terminal bud with numerous over-lapping scales; subterminal buds similar but smaller. Buds contain many budlets; each budlet borne in the axil of a scale. In some species, buds on vigorous shoots may be polycyclic, i.e., 1 or more whorls (cycles) of immature side branches preformed in the bud.

Budlets conical, less than 1 mm long, fleshy, developing into microshoots with bundle-sheath scales and needles; or into

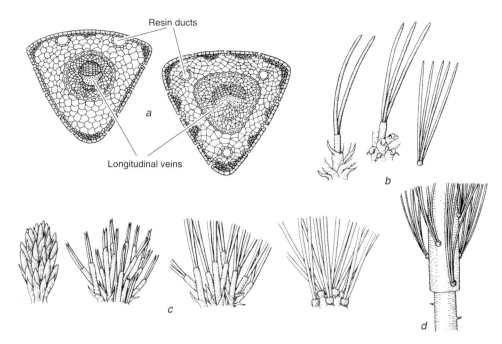

a. Needle cross sections; soft pine needle with 1 longitudinal vein in center (left), hard pine needle with 2 longitudinal veins in center (right). *b.* Needle bundle types; 2-needled, shown with bundle-sheath scale and straight scale leaves (left), 3-needled, shown with bundle-sheath scale and recurved scale leaves (center), 5-needled, bundle-sheath deciduous (right). *c.* Needle development (left to right); 1st year—early spring, late spring, autumn; 2nd year—autumn. *d.* Microshoots on a 2-yr-old twig (bark peeled at base to show woody microshoot stubs).

pollen cones (in a cluster back from the needles), or seed cones and next year's buds (in the subterminal position).

Major Shoots Obvious to the casual observer; bearing a variety of structures including scale leaves, microshoots with needle bundles, pollen cones, seed cones, a terminal bud, and a whorl of subterminal buds, all preformed in the bud during the previous season. Major shoots grow in length and diameter in succeeding seasons. No needles on the basal part of the shoot.

On soft pines, surface of twig smooth with inconspicuous crescent-shaped leaf-cushions just below each microshoot; small circular or oval scars remain when the microshoots abscise. On hard pines, leaf-cushions conspicuous as ridges separated by incised grooves; the forward end of each projecting over and obscuring the scar left by the microshoot after abscission.

Pines may have 3 types of **nodes** on main stems: annual nodes, intermediate nodes formed in the bud, and seasonal nodes resulting from precocious development of buds during the growing season; each type of node is marked by a whorl of branches.

Annual nodes are marked by a faint ring of bud scales or bud-scale scars just ahead of the whorl of branches; the presence of crowded needle bundles behind the node and the absence of needle bundles ahead of the node.

Some species have 1 or more intermediate node(s). **Intermediate nodes** may be recognized by the absence of a ring of bud scales and the uniform arrangement of needle bundles behind them.

Seasonal nodes occur on pines in warmer regions but are rare in Canadian pines; they result from precocious flushes of growth during a growing season. Seasonal nodes resemble annual nodes except that the internode ahead is likely to be quite short.

Under **Twigs** in the species descriptions, features of major shoots are given.

Microshoots Characteristic of pines, but usually inconspicuous. About 1 mm long, set in the axil of a scale leaf, obscured by the projecting end of the leaf-cushion in hard pines.

Microshoots bear bundle-sheath scales and needles; growth in diameter and at the end of the shoot usually ceases after the

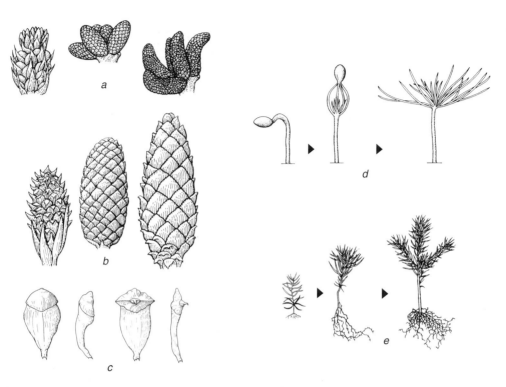

a. Pollen cone development; late winter (left), early spring (center), at time of pollination (right). *b.* Seed cone development; conelet at time of pollination (left), end of 1st
year (center), mature seed cone in 2nd year (right). *c.* Cone scales (outer surface and
profile view); soft pine with smooth, terminal umbo (left), hard pine with dorsal umbo
bearing sharp prickle (right). *d.* Seedling development; emergence of seedling following germination (left), with cotyledons emerging from clinging seed coat (center), with
cotyledons and juvenile leaves (right). e. Seedling development (continued); seedling
at end of 1st year (left), at end of 2nd year (center), at end of 3rd year (right).

1st season; increase in length occurs at the
cambium of the major shoot, which prevents
the microshoot being buried by growth layers
as happens to the base of other side
branches.

If a major shoot is severed, microshoots
immediately behind the stump generate a
winter bud, which will develop into a major
shoot the following spring.

Pollen Cones Catkin-like, yellow, orange, or
red; in clusters toward the base of the new
shoot, usually in the lower part of the crown.
Pollen shed in late spring; cones then wither
and fall away. Pollen cones and seed cones
often on the same tree, but rarely on the
same shoot.

Seed Cones Conelets usually borne in the
upper part of the crown near the tip of new
shoots. Pollination takes place in the spring
when the new shoot has almost finished
elongating and the needles are still enclosed
in the bundle sheath; fertilization does not
take place until about 13 months later. Cones
mature after 2 seasons.

Mature cones woody; scales numerous,
tightly appressed, each with a figured
protuberance called the **umbo**; occur at
nodes, forming a component of a whorl of
side branches.

Seeds Dry, hard, usually winged.

Seedlings With 3–15 cotyledons; new
shoot arises between the cotyledons bearing
primary leaves, narrow, tapering, pointed,
flattened, flexible, clasping the stem. Growth
ends with a rosette of primary leaves, with or
without a terminal bud.

Growth in the 2nd year is expansion of
the rosette (and bud, if present); there is no
prominent annual node marked by bud
scales or by a whorl of side branches. The
new shoot is clothed with primary leaves
which formed in the 1st year (plus scale

jack pine

red pine

pitch pine

Principal branches in regular whorls without lesser branches between.

Pollen cones (above) catkin-like. Seed conelets (below) erect, near the tip of new shoots.

leaves from the bud, if present). Needles in bundles develop along the new shoot in the axils of primary leaves and scale leaves; but are often long and flexible compared with the mature needles. Growth ends with a terminal bud and a whorl of subterminals. Third-year shoots are much like the adult form.

Vegetative Reproduction   By sprouts from latent buds in a few species.

Bark   Smooth when young, with age becoming fissured, with scales or plates.

Wood   Close-grained, soft and uniform in soft pines, hard with prominent latewood in hard pines; longitudinal resin ducts visible with a hand lens.

Size and Form   Vary in size. Eastern white pine is the tallest tree in the eastern forests. On young trees, trunk distinct to the top, crown conical. On mature trees, trunk dissolves in the upper part of a rounded crown. Branches in whorls at nodes; no branches between whorls.

Habitat   Characteristic of sandy well-drained soils, but grow better on richer soils. Generally intolerant of shade, requiring full sunlight for good growth.

Notes   An important source of lumber and wood pulp; improved varieties have been selected and bred to be set out in plantations for wood production. Used extensively in landscaping. Because they retain their needles longer than other conifers, often used for Christmas trees and other decorations. Source of food and shelter for many birds and mammals.

Quick Recognition   Needles in bundles of 2, 3, or 5; seed cones at a node, forming a component of a whorl of side branches; all principal branches in whorls at the nodes of the trunk; no branches between whorls.

Eastern White Pine

Northern white pine, Weymouth pine

Pinus strobus L.

Pin blanc

A soft pine; the tallest tree in eastern Canada.

3

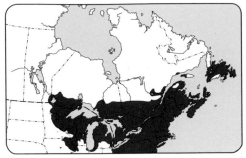

Leaves Needles evergreen, remaining on tree for 1–4 years; in bundles of 5, 5–15 cm long, slender, straight, triangular in cross section, flexible, soft, bluish-green with lines of white dots; edges finely toothed. Bundle-sheath scales deciduous during the 1st season.

Buds Slender, up to 15 mm long, sharp-pointed, with overlapping reddish-brown scales.

Twigs Green and hairy, becoming hairless and orange-brown during the 2nd season. Scars of microshoots round or oval, set in a crescent-shaped leaf-cushion.

Seed Cones Cylindrical, 8–20 cm long, yellowish-green to light brown when mature, pendulous, with a stalk 2 cm long; scales 50–80, usually in 5 spiral rows, thin and rounded at the tip, without prickles. Soon after cones mature, seeds are shed and the cones fall. Good seed crops begin at about 20–30 years and occur every 3–5 years.

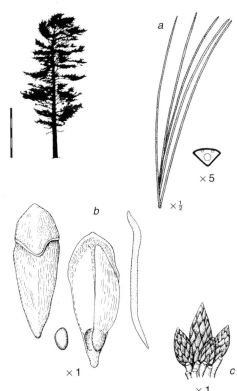

Seeds Reddish-brown, mottled, 5–8 mm long; wing about 20 mm long. Dispersed by wind and seed-eating animals. Require cool moist conditions for some weeks before germination.

Seedlings With 6–11 cotyledons, minutely toothed, 20–30 mm long.

Bark Thin, smooth, grayish-green when young; with age becoming dark grayish-brown and broken into broad scaly ridges 2–5 cm thick, separated by deep longitudinal furrows.

a. Needle bundle and needle cross section. *b.* Cone scale outer surface (left), seed and cone scale inner surface with winged seed (center), cone scale profile (right). *c.* Terminal and subterminal buds.

Wood Soft, light, moderately strong, creamy-white to yellow, straight-grained, easily worked; heartwood moderately decay-resistant.

Size and Form Medium-sized trees, up to 30 m high, 100 cm in diameter, and 200 years old; occasionally larger and older. Young trees: crown conical, with regular whorls of branches. Mature trees, crown irregular, with a few long, stout branches set roughly at right angles to the trunk; branches in upper crown ascending, giving a broadly oval flat-topped outline that often becomes one-sided because of the prevailing wind. Forest-grown trees: trunk often branch-free for lower two-thirds of its height; crown co-lumnar. Root system with 3–5 moderately deep lateral roots, often sinker roots growing down from them.

Closed seed cones yellowish-green, slender.

Young bark (inset) thin, smooth. Mature bark has broad scaly ridges.

Habitat Occurs on a variety of sites, from dry sandy soils and rocky ridges to sphagnum bogs; grows best on moist, sandy loam. Usually mixed with other species. Rapid-growing; thrives in full sunlight; seedlings moderately shade-tolerant; can survive under an open canopy and attain full vigor if the shade is removed within 20 years.

Notes Planted widely for forestry purposes. The most valuable softwood lumber in eastern Canada; used for patterns because of its low shrinkage and uniform texture; also for doors, moldings, trim, siding, paneling, plywood, furniture, and cabinetwork. Reserved in colonial times for Royal Navy shipmasts. Source of food and shelter for wildlife. Used in landscaping.

Attacked by the white pine weevil (*Pissodes strobi*), which kills the leading shoot, resulting in a deformed trunk when a side branch takes over the role of leader; partially shaded trees are less likely to be infested with the weevil. Many trees have been killed by white pine blister rust (*Cronartium ribicola*); can be partially controlled by eliminating the alternate hosts in its life cycle, gooseberry and currant bushes; individual trees can be saved by removing infected branches.

Older thick-barked trees survive most fires and provide a seed supply for the new stand. Fire favors the survival of the species by preparing a favorable seedbed and eliminating competing plants.

The name "Weymouth pine" comes from Lord Weymouth, who planted eastern white pine on large areas of his estate in Wiltshire, England, during the 18th century.

Eastern white pine is the provincial tree of Ontario.

Quick Recognition The only 5-needled pine native to eastern North America.

Western White Pine

Mountain white pine, Idaho white pine, silver pine

Pinus monticola Dougl. ex D. Don

Pin argenté

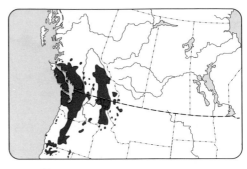

A soft pine of western North America.

Leaves Needles evergreen, remaining on tree for 3–4 years; in bundles of 5, 5–10 cm long, slender, straight, triangular in cross section, flexible, soft, bluish-green with lines of white dots; edges finely toothed. Bundle-sheath scales deciduous during the 1st season.

Buds Slender, up to 10 mm long, blunt-pointed, with overlapping, close-fitting, brownish scales.

Twigs Green and hairy, becoming hairless and brownish during the 2nd season. Scars of microshoots round or oval, set in a crescent-shaped leaf-cushion.

Seed Cones Cylindrical, 10–30 cm long, slightly curved, pendulous, with a stalk 2 cm long; scales 90–160, often reflexed when dry, in spiral rows, thin-tipped, without prickles. Soon after the cones mature, seeds are shed; cones fall during the winter. Good seed crops begin at about 70 years and occur every 3–4 years.

Seeds Brown, mottled, 5–7 mm long; wing 18–26 mm long. Dispersed by wind and seed-eating animals. Require cool moist conditions for some weeks before germination.

Seedlings With 6–10 cotyledons, minutely toothed near their bases, 16–30 mm long.

Bark Thin, smooth, grayish-green when young; with age becoming dark gray to nearly black and broken into small, rectangular to hexagonal, scaly plates 2–4 cm thick, separated by deep longitudinal furrows and horizontal crevices.

Wood Soft, light, moderately strong, non-resinous, creamy-white to yellow; heartwood moderately decay-resistant.

Size and Form Very large trees, up to 50 m high, 150 cm in diameter, and 400 years old. Forest-grown trees: crown slender, columnar, with short whorled branches; trunk branch-free up to 25 m from the ground, with little taper. Open-grown trees: crown sometimes wide and one-sided. Root system wide-spreading, with a few vertical roots.

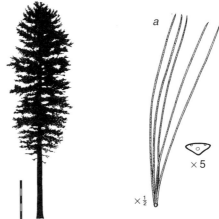

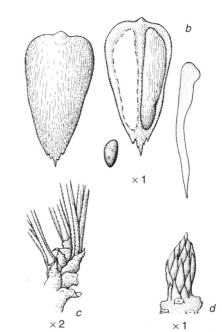

a. Needle bundle and needle cross section. *b.* Cone scale outer surface (left), seed and cone scale inner surface with winged seed (center), cone scale profile (right). *c.* Microshoots in axils of scale leaves. *d.* Terminal bud.

Open seed cones cylindrical.

Young bark (inset) grayish-green. Mature bark has many scaly plates.

Habitat   Thrives on a wide variety of sites, from peat bogs to dry sandy soils and rocky earth; grows best in moist valleys and on gentle slopes. Usually mixed with other species. Moderately shade-tolerant.

Notes   A soft even-grained wood used in carving and for items such as window sashes and frames, doors, patterns, siding, paneling, trim, and wooden matches.

Individual trees susceptible to damage by forest fires, but fire usually favors perpetuation of the species in a given locality by preparing a favorable seedbed and eliminating competing plants.

Susceptible to white pine blister rust (*Cronartium ribicola*); attacked by various bark beetles. Pole blight, a physiological disorder related to drought, results in yellow foliage, dead tops, and the death of many trees.

Named by David Douglas in 1831 during his journey of exploration up the west coast of North America.

Nursery-grown seedlings well-suited for transplanting.

Quick Recognition   The only 5-needled pine that grows at low elevations in the mountains of western Canada; near the coast it reaches higher elevations.

Whitebark Pine
Scrub pine

Pinus albicaulis Engelm.

Pin à blanche écorce

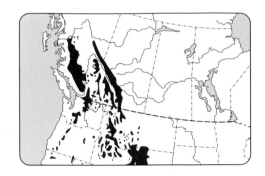

A slow-growing, long-lived soft pine of the mountains of British Columbia and Alberta; grows from 1000 m to the tree line.

Leaves Needles evergreen, remaining on the tree 4–8 years; in bundles of 5, 4–9 cm long, stout, stiff, slightly curved, dark yellowish-green; edges smooth; clustered towards the ends of the twigs. Bundle-sheath scales deciduous during the 1st season.

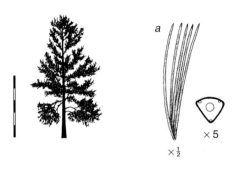

Buds Ovoid, pointed, 5–10 mm long, brown to reddish-brown, with overlapping loose scales.

Twigs Stout, reddish-brown, hairy; becoming gray and smooth during the 2nd season.

Seed Cones Globular to ovoid, 5–8 cm long, stalkless, at right angles to the branch; scales 30–50, thick, tough, pointed, without prickles; cones open only slightly at maturity, but fall and are easily torn apart by birds and squirrels, releasing the seeds. Seed crops at irregular intervals.

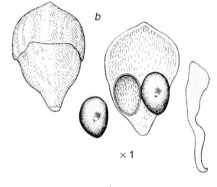

Seeds Large, about 10 mm long, wingless, heavy seed coat; germinate with difficulty.

Seedlings With 8–10 cotyledons, toothless.

Bark Smooth, chalky-white when young; with age broken into narrow brown scaly plates 1 or 2 cm thick; inner bark reddish-brown.

Wood Moderately soft, light; heartwood pale brown; sapwood narrow, white.

Size and Form Varies according to growing conditions. At high elevations near the tree line, a spreading, prostrate shrub, often with a few upright, windblown branches that project above the snow line. At lower elevations, a small, upright multistemmed tree. In favorable locations, a single-stemmed tree up to 20 m high, 50 cm in diameter, and 500 years old, with upswept branches. Root system deep, spreading through the rocky substrate. Windfirm.

a. Needle bundle and needle cross section. *b.* Cone scale outer surface (left), seed and cone scale inner surface with wingless seed (center), cone scale profile (right). *c.* Terminal bud.

Habitat Grows on rocky soils, rock ledges, and cliff faces; requires a moist climate; often occurs in widely scattered clumps. Moderately shade-tolerant.

Seed cones stalkless, globular to ovoid.   Young bark grayish-green.

Trunk varies, often multistemmed.

Notes Forest fires favor survival and regeneration over fir and spruce.

Seriously damaged by mountain pine beetle (*Dendroctonus ponderosae*) and white pine blister rust (*Cronartium ribicola*); some disease-tolerant individuals have been found.

Important for wildlife habitat, watershed protection, and aesthetics in the high mountains.

Quick Recognition Similar to limber pine: both occur at high elevations in the mountains of British Columbia and Alberta; needles in bundles of 5; small trees with irregular form and large branches. Identification is based on cone features [those of limber pine in brackets]: globular to ovoid [cylindrical], 5–8 cm long [8–20 cm], cone scales 30–50 [40–70], thick [slightly thickened at the tip], remain closed [spread open to release the seeds].

Limber Pine

Rocky Mountain white pine

Pinus flexilis James

Pin flexible

A slow-growing, long-lived soft pine of southern British Columbia and Alberta; found above 1000 m to the tree line. Similar to whitebark pine (see Quick Recognition of that species for a comparison); best distinguished by the cones.

Leaves   Needles remaining on the tree for 5–6 years.

Twigs   Greenish-brown, hairy becoming smooth and gray.

Seed Cones   Cylindrical to narrowly ovoid when closed, 8–20 cm long, very short-stalked, more or less at right angles to the stem; scales 40–70, slightly thickened at the tips, spoon-shaped, without a prickle, often with reflexed tips, opening on the tree at maturity to release the seeds; cones shed during the winter. Good seed crops every 2–4 years.

Seedlings   With 6–10 cotyledons, toothless, 25–35 mm long.

Bark   Smooth, pale gray when young; with age becoming thick, rough, dark brown, with wide scaly plates.

Wood   Moderately soft, light; heartwood yellow; sapwood nearly white.

Size and Form   Small trees, up to 12 m high and 60 cm in diameter. Slow-growing, but can live for several hundred years. On mature trees, trunk short, thick, markedly tapered, usually crooked; crown irregular, extending over most of the tree's length. Young branches very tough, flexible, hence the name "limber" pine. Old branches tend to droop, tips upturned. Very old trees sometimes with lower branches longer than the height of the tree.

Habitat   Occurs mainly as single trees or in small open groves on dry rocky exposed sites, but grows on a variety of soils. In southern Alberta, it forms forest outliers on rocky outcrops along the edge of the prairie.

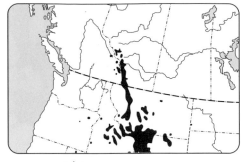

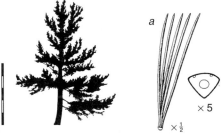

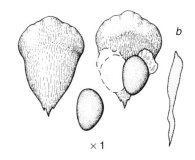

a. Needle bundle and needle cross section. *b.* Cone scale outer surface (left), seed and cone scale inner surface with winged seed (center), cone scale profile (right). *c.* Terminal and subterminal buds.

Notes   Often provides the only tree cover for wildlife on exposed sites at high elevations; useful for the reforestation of snow avalanche paths, for site protection generally, and for increasing the snowpack.

3

Seed cones short-stalked, cylindrical.

Limber Pine

Young bark (inset) pale gray. Trunk usually crooked; mature bark scaly.

Bristlecone Pine*
Colorado bristlecone pine

Pinus aristata Engelm. var. *aristata*

Pin aristé

A small slow-growing soft pine; native to the mountains of Colorado and adjacent states; very long-lived, up to 5000 years old. Irregular in form and picturesque; frequently planted for landscape purposes and for dendrochronological studies. Hardy as far north as Zones C3, NA4.

Needles evergreen, remaining on the tree up to 20 years; in bundles of 5, 2–4 cm long, curved, dark green with a bluish-white cast from resin, tufted toward the tip of the shoot; bundle-sheath scales curling back to form a small rosette, persistent for 2–4 years.

Buds ovoid, up to 8 mm long. **Twigs** light orange, becoming almost black. **Seed cones** ovoid, 4–8 cm long, blunt-tipped, grayish-brown, stalkless; scales with a long slender prickle. **Bark** green, smooth; with age becoming reddish-brown, ridged, fissured. Can grow on poor dry soils, but is intolerant of shade.

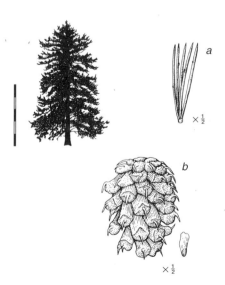

a. Needle bundle. *b.* Mature seed cone (left), winged seed (right).

Ponderosa Pine

Western yellow pine

Pinus ponderosa Dougl. ex P. & C. Laws.

Pin ponderosa

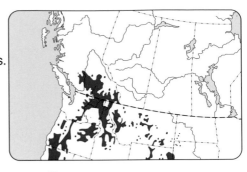

A hard pine; characteristic tree in the southern part of interior British Columbia.

Leaves Needles evergreen, in bundles of 3 (occasionally in 2's or 5's); 12–25 cm long, straight, stiff but flexible, very sharp, dark yellowish-green; edges sharply toothed. Bundle-sheath persistent.

Buds Sharp-pointed, 15–20 mm long, usually resinous.

Twigs Stout, hairless, yellowish-green; changing to orange-brown during the 2nd season.

Seed Cones Cylindrical to narrowly ovoid when closed, 7–15 cm long, lustrous reddish-brown, pendulous, almost stalkless, often in groups of 3; scales thickened at the tips, bearing rigid sharp prickles, opening at maturity to release the seeds. Cones usually fall during the winter, leaving their stalks with a few basal scales attached to the branchlets. Seed crops every 2–5 years.

Seeds Dark brown, mottled, about 7 mm long, with a boat-shaped terminal wing about 20 mm long.

Seedlings With 5–12 cotyledons, toothless, occasionally sparsely toothed on midvein near base, 25–60 mm long.

Bark Dark gray, rough and scaly when young; with age becoming orange-brown and deeply fissured into large, flat, flaky plates 2–10 cm thick.

Wood Uniform in texture, moderately strong and hard; heartwood yellowish to reddish-brown, moderately decay-resistant; sapwood very wide, nearly white to pale yellow.

Size and Form Large trees, up to 35 m high, 100 cm in diameter, and several hundred years old; occasionally 50 m high. Trunk straight, with little taper, often branch-free for most of its length. Crown wide, irregularly cylindrical, flat-topped in old trees; branches stout, lower ones often drooping, upper ones on old crowns ascending. Root

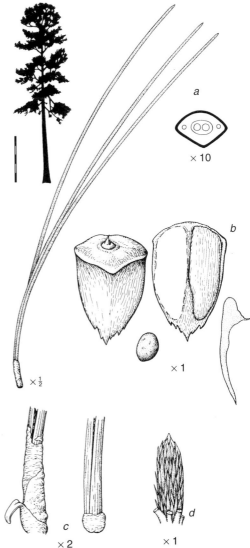

a. Needle bundle and needle cross section. *b.* Cone scale outer surface (left), seed and cone scale inner surface with winged seed (center), cone scale profile (right). *c.* Bundle sheath in 1st year (left), 2nd year (right). *d.* Terminal bud.

Seed cone; scales have sharp prickles.

Young bark rough, scaly.

Mature bark develops large flaky plates.

system very wide-spreading; often with a deep, massive taproot up to 2 m long. Very windfirm.

Habitat Occurs on a wide variety of soils. Grows in pure open stands, especially at lower elevations on areas subject to frequent forest fires; also in mixed stands with Douglas-fir and western larch at elevations up to 1500 m. Intolerant of shade.

Notes Wood is used for window sashes and frames, siding, door moldings, paneling, patterns, cabinetwork, boxes, and crates.

Seeds provide food for birds and small mammals; leaves, twigs, and bark browsed by larger mammals.

Bark beetles can cause severe damage. Fire-resistant because of its thick bark.

Quick Recognition A western species with long needles in bundles of 3.

Pitch Pine

Pinus rigida Mill.

Pin rigide

An eastern hard pine; in Canada restricted to sites along the St. Lawrence River. Dependent on forest fires for reproduction.

Leaves Needles evergreen, in bundles of 3; 7–12 cm long, stiff, twisted, blunt-tipped, yellowish-green; edges sharply toothed. Bundle-sheath persistent.

Buds Sharp pointed, about 1 cm long, reddish-brown, often resinous, with loose, overlapping scales. On vigorous shoots usually polycyclic. Some buds remain dormant in the bark for many years, sprouting forth when the tree is damaged by fire.

Twigs Stout, hairless, greenish to orange or dark brown, ridged and grooved; often in clusters along the trunk. Vigorous shoots usually bear 1 or more loose whorls of side branches at intermediate nodes; terminal and subterminal buds on such shoots often develop precociously during the growing season, forming a seasonal node.

Seed Cones Narrowly ovoid, 5–9 cm long, short-stalked; scales thickened at the tips, bearing a rigid curved sharp spine. Cones may open at maturity; remain closed until opened by fire; or open at irregular intervals. Open cones remain on the tree for many years. Seed production may begin as early as 3 years; good seed crops every 4–9 years.

Seeds 3-angled, 4–5 mm long; wing 15–20 mm.

Seedlings With 4–8 cotyledons, toothless.

Vegetative Reproduction Able to sprout from stumps and to develop new shoots from dormant buds on the trunk if the leaves are killed by fire.

Bark Reddish-brown, smooth, becoming scaly; with age furrowed into large, thick, irregular, flat-topped, dark gray plates.

Wood Coarse-grained, resinous.

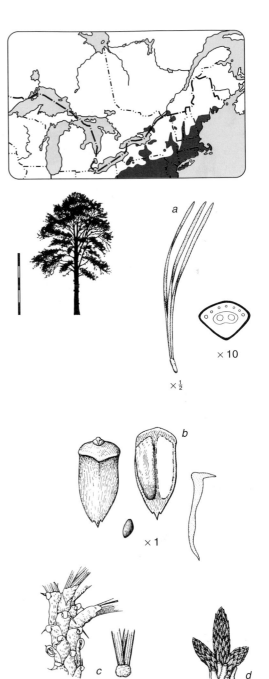

a. Needle bundle and needle cross section. *b.* Cone scale outer surface (left), seed and cone scale inner surface with winged seed (center), cone scale profile (right). *c.* Bundle sheath and scale leaves in 1st year (left), bundle sheath in 2nd year (right). *d.* Terminal and subterminal buds.

Seed cones short-stalked, remain on tree for many years.

Young bark becomes scaly, reddish-brown.

Trunk often bears tufts of needles; mature bark has dark gray plates.

Size and Form   Small trees, up to 20 m high, 30 cm in diameter, and 200 years old. Form is variable. Open-grown trees on exposed sites: crowns irregular; branches many, gnarled, drooping, often dead and covered with clusters of old weathered cones. Forest-grown trees on favorable sites: trunks straight with little taper; crowns regular. Trunks often with clusters of closely packed leafy branchlets, bearing persistent seed cones that appear to be attached directly to the trunk. Root system deep, to 3 m, able to live and grow below the water table.

Habitat   Frequently found on harsh sites such as dry sand-plains, gravelly slopes, rocky ridges, and swamps. Grows in pure stands on areas subject to repeated forest fires; also mixed with other species in open stands. Intolerant of shade.

Notes   Persists in the presence of frequent forest fires that would eliminate other pines because it can produce new shoots from dormant buds in the bark, bear seed cones at an early age, and on certain trees, retain cones with live seeds; seedlings and small trees often with crooks near the base in which dormant buds are protected from fire.
   Useful in reforesting bare sandy land.

Quick Recognition   The only 3-needled pine in eastern Canada; cones with sharp spines; trunks with clusters of branchlets.

Red Pine
Norway pine

Pinus resinosa Ait.

Pin rouge

An eastern hard pine, ranging from south-eastern Manitoba to Newfoundland. A common tree of forest plantations in eastern Canada; uniform in growth and form; little genetic variation.

Leaves   Needles evergreen, in bundles of 2, 10–16 cm long, straight, brittle (breaking in half when bent), pointed, shiny dark green; edges finely and sharply toothed. Bundle-sheath persistent. Dry scale leaves remain at the annual nodes.

Buds   Sharp-pointed, up to 2 cm long, reddish-brown, resinous, with overlapping, loose, hairy scales.

Twigs   Stout, orange to reddish-brown, shiny, grooved and ridged.

Seed Cones   Ovoid, 4–7 cm long, almost stalkless; scales slightly thickened at the tips, without prickles, opening in autumn at maturity to release the seeds. Cones are shed the following year, sometimes leaving a few basal scales on the branchlets. Seed production begins at 15–25 years; good seed crops every 3–7 years.

Seeds   Dull, often mottled, about 5 mm long; wing about 15 mm long.

Seedlings   With 5–10 cotyledons, toothless, 15–25 mm long.

Bark   Reddish to pinkish, scaly; with age, furrowed into broad, flat, scaly plates.

Wood   Relatively light, moderately hard, straight-grained, pale brown to reddish-brown; latewood prominent; sapwood wide.

Size and Form   Medium-sized trees, up to 25 m high, 75 cm in diameter, and 200 years old; occasionally larger and older. Trunk slender, straight, distinct to the tip on young trees; dead branches soon fall leaving trunk clear. Crown conical on young trees, becoming irregular and flat-topped. Upper principal branches upcurved; lower ones horizontally spreading or somewhat drooping, with the foliage crowded towards the tips. Root system moderately deep, wide-spreading; sometimes with a taproot; lateral roots with sinkers. Windfirm.

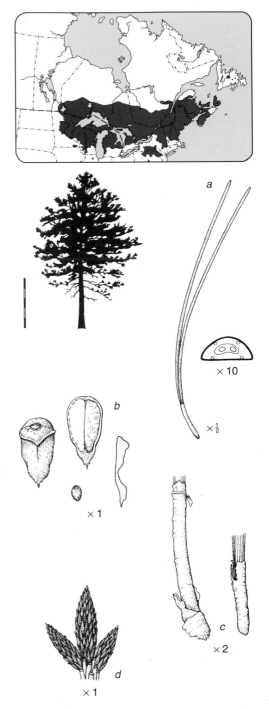

a. Needle bundle and needle cross section. *b.* Cone scale outer surface (left), seed and cone scale inner surface with winged seed (center), cone scale profile (right). *c.* Bundle sheath in 1st year (left), 2nd year (right). *d.* Terminal and subterminal buds.

Seed cones open in autumn, shed
the following year.

Young bark scaly, reddish
to pinkish.

Trunk slender, straight, and clear of branches.

Habitat Occurs most often on sand plains, rock outcrops, and sites where soil fertility is low; in pure stands or mixed with species such as eastern white pine, jack pine, and the aspens. Intolerant of shade.

Notes Wood readily penetrated by preservatives; used for poles and piling. Good structural timber.

Most natural stands originate after a forest fire; fire prepares a seedbed by removing much of the humus, reduces competition from other trees and shrubs, decreases the number of cone-destroying insects, and thins out the overstory.

An attractive tree in recreation areas because of its colorful bark and occurrence in open park-like stands. Useful for windbreaks, snowbreaks, and watershed protection. Does not thrive under urban conditions.

Quick Recognition The only native 2-needled pine in eastern Canada with long needles (needles of Austrian pine as long, but stiffer, very sharp-pointed, bend without breaking apart).

Jack Pine

Banksian pine, gray pine, scrub pine

Pinus banksiana Lamb.
[syn. *P. divaricata* (Ait.) Dum. Cours.]

Pin gris

A hard pine, characteristic of the northern forests; the most widely distributed pine in Canada.

3

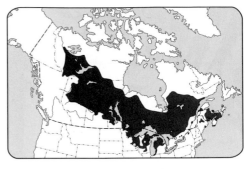

Leaves Needles evergreen, in bundles of 2, 2–4 cm long, straight or slightly twisted, stiff, sharp-pointed, light yellowish-green, spread apart; edges toothed. Bundle-sheath persistent.

Buds Blunt-pointed, up to 15 mm long, pale reddish-brown, resinous. On vigorous shoots often polycyclic.

Twigs Slender, yellowish-green, becoming dark grayish-brown during the 2nd season; ridged and grooved. Vigorous shoots on young trees usually with 1 or more intermediate nodes bearing loose whorls of side branches. Precocious development of terminal and/or subterminal buds may occur during the new growing season.

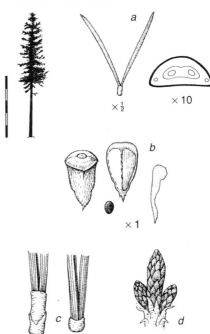

Seed Cones   Variable in shape, oblong to conical, asymmetrical, straight or curved inward, 3–7 cm long, yellowish-brown when mature, stalkless, usually pointing forward, often in clusters of 2 or 3 at nodes; usually remaining closed and persistent on the tree for 10–20 years. Scales 80, thickened at the tips, smooth (sometimes with a minute prickle), held closed by a resin bond; opening when exposed to heat from a wildfire or sometimes from direct sunlight on warm days. Seed crops begin at 5–10 years and occur in most years thereafter.

a. Needle bundle and needle cross section. *b.* Cone scale outer surface (left), seed and cone scale inner surface with winged seed (center), cone scale profile (right). *c.* Bundle sheath in 1st year (left), 2nd year (right). *d.* Terminal and subterminal buds.

Seeds Dark brown or black, 3 mm long, often ridged; wing about 10 mm long.

Seedlings With 3–6 cotyledons, toothless, 12–24 mm long.

Bark Thin, reddish-brown to gray when young becoming dark brown, flaky; with age furrowed into irregular thick plates.

Wood Moderately hard and heavy, weak, light brown.

Size and Form Small trees, up to 20 m high, 30 cm in diameter, and 150 years old; occasionally larger. Open-grown trees: trunk tapered; crown conical, open; branches ascending, arching. Forest-grown trees; slender, straight, with little taper; crown short. On poor soils and rocky sites, the tree is short, often twisted, with long stout branches, some of them dead or dying, giving the crown an unkempt appearance. Root system wide-spreading, moderately deep; often with a taproot.

Seed cones vary in shape; usually point forward on branch.

Young bark becomes flaky, dark brown.

Mature bark thick, with furrows and irregular plates.

Habitat Occurs on poor quality sites such as coarse sands, shallow soils, and rock outcrops, even on permafrost. Grows in pure stands or mixed with other shade-intolerant species such as white birch, trembling aspen, balsam poplar, red pine, and tamarack. Shade-tolerant species such as black spruce, white spruce, and balsam fir often form the understory and succeed as the dominant species in the absence of fire.

After extensive fires, jack pine may occupy all sites from ridge tops to muskegs. Intolerant of shade.

Notes Wood used in general construction; for wood pulp, railway ties, poles, pilings, and mine timbers.

Fire releases live seeds in the persistent cones, makes a favorable seedbed, and removes competing vegetation. Repeated fires at intervals of less than 15 years can, however, eliminate jack pine locally by destroying the seed supply.

Susceptible to rust diseases (*Cronartium* sp.), which induce woody galls that encircle the branches; seldom cause serious damage. Young trees often deformed by browsing of deer and other mammals.

Kirtland's warbler (*Dendroica kirtlandii*), an endangered species, nests only in pure young jack pine stands up to 6 m high and larger than 30 ha.

Fossil evidence indicates that jack pine survived the glacial period in the Appalachian and Ozark mountains.

In Alberta, where the range of jack pine overlaps that of closely related lodgepole pine, hybrids occur, usually with features typical of either parent and some that are intermediate. Controlled crossings are successful; progeny are fertile.

Quick Recognition The only native 2-needled pine in eastern Canada with short needles (Scots pine with longer, bluish, twisted needles, not spread apart in their bundles; see also lodgepole pine).

Lodgepole Pine

Rocky Mountain lodgepole pine, black pine

Pinus contorta Dougl. ex Loud.
var. *latifolia* Engelm.
[syn. *P. murrayana* Balf.]

Pin tordu latifolié

In Canada, the species *Pinus contorta*, a hard pine, has 2 varieties: lodgepole pine, a tall straight inland tree distributed from Yukon through interior British Columbia, into western Alberta, and southward; and shore pine, a short scrubby tree confined to a relatively narrow strip along the Pacific coast.

Leaves   Needles evergreen, in bundles of 2, 3–7 cm long, usually twisted, stiff, very sharp pointed, dark green to yellowish-green, not spread apart; edges sharply toothed. Bundle- sheath persistent.

Buds   Blunt-pointed, up to 15 mm long, reddish-brown, resinous. On vigorous shoots often polycyclic.

Twigs   Orange-brown, becoming reddish-brown or very dark brown during the 2nd season; ridged and grooved. Vigorous shoots on young trees usually with 1 or more intermediate nodes bearing loose whorls of side branches. Precocious development of terminal and/or subterminal buds may occur during the current growing season.

Seed Cones   Short-cylindrical to ovoid, sometimes asymmetrical, 3–6 cm long, purplish-brown, stalkless, at right angles to the branch or pointing back, in small clusters at the nodes, usually closed and on the tree for 10–20 years. Scales thickened at the tips, with a curved prickle, usually held closed by a resin bond; opening when exposed to the heat from a wildfire or from direct sunlight. Seed production begins at 5–10 years; good crops every 1–3 years.

Seeds   Brownish, often mottled and ridged on one side, 3 mm long; wing about 10 mm long.

Bark   Relatively thin, less than 2 cm thick, orange-brown to gray, with fine scales.

Wood   Soft and light to moderately hard and heavy, light yellow to yellowish-brown, tangential surface often prominently dimpled.

Size and Form   Medium-sized trees, up to 30 m high, 60 cm in diameter, and 200 years old. Trunk often straight with little taper,

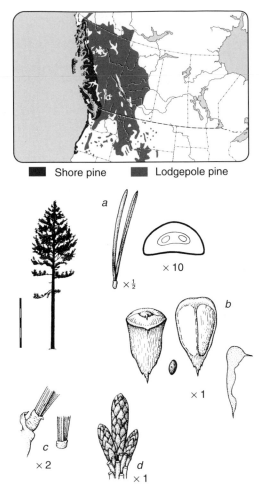

■ Shore pine   ■ Lodgepole pine

a. Needle bundle and needle cross section. *b.* Cone scale outer surface (left), seed and cone scale inner surface with winged seed (center), cone scale profile (right). *c.* Bundle sheath in 1st year (left), 2nd year (right). *d.* Terminal and subterminal buds.

especially in dense stands where the live crown is small. Crown narrowly conical; branches slender, short. Root system a taproot and several spreading lateral roots with vertical sinkers.

Habitat   Occurs on a wide range of soils and sites, including wet depressions. Able to colonize recently burned areas mainly because of an abundant seed supply in the closed cones. Most stands are of fire origin. Large pure stands common; their density sometimes restricting normal growth.

Notes   Important for timber; used in general construction, for wood pulp, and, after treatment with preservatives, for railway ties, poles, and mine timbers.
      Hybridizes with jack pine.

Seed cones at right angles to the branch or pointing back.

Young bark at first smooth, becomes scaly.

Mature bark thin, with many fine scales.

"Lodgepole" derives from use of this tree for poles to support the lodges or tepees of native people. The Cypress Hills in Alberta and Saskatchewan were named after this pine, once commonly called "cypress".

Various bark beetles (*Dendroctonus* sp.) are serious pests; comandra blister rust (*Cronartium comandrae*) forms cankers on the stems that may kill the tree.

> **Quick Recognition** A western species with 2 short needles per bundle. Contrasting features of jack pine in brackets. Needles strongly [moderately] twisted, roughly parallel [spread apart]. Cones set at right angles to the branch or pointing back [pointing forward], with [without] a prickle on the scale. Bark thin [thick], with small thin scales [furrowed with thick scales]. Crown spire-like [conical or irregular].

Shore Pine
Coast pine, beach pine

Pinus contorta Dougl. ex Loud. var. *contorta*

Pin tordu

Features differing from those of lodgepole pine follow. A small (up to 15 m) branchy tree. **Needles** shorter, darker green, stiffer. **Seed cones** reflexed, opening at maturity but sometimes remaining on the tree. **Seeds** larger (3–5 mm long); slower to germinate. **Bark** thicker, becoming deeply furrowed into flat, thick, coarse, dark reddish-brown plates. **Trunk** usually with more taper, often crooked; **crown** often deformed by prevailing winds. Found up to 600 m; on rocky ridges, coastal sand dunes, and in bogs. Not so dependent on fire.

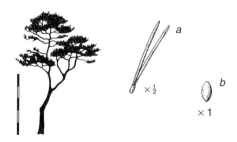

a. Needle bundle. b. Seed.

Scots Pine**

Scotch pine

Pinus sylvestris L.

Pin sylvestre

The most widely distributed of the world's pines; found from western Europe to eastern Asia, from the Arctic Circle to the Mediterranean Sea. A hard pine; one of the first species to be introduced to North America. Frequently planted for landscape purposes, erosion control, and shelterbelts. Hardy as far north as Zones C2, NA2, depending on seed source.

Leaves Needles evergreen, in bundles of 2, 4–8 cm long, twisted, stiff, sharp-pointed, bluish-green to grayish-green; edges finely toothed. Bundle-sheath persistent, 5–8 mm long.

Buds Ovoid, sharp-pointed, 6–12 mm long, reddish-brown, usually non-resinous; tips of some lower scales loose.

Twigs Greenish-brown, becoming grayish-brown in the 2nd season; hairless, ridged and grooved.

Seed Cones Conical to ovoid, often asymmetrical, 2.5–7 cm long, usually in clusters of 2 or 3, pointing back along the stem; scales woody, thick, tips raised and 4-sided, usually without a prickle, bright green in the 2nd spring, maturing and turning purplish-brown by the end of the season. Seeds released slowly during the winter and following spring. Seed production begins at 10–15 years; good crops every 3–6 years.

Seeds Very dark brown, 2–4 mm long; wing brown, 10–15 mm long, separates from the seed and ends in 2 claw-like projections.

Seedlings With 3–8 (often 6) cotyledons, toothless,

Bark On the upper part of the tree, orange-red, smooth, peeling in papery flakes and strips; on the lower part, grayish- to reddish-brown, deeply fissured into irregular, longitudinal, loose scaly plates revealing the brownish-red inner bark.

Size and Form Medium-sized trees in North America, up to 30 m high. Trunk varying from straight and slender with few branches to short and crooked with large branches, depending on seed source, growing conditions, and activity of pests and diseases. Crown nearly conical, becoming rounded or flat-topped with age. Root system moderately deep and wide-spreading, especially on sandy soils.

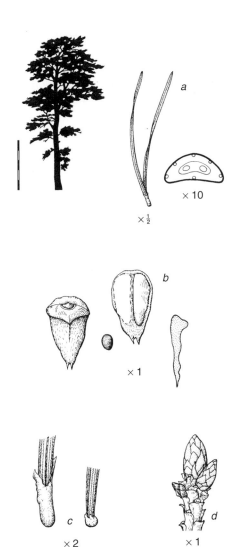

a. Needle bundle and needle cross section. *b.* Cone scale outer surface (left), seed and cone scale inner surface with winged seed (center), cone scale profile (right). *c.* Bundle sheath in 1st year (left), 2nd year (right). *d.* Terminal and subterminal buds.

Seed cone scales raised, 4-sided.

Young bark has bright
orange-red papery flakes.

Mature bark fissured, with loose scaly plates.

Notes In Europe, a tall straight tree with wood of excellent quality. In North America, trunks are seldom straight and the wood quality is poor because of seed source chosen by early settlers. Scots pine can succeed in most locations in the forested parts of Canada and the northern United States, if from a suitable seed source. Intolerant of shade.

A common Christmas tree because of the blue-green color, stiff branches, short densely-spaced needles, and excellent needle retention. Popular with growers because of rapid growth, good response to shaping, and abundant seeds with good storing qualities. However, subject to a great many pests and diseases.

> **Quick Recognition** Needles twisted, blue-green in bundles of 2; bark orange. See jack pine and mugho pine.

Austrian Pine*
European black pine

Pinus nigra Arnold

Pin noir d'Autriche

One of the most common ornamental trees; a hard pine native to southern Europe; many varieties exist, usually named by place of origin. Planted in cities and along highways because of its attractive form, dark green foliage, and tolerance of salt spray, air pollution, and dry soil. Hardy as far north as Zones C4, NA4.

Leaves Needles evergreen, in bundles of 2, 8–16 cm long, straight, stiff (but not breaking cleanly in half when bent), sharp, dark green; edges finely toothed; bundle-sheath persistent, about 10 mm long.

Buds Cylindrical, long-pointed, pale brown, usually white with resin; scale tips loose.

Twigs stout, yellowish-green to greenish-brown, hairless, ridged and grooved.

Seed Cones Ovoid, 5–8 cm long, light shiny brown, stalkless, in clusters of 2–4, at right angles to the branch; scales with a ridge ending in a small prickle.

Seeds Reddish brown, mottled, 6–7 mm long, wing 18–20 mm long.

Seedlings With 6–8 cotyledons, toothless, 25–45 mm long.

Bark dark brown to dark gray, deeply furrowed.

Size and Form Generally under 30 m high in Canada; crown broadly conical with regular whorls of branches, becoming irregular and umbrella-shaped with age.

Notes Easily grown from seed; transplants well. Succeeds on a great variety of soils and sites if a suitable seed source is selected. Intolerant of shade.

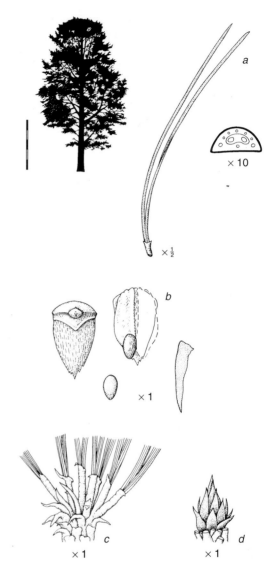

a. Needle bundle and needle cross section. *b.* Cone scale outer surface (left), seed and cone scale inner surface with winged seed (center), cone scale profile (right). *c.* Base of microshoots with bundle sheaths and scale leaves. *d.* Terminal bud.

Seed cones in clusters of 2–4, at right angles to branch.

Austrian Pine

Young bark has yellowish flakes.

Mature bark dark, deeply furrowed.

Mugho Pine*
Swiss mountain pine

Pinus mugo Turra

Pin mugo

Native to the mountains of southern Europe; usually a shrub with upswept irregular branches; some
varieties become small trees. Widely used for
foundation planting in landscape work; useful for
stabilizing steep slopes. Very hardy, surviving as
far north as Zones C2, NA2.

   Needles in bundles of 2, 3–6 cm long,
curved, somewhat twisted, stiff, pointed, dark
green; edges finely toothed; bundle-sheath persistent, up to 10 mm long. **Buds** cylindrical, about
1 cm long, reddish-brown, resinous; scale tips
pressed against the bud. **Twigs** green, becoming
brown in the 2nd season, hairless, with prominent
ridges and grooves. **Bark** dark gray, scaly. **Seed
cones** ovoid, 3–6 cm long, shiny yellowish-brown
to dark brown, stalkless or short-stalked, in small
clusters at annual nodes, almost at right angles to
the branch; scales with a dark brown ring surrounding the prickle at the tip; on lowest scales,
tips turning downward.

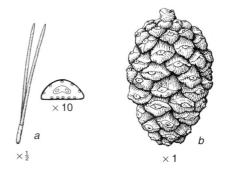

a. Needle bundle and needle cross section.
b. Mature seed cone.

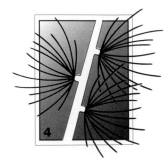

Group 4.

Needles deciduous (or evergreen), in tufts of 10 or more on dwarf shoots, also single on long shoots; seeds in cones

Group 4 contains two genera of the pine family — the larches or tamaracks (*Larix*) and the "true" cedars (*Cedrus*). Species in both genera are characterized by the presence of dwarf shoots bearing a tuft of 10–60 needles; they also have needles on the long (extension) shoots, singly placed as in spruce and fir. The cones are borne on the dwarf shoots.

Larch needles are deciduous and very flexible; the seed cones are ovoid and remain intact on the tree for some months after maturing. Cedar needles are evergreen, stiff, and sharp-pointed; the seed cones are oblong and disintegrate when mature.

Larches/ Tamaracks

Genus *Larix*
Pinaceae: Pine Family

Les mélèzes

The larch genus contains about 10 species distributed throughout the cooler parts of the Northern Hemisphere, from the arctic tree line southward, and from the alpine tree line to sea level; 3 species are native to Canada and North America, the remainder to Europe and Asia. Larches occur in every province and territory of Canada. They are notable for being deciduous conifers.

Leaves Deciduous, 20–50 mm long, needle-like, flexible, stalkless, pale green, with 2 lines of white dots on the lower surface, less conspicuous dots on the upper surface; 2 resin ducts visible near outer margin in cross section. Needles spirally arranged; those on long (or extension) shoots spaced singly along the shoot, at the forward end of a leaf-cushion; also in a cluster at the shoot base; those on dwarf shoots very close together in tufts of 10–60, length variable. Needles on vigorous shoots usually longer than average. Appear earlier in spring than most species; turn bright yellow in autumn.

Buds Terminal bud small, short, rounded, 2–3 mm in diameter; with many pointed scales, non-resinous in most species. Lateral buds smaller, in the axils of some leaves. Each dwarf shoot ends in a terminal bud which may develop into a dwarf shoot, a pollen cone, or a seed cone.

Twigs Moderately stiff and slender; ridged and grooved with leaf-cushions; on leafless branchlets, leaf-cushions terminated by a leaf scar set at right angles to the twig.

Dwarf branches prominent, marked with annual circles of leaf scars; live for several years bearing annual crops of needles; or a seed cone or a pollen cone; can develop into long shoots.

Spring flush of growth (including all dwarf shoots) preformed in a bud; subsequent growth of long shoots neoformed during the growing season. Most side shoots develop from buds on previous year's twig, but sometimes neoformed on vigorous leading shoots. Growth ends for the season with the formation of a terminal bud.

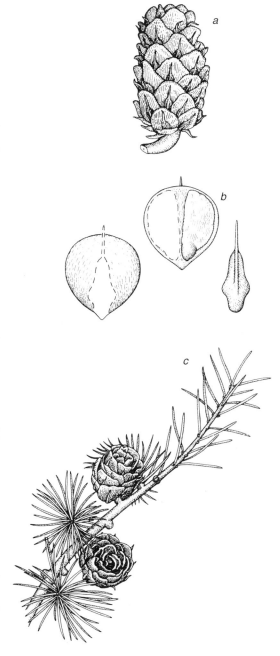

a. Mature seed cone. *b*. Cone scale; outer surface showing bract position (left), inner surface with winged seed (center), bract (right). *c*. Branch, showing singly spaced needles on long shoot, needles in tufts on short shoots.

Pollen Cones Globular or elongated, yellow, catkin-like, at the end of shoots; often on the same branch as seed cones but more abundant in the lower part of the crown. Appear in early spring; wither and fall away after pollen is shed.

Seed Cones Conelets (at pollination) red, green, or yellowish, soft, erect; set in a small cluster of leaves at the tip of a dwarf shoot; often on the same branch as a pollen cone but more abundant in the upper part of the crown; occasionally borne laterally on a long shoot. Scales 20–60; bracts separate, trident-shaped, longer than the scale. Mature in 1 season.
    Mature cones 1–5 cm long, usually erect, remaining on tree for several years.

Seeds Small, 2–3 mm long; wing longer than the seed.

Seedlings Newly germinated seedlings 2–3 cm tall, with a slender stalk surmounted by a whorl of 4–6 linear, curved, green, cotyledons, each about 2 cm long. A new shoot with spirally arranged needles soon forms above the cotyledons; needles sometimes remain green into the 2nd growing season.

Bark Scaly, gray to brown.

Wood Hard, strong, durable, commonly with a spiral grain; latewood prominent; resin ducts visible with a hand lens.

Size and Form Tall slender trees. Trunks straight to sinuous, distinct through the narrow open crown to the tip of the tree. Principal branches slender, irregularly placed along the trunk; lesser branches in between.

Habitat Tolerate cold continental climates; grow on a variety of sites, but best on upland sites. Intolerant of shade.

Notes Larch wood is used for general construction, crates, and pulpwood; especially valuable for structures in contact with water, such as bridges, well cribbing, foundations, and boat timbers. One of the best conifers for firewood. Bundles of branchlets can be fashioned into duck decoys and other objects. Leaves, twigs, bark, and seeds are a source of food for wildlife.
    Amenable to hybridization and selection; among the first trees to be tested in this way for forestry purposes.
    Frequently planted for landscape purposes because of their pleasing form and soft green foliage that turns brilliant yellow in autumn.

Quick Recognition In summer, tufts of soft needles borne on dwarf shoots. In winter, leafless with knobby dwarf shoots on slender branchlets and upright persistent cones.

Photos of pollen cones and seed conelets (opposite page) western larch.

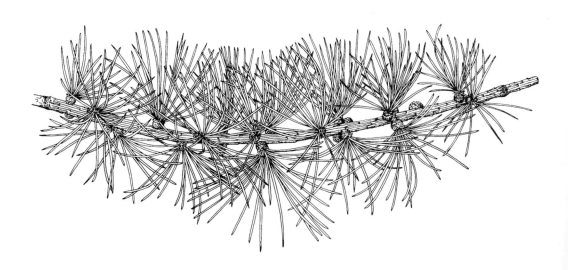

Leaves turn brilliant yellow in autumn.

Larches are leafless in winter.

Pollen cones catkin-like; often on same branch as seed conelets.

Seed conelets soft and erect; at tips of dwarf shoots.

Western Larch
Western tamarack

Larix occidentalis Nutt.

Mélèze de l'Ouest

The world's largest larch; occurs in south-eastern British Columbia, into Alberta.

Leaves Flattened above, keeled below, triangular in cross section, 3–5 cm long, shiny pale green, 15–30 per tuft.

Buds Terminal bud dark brown with fringed scales, downy on dwarf shoots.

Twigs Hairy at first, becoming hairless and orange-brown during the summer.

Seed Cones Conelets 10–20 mm long, red. Mature cones ovoid, 3–5 cm long, reddish-brown; about 30 scales, hairy on the lower side, tips curving toward cone base when the cone is open; bract tip extends beyond the scale. Seed production begins at 15–25 years, abundant by 50 years, often continuing for 200 years and more. Seeds produced in most years, abundantly every 4 or 5 years.

Seeds 5 mm long; wing 8 mm long.

Bark Reddish-brown, scaly when young, becoming thick (up to 15 cm), deeply furrowed with flat flaky ridges.

Wood Heavy, hard, strong; heartwood brown, moderately resistant to decay; sapwood yellowish-white, narrow.

Size and Form Very large trees, up to 70 m high, 200 cm in diameter, and 400 years old. Trunk usually branch-free over much of its length; crown short, narrow, pyramidal. Principal branches usually horizontal but often drooping in the lower crown of open-grown trees. Root system deep, wide-spreading.

Habitat Grows at elevations between 400 and 1500 m, on deep, well-drained, coarsely textured, moist soils, where small pure stands may form; usually mixed with Douglas-fir, western white pine, lodgepole pine, Engelmann spruce, subalpine fir, western hemlock, ponderosa pine.

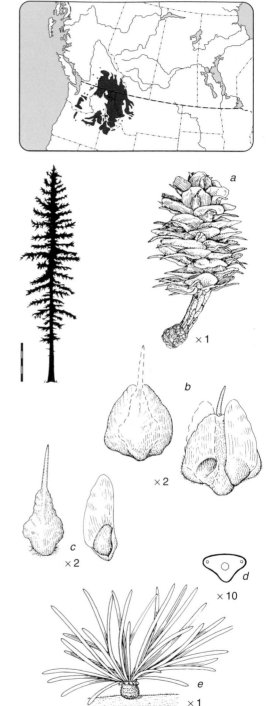

a. Seed cone. b. Cone scale; outer surface showing bract position (left), inner surface with winged seed (right). c. Bract (left); winged seed inner surface (right). d. Needle cross section. e. Dwarf shoot with tuft of needles.

Seed cones have straight bracts extending beyond the scales.

Young bark scaly.

Mature bark thick, deeply furrowed.

Notes The most important native larch for pulpwood production and for lumber; used for building construction, piling, flooring, interior and exterior finishing. Old trees often fire-scarred at the base.

Quick Recognition The largest larch; a western species, usually occurring below 1500 m; twigs slightly hairy; cone bracts projecting beyond the scales. Compare subalpine larch (at higher elevations; small irregular tree with densely hairy twigs) and tamarack (shorter needles, smaller cones with hidden bracts).

Subalpine Larch
Alpine larch, timberline larch

Larix lyallii Parl.

Mélèze subalpin

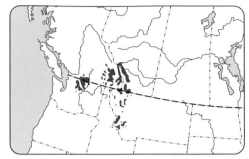

Occurs in the mountains of southern British Columbia and Alberta at elevations of 1200–2200 m, higher than any other tree; often forming the alpine tree line.

Leaves Somewhat stiff, 4-sided in cross section, 4–5 cm long, pale bluish-green, 30–40 per tuft; sometimes remaining green for 2 seasons on trees up to 20 years old.

Buds Scales with a dense fringe of hairs sometimes hiding the bud.

Twigs Stout, tough, densely hairy. Dwarf shoots often several centimeters long, longer than those of other larches; dwarf shoots occur as the new growth on long shoots much more often than in other larches.

Seed Cones Conelets 10–15 mm long, deep purplish-red. Mature cones ovoid, 4–5 cm long, purplish-brown, standing out from the twig in all directions; scales hairy, curving toward the base of the cone; bracts fringed, extending beyond the scales, with tips curving toward base of cone. Seed crops are infrequent; seedlings are rarely found.

Seeds About 3 mm long; wing 6 mm long.

Bark Thin, smooth, gray to yellowish-gray when young; becoming thicker (2–3 cm), with irregularly shaped reddish scaly plates.

Size and Form Small trees, up to 12 m high, 50 cm in diameter, and several hundred years old. Trunk short, sturdy, and tapering rapidly; crown ragged; principal branches irregularly spaced, long, gnarled, thick near the trunk, wide-spreading, often drooping with upturned tips; needles mostly confined to the end of the branch. Dead branch stubs and parts of branches scattered along the trunk. Size and form depend on the growing conditions; stunted under severe conditions, large and handsome under favorable conditions. Deep-rooted, windfirm.

Habitat Grows on acidic gravelly soils in small, open, pure stands above 1500 m to the alpine tree line; mixed with subalpine fir, Engelmann spruce, mountain hemlock, whitebark pine at lower elevations.

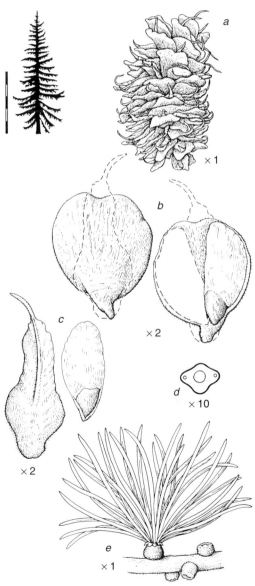

a. Seed cone. b. Cone scale; outer surface showing bract position (left), inner surface with winged seed (right). c. Bract (left); winged seed inner surface (right). d. Needle cross section. e. Dwarf shoot with tuft of needles.

Seed cones; bract tips curve backwards towards base of cone.

Young bark smooth, gray.

Mature bark has irregular scaly plates.

Notes A valuable tree for watershed protection, avalanche control, wildlife habitat, and mountain scenery.

Quick Recognition Small western tree with irregular form, pale green foliage, and densely hairy twigs; occurring above 1500 m.

Tamarack

Hackmatack, eastern larch, American larch, Alaska larch, juniper

Larix laricina (Du Roi) K. Koch

Mélèze laricin

Occurs in every province and territory of Canada.

Leaves Flattened above, keeled below, or triangular or 4-sided, 2–5 cm long, bluish-green, 15–60 per tuft.

Buds Terminal bud dark red or brown, smooth; on dwarf shoots, surrounded by a circle of hairs.

Twigs Hairless, orange-brown to pinkish.

Seed Cones Conelets 5–10 mm long, red, pink, or yellowish-green; occasionally a leafy shoot develops at the tip. Mature cones broadly ovoid, 1–2 cm long, light brown, on stout, short, curved stalks; scales up to 20, smooth; bracts much shorter than the scales, visible only at the base of the cone. Cones begin to open in mid-August; seeds shed during the following months. Seed production begins at about 10 years, peaking at about 75 years. Seeds produced in most years, abundantly every 3–6 years.

Seeds 3 mm long; wing 6 mm long.

Bark Thin, smooth, gray when young, becoming reddish-brown, scaly. Newly exposed bark reddish-purple.

Size and Form Medium-sized trees, up to 25 m high, 40 cm in diameter, and 150 years old. Trunk slender, straight or sinuous; crown narrowly conical, open, becoming irregular with age. Principal branches horizontal or sometimes ascending. Root system shallow, wide-spreading. In nutrient-poor bogs, and near the tree line in the far north and on mountain slopes, trees often stunted with short needles and narrow cone scales.

Habitat Found mainly on cold, wet, poorly drained sites such as sphagnum bogs and muskeg; mixed with black spruce and/or eastern white-cedar. Grows better on moist, well-drained, light soils; mixed with black spruce, white spruce, trembling aspen, and white birch. May occur in pure stands in a narrow band around bogs.

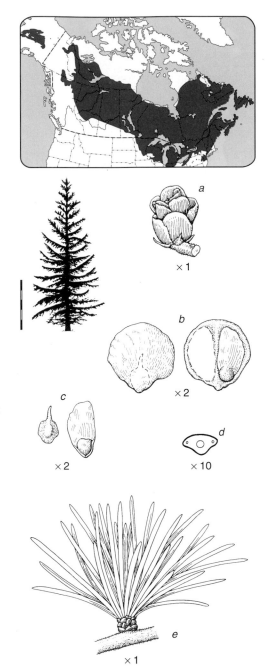

a. Seed cone. *b.* Cone scale; outer surface showing bract position (left), inner surface with winged seed (right). *c.* Bract (left); winged seed inner surface (right). *d.* Needle cross section. *e.* Dwarf shoot with tuft of needles.

Seed cones broadly ovoid, bracts shorter than scales.

Young bark smooth, gray, becomes scaly.

Trunk usually slender and straight.

Notes From time to time tamarack forests have been severely damaged by larvae of the larch sawfly (*Pristiphora erichsonii*), which feed on the foliage. Individual trees are frequently damaged by porcupines (*Erethizon dorsatum*), which eat the bark.

Quick Recognition The only native larch in eastern North America. Needles short and cones small compared with European and Japanese larches.

European Larch**

Larix decidua Mill.
[syn. *L. europaea* DC.]

Mélèze d'Europe

Medium-sized trees, up to 30 m high and 100 cm in diameter; native to Europe; frequently planted in eastern North America for forestry and ornamental purposes; naturalized locally. Hardy as far north as Zones C3, NA2.
   Needles flattened above, keeled below, 2–5 cm long, bright green, 30–55 per tuft. **Buds** light brown, slightly resinous. **Twigs** pale yellow, hairless; dwarf branches darkened by bud-scale remnants. **Seed cones** ovoid, 2–4 cm long, reddish-brown, stalked; 40–60 scales, hairy, broad, rounded, curving slightly inward; bracts about one-half the length of the scales. Vigorous young trees sometimes grow 1 m per year. Principal **branches** long, sweeping downward; secondary branches drooping.

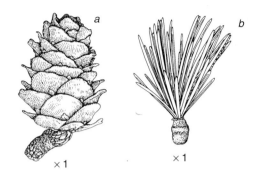

a. Seed cone. b. Dwarf shoot with tuft of needles.

Japanese Larch*

Larix kaempferi (Lamb.) Carrière
[syn. *L. leptolepis* (Siebold & Zucc.) Gord.]

Mélèze du Japon

Medium-sized trees, up to 30 m high; native to Japan, where it occurs in a few mountainous locations with a mild, humid climate. Frequently planted in Canada as an ornamental and some-times for forestry purposes because it grows better than European or native larch in mild maritime climates, especially during the first 20 years. Hardy as far north as Zones C2, NA4.
   Needles flattened above, keeled below, 2–5 cm long, broad, grayish-green to bluish-green on the upper surface, with distinct lines of white beneath; 40–50 per tuft. **Buds** ovoid, reddish-brown, resinous; scales fringed with hairs, tips curving outward. **Twigs** reddish-violet or orange-red, sometimes hairy; branchlets orange-brown. **Seed cones** globular, 2–5 cm long, light brown; about 30 scales, soft, fringed, tips curving toward the cone base; bracts less than one-half as long as the scales; open cones form a rosette. Principal **branches** long, horizontal, upturned at the tips. Requires well-drained soil with plenty of moisture.
   Hybrids between European larch and Japanese larch were first noticed about 1904 in a nursery at Dunkeld, Scotland. Since then, such hybrids (*L. ×eurolepis* Henry) have been produced in abun-dance under controlled conditions; most of their morphological features are intermediate between the parents. These hybrids are useful for forestry purposes because some individuals inherit the good stem form of European larch and the disease

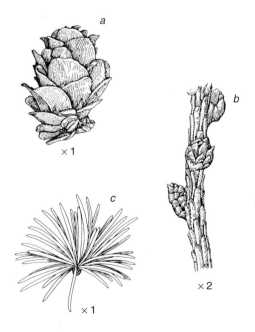

a. Seed cone. b. Twig and buds. c. Dwarf shoot with tuft of needles.

resistance of Japanese larch. In addition, the hy-brids sometimes grow more rapidly than either parent, a phenomenon called heterosis.

4

Siberian Larch*

Larix sibirica Ledeb.
[syn. *L. russica* (Endl.) Sab.]

Mélèze de Sibérie

Medium-sized trees, up to 30 m high; native to northeastern Russia and western Siberia. A hardy species, leafing out early in the spring and continuing to grow until early autumn. Frequently planted in the Prairie provinces. Hardy as far north as Zones C2, NA2.

 Needles keeled below, 2−4 cm long, upper surface green to grayish-blue, with lines of white dots below; 15−30 per tuft; appear earlier in spring than other larches. **Buds** ovoid, brown, darker at the base, resinous. **Twigs** yellowish-gray, hairy in spring, becoming smooth and shiny in summer. **Seed cones** ovoid, tapered to the tip, 2.5−4 cm, on short stalks; about 30 scales, leathery, hairy, wavy-margined, curving inward; bracts shorter than scales. Principal **branches** in upper crown ascending, becoming horizontal with age.

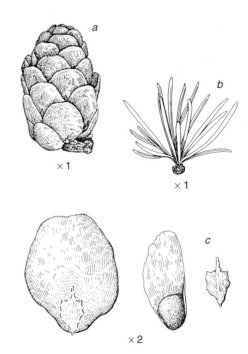

a. Seed cone. *b.* Dwarf shoot with tuft of needles. *c.* Cone scale outer surface showing bract position (left), winged seed inner surface (center), and bract (right).

Dahurian Larch*

Larix gmelinii (Rupr.) Kuzen.

Mélèze de Dahurie

Medium-sized trees, up to 30 m high; native to eastern Siberia; planted in the Prairie provinces. Hardy as far north as Zone C2, NA2.

 Needles flat above, keeled below, about 3 cm long, bright green, with distinct lines of white dots; those on dwarf shoots in a cup-like formation; remaining green late in the season. **Buds** non-resinous, yellowish-brown, darker at the base. **Twigs** hairy, yellowish-brown, reddish in winter. **Seed cones** 2−3 cm long; 20−40 scales; bracts shorter than scales. Principal **branches** long, secondary branches drooping.

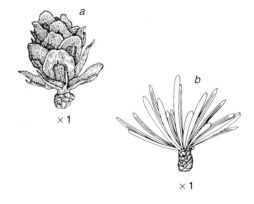

a. Seed cone. *b.* Dwarf shoot with tuft of needles.

Cedars
Genus *Cedrus*
Pinaceae: Pine Family

Les cèdres

About 70 different genera have been called cedar. *Cedrus* is considered the "true" cedar. It is a small genus with 4 species; none is native to North America. The many cultivars are highly regarded as ornamentals and planted in the milder parts of British Columbia.

Leaves Evergreen, remaining on tree for 3–6 years; needle-like, stiff, pointed; lines of white dots on all surfaces; generally triangular in cross section, with a midvein and 2 resin ducts. Needles on long shoots, singly placed; on dwarf shoots, in tufts of 10–60.

Buds Very small, ovoid, brown.

Twigs Downy.

Pollen Cones Similar to the seed cones, but may occur on separate trees; pollen shed in summer or autumn.

Seed Cones Large, oblong, resinous, upright; mature in 2 or 3 years, and then disintegrate, leaving the cone axis terminating the dwarf shoot.

Bark Dark gray, smooth on when young, fissured and scaly with age.

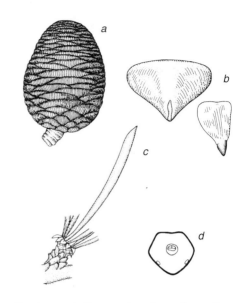

a. Seed cone. *b.* Cone scale outer surface with bract (left); winged seed (right). *c.* Needle and needle bases on dwarf shoot. *d.* Needle cross section.

Size and Form Very large trees, up to 50 m high. Crown broad, irregular; rounded or flat in old trees. Principal branches long, broad, with drooping tips.

Notes Cedar wood has been prized since antiquity; sweet-scented, oily, durable, uniform, easily worked.

Differences between the species are small and are given in the key on page **415**.

Deodar Cedar*
Cedrus deodara (Rox.) Loud.

Cèdre de l'Himalaya

Native to the Himalayas. Hardy in Canada as far north as Zones C8, NA7.

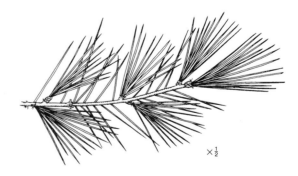

$\times \frac{1}{2}$

Leafy shoot; leaves up to 5 cm long.

Atlas cedar

Cedars will grow in the milder parts of British Columbia.

Atlas Cedar*

Cedrus atlantica (Endl.) G. Manetti
ex Carrière

Cèdre de l'Atlas

Native to the Atlas Mountains of North Africa.
Hardy in Canada as far north as Zones NA6, C7.

$\times \frac{1}{2}$

Leafy shoot; leaves usually less than 2.6 cm long.

Cedar-of-Lebanon*

Cedrus libani A. Rich.
[syn. *C. libanotica* Link]

Cèdre du Liban

Native to Asia Minor; somewhat hardier than Atlas
and deodar cedar. Hardy in Canada as far north as
Zones C6, NA5.

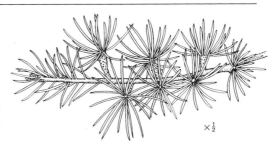

$\times \frac{1}{2}$

Leafy shoot; leaves 2.6–3 cm long.

Group 5.

Needles evergreen, single, flat or 4-sided; seeds in cones

Group 5 comprises four genera of the pine family: fir (*Abies*), spruce (*Picea*), Douglas-fir (*Pseudotsuga*), and hemlock (*Tsuga*). Spruce needles are four-sided; needles of species in the other three genera are flattened, with a prominent midvein. Each needle is singly placed on the twig; successive needles are arranged in spirals around the twig. On upright twigs, the spiral arrangement is quite evident — the needles spread out all around the twig. However, on horizontal twigs, the needles may appear to be two-ranked, or concentrated on the upper side of the twig often with a parting (as with hair) down the middle; such modifications are brought about by twisting and bending at the needle base.

White dots can be seen on most needles; they are exudates of wax or resin from the stomates, microscopic pores in the leaf surface. The number of lines of dots and the surfaces on which they occur are often helpful in identification of species.

All trees in this group have terminal buds and lateral buds scattered along the twig. A bud may develop into a leafy shoot, a seed cone, a pollen cone, or may remain dormant; differentiation occurs in the previous season. As new shoots develop, the terminal shoot is the longest, and side shoots are progressively shorter behind it.

Pollen cones and seed cones are on the same tree (monoecious).

Firs

Genus *Abies*
Pinaceae: Pine Family

Les sapins

Firs are found in the cooler parts of the North Temperate Zone. Worldwide about 40 species are recognized; 9 are native to North America, 4 to Canada. Their single straight trunks, distinct to the tip of spire-like crowns, make them stately trees.

Cone features are essential for the identification of fir species; however, cones are at the top of the tree and disintegrate while still attached; sometimes, cone scales or bracts can be found on the ground beneath the tree. In the following descriptions, features under **Leaves** pertain to needles on well-developed horizontal branchlets, without cones, unless otherwise noted.

Leaves Evergreen, remaining on the tree up to 20 years; flat, flexible, 15−75 mm long; tip blunt or notched; stalkless, but narrowed and then flared to a rounded base at the point of attachment; longitudinal lines of white dots on the lower surface or on both surfaces. Needles spirally arranged along the twig, but on horizontal twigs, bent and twisted at the base so as to appear in 2 ranks or crowded on the upper side and parted down the middle. Midvein prominent. Resin ducts (2) evident in cross section; their position, near or remote from the lower surface, is useful in identifying species.

Needles on the leading shoot and adjacent lateral branches at the tree top are stiff, pointed, and spread out in all directions, thus resembling spruce needles; those on horizontal branches bearing seed cones (usually on upper branches) are also stiff and pointed, but curved upwards.

Buds Terminal bud and the subterminal buds form a compact group; each subterminal bud is in the axil of a scale at the base of the terminal bud; hence, there are no needles between the terminal bud and the subterminals. Lateral buds are spaced along the twig; rounded, less than 7 mm long, usually resinous.

Twigs Stout, somewhat brittle, smooth; leaf scars (on the branchlets) flat, circular. New shoot and its needles preformed in the bud; the longest new side shoots develop at the annual node from the subterminal buds; those farther down the internode are progressively shorter. Annual nodes marked by persistent bud scales.

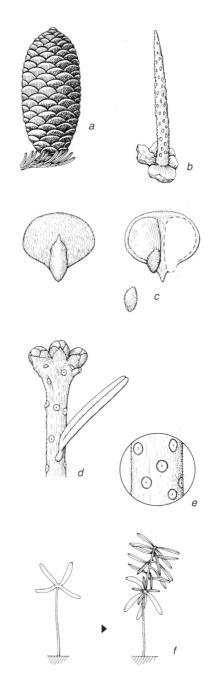

a. Mature seed cone. b. Cone axis after scales shed. c. Cone scale; outer surface with bract (left), inner surface with winged seed (right), seed (below). d. Twig terminal showing terminal and subterminal buds, leaf scars, and leaf attachment. e. Detail of flat, circular leaf scars. f. Seedling development; cotelydon stage (left), leaf stage (right).

Pollen Cones Soft, catkin-like, borne in the middle part of the crown, in the leaf axils of the previous year's twig, pendulous. Pollination takes place in late spring; cones wither and fall away soon after, leaving persistent bud scales on the branchlet.

Seed Cones Conelets erect, borne in the upper part of the crown, in the leaf axils of the previous year's twig. At pollination, the bract below each cone scale is larger than the scale.

   Mature cones cylindrical or barrel-shaped, erect, often resin-coated. In most species, the scales outgrow the bracts; when the seeds ripen at the end of the 1st season, bracts and scales fall away leaving a persistent erect cone axis.

Seeds In pairs on the upper side of the cone scales, often resinous, winged at one end; wing does not easily separate from the seed. Often more than half of the seeds are empty. Seeds require moist chilling for several weeks to ensure prompt germination.

Seedlings Newly germinated seedlings 2–4 cm tall, with a stalk surmounted by a whorl of 3–7 upcurved, green cotyledons, 2–3 cm long with blunt tips and lines of white dots on the upper surface. A terminal bud or an upright leafy shoot ending in a terminal bud develops above the cotyledons in the 1st season. Root grows vigorously, even through humus layers.

Vegetative Reproduction Layering (rooting of attached branches) occurs occasionally.

Bark Smooth, thin, with conspicuous raised resin blisters when young; becoming scaly, or thick and furrowed with age.

Wood Properties similar among the firs: light, straight-grained, soft, pale, no distinct heartwood, little latewood, no resin ducts.

Size and Form Medium-sized to very large trees. Trunk distinct to the top of the tree; leading shoot upright. Crown dense, regular, slender, spire-like. Principal branches horizontal, in regular whorls at the annual nodes; each forms a flat branch complex that droops slightly at the tip; smaller branches occur between the whorls. Some dead branches retained along the trunk below the live crown.

Habitat Found in northern regions and at higher elevations; occurs in pure stands and mixed with spruces, birches, and aspens; very shade-tolerant.

Notes Widely used for wood pulp and lumber. Important to wildlife for food and shelter. Excellent for Christmas trees because the needles are fragrant and remain on the tree for some weeks after harvesting; also firs respond well to shearing–cutting back new shoots to improve tree form. Resin from the bark blisters (marketed as Canada balsam) is used in preparing microscope slides because its refractive index is compatible with glass. Leaf oils are used in medicinal products.

   Firs can survive for many years in the shade of other trees and then grow well when those trees are removed. Sensitive to drought. Susceptible to heart rot and various pests including balsam woolly adelgid (*Adelges piceae*), eastern spruce budworm (*Choristoneura fumiferana*), and hemlock looper (*Lambdina fiscellaria fiscellaria*).

Quick Recognition [See spruce genus for contrasting features.] Prominent resin blisters on bark. Needles flat, flexible, tip blunt or notched; twigs smooth; buds rounded, resinous; subterminal buds form a compact group touching terminal bud. Cones erect, scales deciduous, axis persistent.

Seed conelets erect, borne in upper part of crown.

Principal branches in whorls, with lesser branches in between.

Pollen cones pendulous, borne in middle part of crown.

All photos balsam fir.

Cone axes persist after seeds, scales, and bracts are shed.

Balsam Fir
Canada balsam

Abies balsamea (L.)
Mill.

Sapin baumier

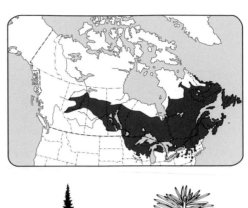

A distinctive tree of the northern forests of central and eastern Canada because its regular crown gradually tapers to a spire-like top.

Leaves 15–25 mm long, tip rounded or notched; upper surface shiny dark green, often with a few white dots toward the tip; 10–12 lines of white dots beneath. Needles arranged in 2 ranks, shorter on the upper side of the twig. Resin ducts small, remote from the surface. Strong odor when crushed.

Buds Broadly ovoid, 5 mm long, resinous.

Twigs Greenish-gray, smooth, hairy.

Seed Cones Erect, barrel-shaped, 4–10 cm long, grayish-brown, resinous; bracts with rounded shoulders and a needle-shaped tip, sometimes longer than the scales. Cones break up from early September, leaving the bare axis on the tree for several years. Seeds abundant. Good seed years occur every 2–4 years.

Seeds 3–6 mm long, resinous; wing 10–15 mm long, purple to brown, firmly attached to seed coat.

Bark Grayish, smooth, with raised resin blisters when young; with age, breaking into irregular brownish scales.

Wood Light, soft, weak, somewhat brittle, odorless, white, little contrast between earlywood and latewood, or heartwood and sapwood.

Size and Form Medium-sized trees, up to 25 m high, 70 cm in diameter, and 150 years old. Trunk slightly tapered below the crown. Dead branches persist for several years; on open-grown trees, the lower branches remain alive and green foliage extends to the ground. Root system shallow.

Habitat Adaptable to a variety of soils and climates; grows in pure stands or mixed with trembling aspen, white birch, white spruce, black spruce, red spruce, and eastern hemlock.

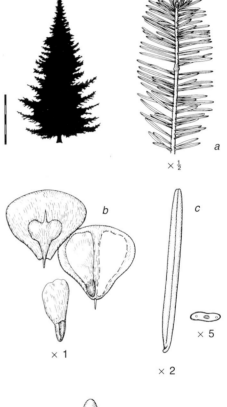

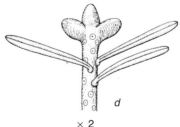

a. Twig and needles. *b.* Cone scale; outer surface with bract (left), inner surface with winged seed (right), winged seed inner surface (below). *c.* Needle and needle cross section. *d.* Twig terminal.

Seed cones often in dense groups.

Young bark smooth, with
raised resin blisters.

Irregular scales beginning to show at base of trunk.

Notes Widely used as a source of wood pulp and lumber. Favored for Christmas trees because the needles stay on for a long time after cutting.

Every few decades for at least a century, balsam fir stands have been ravaged by eastern spruce budworm (*Choristoneura fumiferana*); the control of the budworm by chemical spraying has generated a controversy concerning the effects of these pesticides on the environment. Damaged stands are usually followed by another stand of balsam fir. Balsam fir is also susceptible to butt rot, balsam woolly adelgid (*Adelges piceae*), and hemlock looper (*Lambdina fiscellaria fiscellaria*).

Balsam fir seedlings often predominate under a mixed stand: the seeds, which are abundant, germinate and survive in humus layers; the seedlings are very shade-tolerant. Young stands of balsam fir in eastern

Canada are often so dense that walking through them is difficult.

Forest fire prevents the regeneration of balsam fir.

The species occurs less frequently and its vigor declines from east to west, probably because of increasing dryness.

Trees with long protruding bracts on the seed cones are common in the coastal regions of eastern Canada and have been named as a variety: bracted balsam fir (*Abies balsamea* var. *phanerolepis* Fernald).

> **Quick Recognition** The only fir native to eastern Canada; distinguished from spruce by its flat blunt needles, smooth twigs, and spire-like crown; and from eastern hemlock by its stout twigs and regular branching pattern.

Subalpine Fir
Alpine fir

Abies lasiocarpa (Hook.) Nutt.

Sapin subalpin

The smallest of the western firs. A common tree in central British Columbia and northward into Yukon.

Leaves 25–40 mm long, length variable on same twig; tip rounded or notched; grayish-green to light bluish-green; lines of white dots, more prominent on the lower surface. Needles crowded, curving upwards from the sides of the twig and standing almost erect; those below appressed; occasionally arranged in 2 ranks. Resin ducts large, remote from the surface.

Buds Broadly ovoid, 3–6 mm long, light brown, resinous.

Twigs Stout, brownish, becoming ash-gray, hairy, retaining hair for several years.

Seed Cones Erect, barrel-shaped, 4–10 cm long, grayish-brown, resinous; scales wide; bracts shorter than the scales, wide, with an abruptly pointed needle-shaped tip. Cones break up from early September, leaving the bare axis on the tree for several years. Seeds abundant. Good seed years occur about every 3 years.

Seeds About 6 mm long, resinous; wing 10–18 mm long, purple to brown, firmly attached to seed coat.

Vegetative Reproduction Open-grown trees frequently reproduce by layering, especially at higher elevations.

Bark Smooth, ash-gray, blotched with raised resin blisters when young; with age, breaking into irregular grayish-brown scales.

Wood Light, soft, relatively weak, odorless, pale, no distinct heartwood.

Size and Form Medium-sized to large trees, up to 30 m high, 75 cm in diameter, and 200 years old; may be shrubby on exposed ridges at the tree line. Trunk cylindrical; crown narrow, dense, with a spire-like top. Branches short, drooping; lower ones of open-grown trees sometimes touching the ground. Principal branches frond-like when the lateral branches are branchless. Root system shallow, wide-spreading.

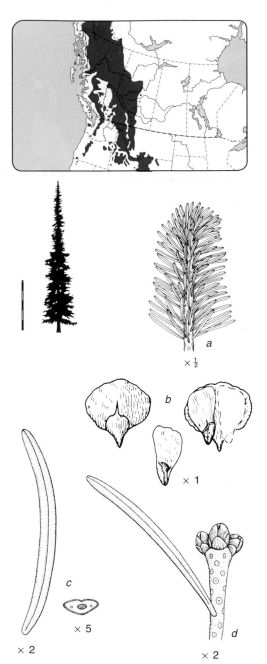

a. Twig and needles. *b.* Cone scale; outer surface with bract (left), inner surface with winged seed (right), winged seed inner surface (below). *c.* Needle and needle cross section. *d.* Twig terminal.

Seed cones barrel-shaped, grayish-brown.

Resin blisters prominent on young bark.

Trunk only slightly tapered; mature bark scaly, grayish-brown.

Habitat High-altitude tree (600–2000 m) in southern part of range; uncommon near Pacific coast. Found on a variety of soils; exists in pure stands or mixed with Engelmann spruce, mountain hemlock, sub-alpine larch, lodgepole pine, white birch, and trembling aspen. Frequently colonizes newly exposed ground; also capable of reproducing on humus layers under a forest cover. Shade-tolerant.

Notes Used for wood pulp and lumber. Seeds are a favorite food of squirrels.

> **Quick Recognition** Bract tip broad, sharp.

Amabilis Fir
Pacific silver fir

Abies amabilis (Dougl. ex Loud.)
Dougl. ex J. Forbes

Sapin gracieux

Commonly found in the subalpine forests of western British Columbia and as far south as northern California.

Leaves 20–30 mm long, tip notched; upper surface shiny dark green, grooved, often with a few white dots toward the tip; several lines of white dots beneath. Needles horizontally spreading and curved upwards on the lower surface and sides of the twig; shorter, pressed along the twig, and pointed forward on the upper surface. Resin ducts small, near lower surface. Odor of oranges when crushed.

Buds Rounded, 3–7 mm long, dark bluish-brown resinous.

Twigs Stout, dark yellowish-brown to grayish-brown, minutely hairy.

Seed Cones Ovoid-conical, 9–14 cm long, brown; scales as wide as they are long; bracts shorter than the scales, with shoulders sloping away from the broad base of a long wedge-shaped tip. Cones ripen in late August; seeds shed a few weeks later.

Seeds 8–16 mm long; wings 25–40 mm long.

Bark Smooth, light gray, blotched with white patches, becoming scaly and grooved at the base of mature trees.

Wood Light, soft, weak, creamy-white to yellowish-brown, no distinct heartwood.

Size and Form Large trees, up to 40 m high, 90 cm in diameter, and 300 years old; occasionally larger. Slow-growing; a tree 60 cm in diameter can be 200 years old. Trunk slender; crown slender, conical. Principal branches mostly horizontal; lower ones drooping, persistent after they die. Root system moderately deep, wide-spreading.

Habitat Found on a variety of soils; occasionally grows in pure stands, but more often mixed with Sitka spruce, western hemlock, Douglas-fir, and western redcedar. Very shade-tolerant.

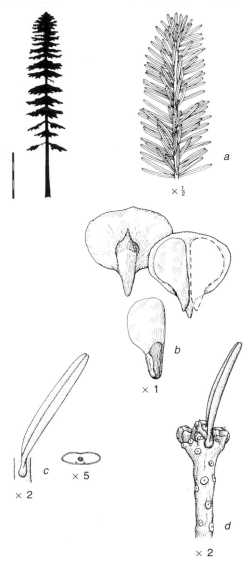

a. Twig and needles. *b.* Cone scale; outer surface with bract (left), inner surface with winged seed (right), winged seed inner surface (below). *c.* Needle and needle cross section. *d.* Twig terminal.

Seed cones ovoid-conical; scales as wide as long.

Young bark smooth, light gray.

Mature bark scaly and grooved.

Notes Used mainly for wood pulp and lumber. Because of its symmetrical shape and shiny foliage, it is planted as an ornamental (Zones C6, NA5).

Quick Recognition Needles on upper side of twig point forward; with white dots only on the lower surface. Seed cone bracts wide with sloping shoulders.

Grand Fir
Lowland fir

Abies grandis (Dougl. ex D. Don) Lindl.

Sapin grandissime

Largest of the firs; occurs at relatively low elevations in the coastal and interior forests of southern British Columbia.

Leaves Relatively stiff, 20–35 mm long, shorter on the upper side of the twig; tip notched; upper surface shiny dark green, grooved, sometimes with a few white dots at the tip; several lines of white dots beneath. Needles horizontally spreading in 2 distinct ranks. Resin ducts near lower surface. Odor of oranges when crushed.

Buds Broadly ovoid, 5 mm long, blunt-tipped, purple or bluish-brown, resinous.

Twigs Slender, olive-green to dark orange-brown, minutely hairy.

Seed Cones Cylindrical to narrowly ovoid, with a blunt or sunken tip, 5–12 cm long, green to purplish; scales much wider (25–30 mm) than long; bracts shorter than the scales, with broad pointed shoulders sloping inward to the base of the tip, a small tooth no higher than the tops of the shoulders. Seed production less prolific than other firs.

Seeds Light brown or tan, 6–10 mm long; wing 12–20 mm long.

Bark Smooth, grayish-brown, with resin blisters and white blotches when young; with age, becoming deep brown, thick and scaly, separating into dark gray flat ridges.

Wood Light, soft, relatively weak, odorless, light brown.

Size and Form Large trees; in coastal forests, up to 40 m high, 90 cm in diameter, and 250 years old; some trees on Vancouver Island up to 75 m high. In interior British Columbia, trees seldom reach 40 m high. On old trees, trunk slightly tapered, long compared with the length of the crown; crown cylindrical or oval, rounded at the top; principal branches horizontal, lower ones drooping, with upturned tips. On younger, open-grown trees, trunk is completely hidden by branches that extend to the ground; crown pointed at top. Roots deep, wide-spreading.

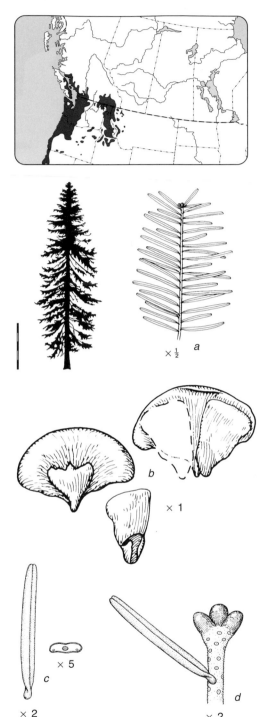

a. Twig and needles. b. Cone scale; outer surface with bract (left), inner surface with winged seed (right), winged seed, inner surface (below). c. Needle and needle cross section. d. Twig terminal.

Seed cones cylindrical; scales much wider than long.

Young bark grayish-brown.

Mature bark has dark gray flat ridges.

Habitat Grows best on deep, well-drained, alluvial soils; occasionally found in pure stands; often mixed with such species as western redcedar, black cottonwood, western hemlock, Douglas-fir, red alder, and Sitka spruce. Moderately shade-tolerant.

Notes Used for lumber, wood pulp, plywood.

Quick Recognition A lowland species. Needles long and blunt, lines of white dots on lower surface; resin ducts near lower surface. Cone scales very broad; bract tip short.

Noble Fir*

Abies procera Rehd.

Sapin noble

Very large trees of the Rocky Mountains, range stops just short of the southern border of British Columbia; planted in Canada as an ornamental. Hardy as far north as Zones C6, NA5.

Needles very narrow, 20–30 mm long, comparatively thick, tip rounded, grooved on the upper surface, lines of white dots on both surfaces; arranged along both sides of the twig; upper needles pressed against the twig for part of their length and then curved upwards; resin ducts near lower surface.

Buds globular, small, slightly resinous, with the outer scales separating from the bud. **Twigs** reddish-brown, hairy.

Seed cones cylindrical, 11–18 cm long; bracts prominent, long-pointed, light yellow to light green, longer than the scales, reflexed so as to hide them.

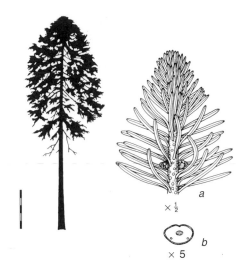

a. Twig and needles. *b.* Needle cross section.

Red Fir*

California red fir

Abies magnifica A. Murr.

Sapin rouge

Very large trees, native to the Rocky Mountains south of Canada; frequently planted in British Columbia. Hardy as far north as Zone C6, NA5.

Needles 20–40 mm long, somewhat 4-sided, no groove on the upper surface, lines of white dots on all sides; arranged along both sides and the upper surface of the twig; upper needles pressed against the twig for part of their length and then curved upwards; resin ducts near lower surface.

Buds globular, small, slightly resinous, with the outer scales separating from the bud. **Twigs** brown, hairy.

Seed cones cylindrical, 15–22 cm long; typically with the bracts shorter than the scales. In the variety Shasta red fir (*A. magnifica* var. *shastensis*), bracts extend beyond the scales, giving the cone a bristly appearance.

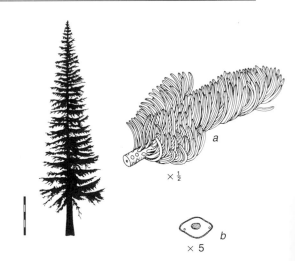

a. Twig and needles. *b.* Needle cross section.

5

Nordmann Fir*
Abies nordmanniana (Steven) Spach

Sapin de Nordmann

Very large trees native to the Caucasus Mountains; occasionally planted as ornamentals in Canada. Hardy as far north as Zone C6, NA4.

Needles closely spaced, 25–35 mm long, blunt-tipped, dark green on the upper surface, lighter beneath with lines of white dots; horizontally spreading on the sides of the twig; upper needles shorter and directed forward, thus covering the upper twig surface. **Buds** ovoid, non-resinous, with appressed scales. **Twigs** greenish-yellow, slightly hairy, becoming hairless. Seed cones cylindrical, 12–20 cm long; bract tips showing.

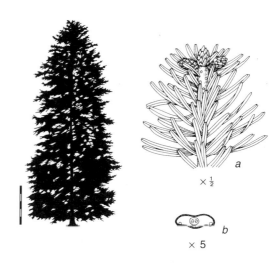

× ½

× 5

a. Twig and needles. *b.* Needle cross section.

Spanish Fir*
Abies pinsapo Boiss.

Sapin d'Espagne

Medium-sized attractive trees, with branches in regular whorls forming a symmetrical crown. Native to the mountains of southern Spain; planted as an ornamental in Canada. Hardy as far north as Zone C7, NA6.

Needles stiff, 8–15 mm long, lines of white dots on both surfaces; base not twisted as much as in most other firs, thus arrangement resembles spruce. **Buds** ovoid, slightly resinous. **Twigs** brownish-green, hairless. **Seed cones** cylindrical, 9–18 cm long.

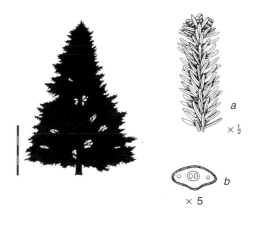

× ½

× 5

a. Twig and needles. *b.* Needle cross section.

White Fir*
Colorado fir

Abies concolor (Gord. & Glend.)
Lindl.

Sapin argenté

Very large trees, native to mountainous regions of
the western United States; commonly planted in
Canada because of the attractive needles, hardiness, fast growth, and tolerance of drought and
shade. Hybridizes with grand fir. Hardy as far north
as Zones C4, NA3.

Needles 40–70 mm long, blunt or somewhat
pointed; intensely whitish-green owing to dense
lines of white dots on both sides; horizontally
spreading in 2 distinct ranks on horizontal branches; side needles sickle-shaped and curved
upwards; upper needles pointed forward; resin
ducts near lower surface; odor of oranges when
crushed.

Buds broadly conical, large, yellowish-brown,
resinous. **Twigs** olive-green, becoming grayish
with age, hairless. **Bark** Rough, light gray.

Seed Cones cylindrical, 7–12 cm long, green
or purple, becoming brown; scales 25 mm wide;
bracts shorter than the scales, with a small point.

5

a. Twig and needles. *b.* Needle cross section.

Spruces

Genus *Picea*
Pinaceae: Pine Family

Les épinettes

Spruces are distributed through the cooler parts of the North Temperate Zone and at higher elevations farther south. Worldwide about 40 species are recognized; of the 7 species native to North America, 5 are found in Canada: red spruce in the east, Sitka and Engelmann spruces in the west, and 2 transcontinental species, white and black spruces. Several non-native species have been introduced, including Norway, Colorado, and Serbian spruces.

Identifying spruces is often difficult. Wide-ranging species vary from one part of their range to another. Where the ranges of white, Engelmann, and Sitka spruces overlap, intermediate forms are common. The same is true for red and black spruces in the east. Cone features are often the most reliable guide in identifying species. In the following descriptions, features under **Leaves** pertain to needles on well-developed horizontal branchlets, unless otherwise noted.

Leaves Evergreen, remaining on the tree for 7–10 years; 4-sided (flattened in Sitka and Serbian spruces), stiff, 6–30 mm long, pointed; lines of white dots on the leaf surfaces; stalkless, but jointed at the base and attached to a leaf-peg. Needles spirally arranged along the twig, but often bent and twisted at the base causing them to spread from both sides of the twig and/or be crowded on the upper side, especially on horizontal twigs. Midvein and 2 resin ducts visible in cross section in most species.

Buds Conical, less than 5 mm long, many overlapping scales, non-resinous; terminal bud prominent, with a small cluster of lateral buds usually occurring just below it, separated by a few needles; other smaller lateral buds spaced along the twig. Reproductive buds may be terminal or lateral.

Twigs Stiff, tough, characterized by leaf-pegs that remain on the branchlet after the needles fall; each leaf-peg is a projection from the forward end of a leaf-cushion, a ridge of bark about 1 cm long, separated by incised grooves. New shoot and its needles preformed in the bud. Preformed shoot growth ceases in early summer; however, the leading shoot on vigorous young trees may

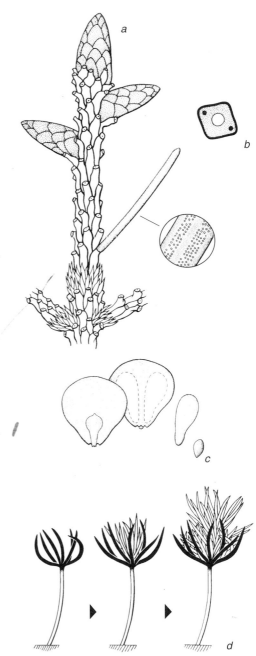

a. Twig showing terminal and lateral buds, leaf-cushions, leaf-pegs, and leaf attachment; detail of leaf surface with lines of white stomatal dots. b. Needle cross section. c. Cone scale; outer surface with bract (left), inner surface (center), winged seed and seed (right). d. Seedling development; cotyledon stage (left), leaf stage (center), end of 1st growing season (right).

continue to elongate by neoformed growth and may bear side shoots.

Pollen Cones Cylindrical, fleshy, catkin-like, 10–20 mm long, erect; borne in the middle and lower part of the crown, in terminal and lateral positions on the previous year's twig. Pollination takes place in spring as new leafy shoots emerge from their buds; cones then wither and fall away.

Seed Cones Borne in the upper part of the crown, laterally and at the tips of previous year's twigs. At pollination, conelets 10–20 mm long, soft, fleshly, green or purplish-red, erect; scales spread apart and often reflexed to permit the entry of pollen. After a few days, the scales come together, and the cone bends downward. Full-size by mid-July; mature in late summer or early autumn.

Mature cones cylindrical to ovoid, usually pendulous. Scales numerous, leathery; bracts small. Seeds are released over several months; cones fall intact before or during the next growing season; however, cones and seeds of black spruce persist for years. Seed production may begin before the 10th year, with good crops every few years thereafter.

Seeds Small, dry, winged, wind-dispersed. Wing clasps the seed; easily detached. Among the easiest seeds to store; viable for many years when kept dry in airtight containers at subfreezing temperatures. Germination is normally prompt, but may be aided by moist prechilling for several weeks at temperatures just above freezing.

Seedlings Newly germinated seedlings 2–4 cm tall, with a stalk surmounted by a whorl of 3–10 upcurved, green cotyledons 1–3 cm long. In a week or two, an upright leafy shoot develops above the cotyledons; young needles soft, flexible.

Vegetative Reproduction Layering, the rooting of attached branches, occurs frequently in black spruce and can occur in any spruce species.

Detached stem tips can be rooted; those from younger trees root more readily. Ornamental cultivars are reproduced by grafting.

Bark Scaly, thin.

Wood Properties similar among the spruces: light, straight-grained, soft, moderately strong, resilient, pale; no distinct heartwood; latewood distinct from earlywood; low in resin. Longitudinal resin ducts visible with a hand lens.

Size and Form Size varies from the huge Sitka spruce on the Pacific coast to the stunted Engelmann spruce at the alpine tree line and the small black spruce in northern bogs. Trunk straight, distinct to the top of the tree; leading shoot upright. Crown dense, conical, may extend well down the trunk. Principal branches generally straight and horizontal, in loose whorls at the annual nodes; lesser branches between the whorls. Dead branches may remain on the trunk for many years. Root system shallow.

Habitat Usually found on well-drained sites. A spruce species will usually be found at arctic and alpine tree lines, where it may be growing on permafrost. Most species are shade-tolerant.

Notes A source of wood pulp. First among Canadian species in volume of lumber produced. Used in general construction, mill work, interior finishings, and for plywood, boxes, and musical instruments. Because it has little taste or odor, a preferred material for food containers. Commonly planted in reforestation programs and for landscape purposes; often used for Christmas trees.

Important to wildlife; deer and rabbits eat the shoots; porcupines the bark; birds and small rodents the buds and seeds; many animals find shelter under or within the dense crowns.

Much genetic improvement work on spruce has been carried out in Canada and elsewhere for forestry purposes. Promising strains have been selected, and many hybrids have been produced.

Individual trees are susceptible to fire damage because of thin bark; the regeneration of some species by seeds, however, is favored by fire, which removes competing vegetation and prepares a favorable seedbed. Subject to windthrow because of the shallow root system.

Principal branches in whorls, with many lesser branches in between.

Seed conelets at first erect, pendulous after pollination.

Pollen cones borne terminally and laterally on twigs.

All photos white spruce.

Quick Recognition [Contrasting features of *Abies* in brackets]. Leaf-pegs and the leaf-cushions associated with them on the twigs and branchlets are a prominent feature [twigs and branchlets smooth, with flat round leaf scars]. Needles 4-sided [flat, with upper and lower surfaces and a prominent midvein], stiff [flexible], pointed [usually rounded or notched], borne on leaf-pegs [attached directly to the shoot]. Closest subterminal buds separated from the terminal by a few needles [no needles between the terminal bud and the cluster of subterminals]. Cones cylindrical or ovoid, pendulous [erect], often at the tip of a branchlet [always attached at the side]; remain intact after falling [disintegrate when mature, leaving an upright stalk attached to the branchlet]; scales numerous. Crown conical; principal branches in loose whorls near annual nodes; many lesser branches between the whorls. Bark scaly [smooth with blisters].

Sitka Spruce
Coast spruce, tideland spruce

Picea sitchensis (Bong.) Carrière

Épinette de Sitka

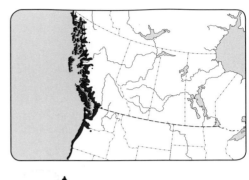

The tallest conifer in Canada; a coastal species, found in Alaska, British Columbia, and southward into the United States.

Leaves Straight, flattened, keeled below, 20–30 mm long, sharp-pointed; yellowish-green on the upper surface, prominent lines of white dots beneath; needles tending to radiate at right angles from the twig.

Buds Conical to dome-shaped, resinous; scales appressed, outer scales blunt-tipped, shorter than the bud.

Twigs Light gray to yellowish-brown, lighter than the buds, hairless.

Seed Cones Broadly cylindrical, 5–10 cm long. Scales yellow to light brown, thin, brittle, loose-fitting, elongated, broadest near the middle; outer margin, wavy, irregularly toothed; bracts visible between open scales. Cones open in late autumn, shed during the succeeding months.

Seeds Reddish-brown, 2–3 mm long; wing 5–8 mm.

Bark Thin, broken into large, loose, reddish-brown scales; newly exposed bark rusty-gray.

Wood Light, soft, resilient, relatively strong; heartwood light pinkish-brown with gradual transition into a creamy-white sapwood.

Size and Form Very large trees, up to 55 m high, 200 cm in diameter, and 700–800 years old; the Carmanah Giant, a 95-m Sitka spruce on Vancouver Island, is reputed to be the tallest tree in Canada. Trunk massive, often buttressed at the base. Crown rather open. Principal branches horizontal; some secondary branches drooping; new shoots may develop along the trunk. Root system shallow, wide-spreading.

Habitat Usually occurs in the fog belt of the Pacific coast, along inlets and borders of streams inland for about 150 km to elevations of 500 m. Most abundant on Vancouver Island and the Queen Charlotte Islands, and in northern coastal forests on deep well-drained alluvial gravel. Grows in pure stands, more often mixed with western hemlock, Douglas-fir, western redcedar, yellow-cedar, grand fir, red alder, and black cottonwood.

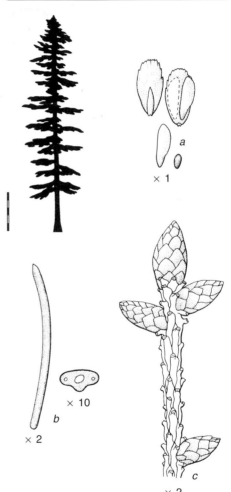

a. Cone scale; outer surface with bract (left), inner surface with winged seed (center), winged seed and seed (below). *b.* Needle and needle cross section. *c.* Twig terminal.

Seed cone; scales loose-fitting, brittle, irregularly toothed.

Young bark smooth, soon becomes scaly.

Trunk massive, often buttressed at base.

Notes Because of its great size, yields a large proportion of clear lumber, source of wood pulp. During the Second World War, aircraft, including the famed Mosquito bomber, were made from Sitka spruce.

In the northern inland parts of its range, where Sitka spruce occurs with white spruce, hybrids are common (*P.* ×*lutzii* Little). In southern British Columbia, Sitka spruce occasionally hybridizes with Engelmann spruce. Widely used as a reforestation species in western Europe.

Quick Recognition Needles flattened, sharp, lines of white dots prominent on the lower surface. Cone scales loose-fitting, pale, broadest near the middle, outer margin toothed.

Engelmann Spruce

Mountain spruce, Columbian spruce, silver spruce, white spruce

Picea engelmannii Parry ex Engelm.

Épinette d'Engelmann

Found in the mountains of interior British Columbia, adjacent parts of Alberta, and southward into the United States.

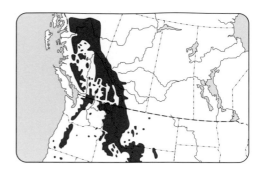

Leaves Curved, somewhat flexible, 15–25 mm long, tip pointed or blunt; bluish-green but often coated with a whitish powder; lines of white dots on all sides; upper needles on a horizontal twig tending to point toward the tip. Aromatic when crushed.

Buds Conical, non-resinous; scales light brown, rounded; outer scales long-pointed, slightly reflexed.

Twigs Grayish to light brown, somewhat hairy.

Seed Cones Cylindrical to narrowly ovoid, 3–7 cm long, shiny. Scales yellowish-brown, thin, flexible, loose-fitting, tapered at both ends, irregularly toothed and often split at the tip; bracts comparatively prominent. Cones open in autumn, shed during the winter or following spring, often retaining some seeds.

Seeds Dark brown, 2–4 mm long; wing 5–8 mm.

Bark Thin, broken into large, loose, coarse, rounded, brownish scales; newly exposed bark silvery-white, resinous.

Size and Form Large trees, up to 35 m high, 90 cm in diameter, and 300 years old; 55-m-high trees have been reported. Crown dense, symmetrical, narrow, spire-like; lower branches often sloping downward; secondary branches may hang vertically.

Habitat Usually grows on mountain slopes at elevations of between 1000 and 2000 m, also along streams at lower elevations. Occurs in pure stands, but more often mixed with subalpine fir, western hemlock, western larch, lodgepole pine, birches, and aspen. Frequently forms part of the alpine tree line. Shade-tolerant.

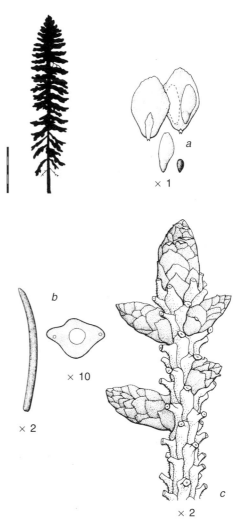

a. Cone scale; outer surface with bract (left), inner surface with winged seed (right), winged seed and seed (below). *b*. Needle and needle cross section. *c*. Twig terminal.

5

Seed cone; scales loose-fitting, often split at tip.

Young bark thin, scaly.

Mature bark has loose, brownish scales.

Notes An important forestry species in interior British Columbia; used for wood pulp and lumber. Hybrids occur where the range of Engelmann overlaps that of white, Sitka, and Colorado spruce.

Quick Recognition Difficult to separate from white spruce. Twigs somewhat hairy. Cone scales loose-fitting, flexible, broadest near the middle, tapered to a toothed, split, blunt tip. Needles aromatic when bruised.

White Spruce

Cat spruce, skunk spruce, pasture
spruce, Canadian spruce

Picea glauca (Moench) Voss

Épinette blanche

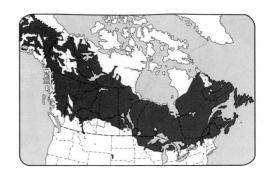

Found in all forested regions of Canada
except on the Pacific coast.

Leaves Straight, stiff, 15–22 mm long, tip
pointed but not sharp, green to bluish-green,
lines of white dots on all sides. Unpleasantly
pungent when crushed.

Buds Ovoid, 6 mm long, blunt-pointed, non-
resinous; scales tight-fitting, margins ragged,
curled out; outer scales shorter than the bud.

Twigs Shiny, light greenish-gray, often
tinged with orange or purple, hairless; leaf-
cushions rounded, grooves open. Twigs of
seedlings may be hairy.

Seed Cones Slender, cylindrical, 3–6 cm
long, blunt-tipped, stalkless. Scales light
brown, thin, tough, flexible, close-fitting;
outer margin rounded, smooth. Cones open
in late summer; seeds released from late
summer to spring. Mature open cones easily
compressed but scales do not break.

Seeds 2–4 mm long; wing 4–8 mm long.

Bark Smooth, thin, light gray when young,
darker gray and scaly with age; newly exposed bark salmon pink, silvery.

Size and Form Medium-sized trees, up to
25 m high, 60 cm in diameter, and 200 years
old; occasionally larger. Crown broadly conical, ragged, irregular, densely foliated, but
spire-like in northern parts of the range.
Principal branches bushy, generally horizontal, but sometimes sloping downward in
the lower part of the crown; tips gradually
upturned. Root system shallow, with many
tough, pliable, wide-spreading branch roots.

Habitat Common in northern forests; occurs
on a variety of soils and under a wide range
of climatic conditions; associated with
trembling aspen, white birch, black spruce,
and balsam fir. In eastern Canada, invades
abandoned farmland. Shade-tolerant; after
being suppressed, recovers well when exposed to more light. Often found at the arctic
tree line.

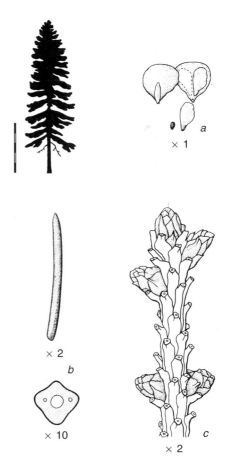

a. Cone scale; outer surface with bract (left), inner
surface with winged seed (right), seed and winged
seed (below). *b.* Needle and needle cross section.
c. Twig terminal.

5

Seed cones; scales close-fitting, flexible, margins smooth.

Young bark smooth, light gray.

Mature bark dark gray and scaly.

Notes Important in Canada for the production of wood pulp and lumber. Very hardy, frequently planted for landscape and forestry purposes.

Considerable genetic diversity within the species, as indicated by the variation in form and by its wide range of sites. For example, a population found in northern Alberta, Yukon, and Alaska (Porsild spruce, *Picea glauca* var. *porsildii* Raup) has smooth bark covered with resin blisters similar to balsam fir.

Where white spruce grows with Sitka and Engelmann spruce, natural hybrids occur.

On floodplains of some northern rivers, tiers of adventitious roots may develop along the stem in successive layers of sediment, giving rise to exceptionally large trees.

White spruce is the provincial tree of Manitoba.

Quick Recognition Needles bluish-green. Twigs shiny, light greenish-gray, tinged with orange or purple, hairless; leaf-cushions rounded, grooves open. Newly exposed bark salmon pink. Cones cylindrical, with flexible, tight-fitting, rounded, smooth-edged scales.

Red Spruce
Eastern spruce, yellow spruce,
he-balsam

Picea rubens Sarg.
[syn. *P. rubra* Link]

Épinette rouge

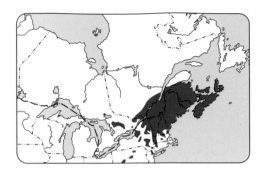

The common spruce in the Maritime provinces, and southward into the Appalachian
Mountains of the United States; present but
uncommon in south-central Ontario and in
Quebec north of the St. Lawrence River.

Leaves Curved, 10–16 mm long, blunt-
pointed, shiny yellowish-green, lines of faint
white dots usually present on all sides;
needles pointing forward, often pressed
close to the twig.

Buds Conical, chestnut-brown, shiny or
slightly resinous; outer scales hairy, narrow,
with long slender points projecting beyond
the tip of the bud.

Twigs Shiny, yellow or yellowish-orange,
with a brown stain at the base of the leaf-
peg. Hairs on twig narrowly conical. In the
2nd year, leaf-cushions rounded, grooves V-
shaped; twig color reddish-brown similar to
Norway spruce.

Seed Cones Ovoid, 3–5 cm long, rich
chocolate-brown, blunt-pointed, abruptly
tapered to a very short stalk. Scales stiff,
lightly striated; margin firm, smooth or slightly
rough. Cones open in autumn, usually shed
during the following year. Open cones
broadly ovoid, with wide-spreading scales
easily separated from the axis. Good seed
crops every 2–11 years.

Seeds Dark brown, 2 mm long; wing
3–5 mm.

Bark Reddish-brown, shredded when
young, separating into to reddish-black
scales or plates with age; at maturity dark
and furrowed, resembling old white pine.
Newly exposed bark dull yellow or reddish-
brown.

Size and Form Medium-sized trees, up to
25 m high, 60 cm in diameter, and 300 years
old, occasionally larger. Crown broadly conical, rather open. Principal branches about
3 m long in the lower part of the crown, with
a flat non-bushy appearance, sloping downward; tips abruptly upturned.

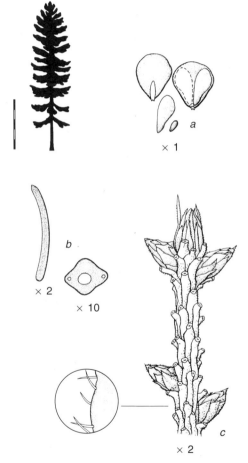

a. Cone scale; outer surface with bract (left), inner
surface with winged seed (right), winged seed and
seed (below). *b.* Needle and needle cross section.
c. Twig terminal with detail of conical twig hairs.

5

Seed cones; scales close-fitting, firm, margin smooth or slightly rough.

Young bark shredded, reddish-brown.

Mature bark dark and furrowed.

Habitat Typically a component of mature forests on moist upland sites, mixed with eastern white pine, balsam fir, eastern hemlock, yellow birch, and sugar maple. In Ontario and Quebec found on cool moist sites such as north-facing slopes and lake shores. Very shade-tolerant, surviving as an understory tree for many years.

Notes Where the ranges of red and black spruce overlap, hybrids are common, with features intermediate between the parents.

Susceptible to damage by windthrow. Red spruce is the provincial tree of Nova Scotia.

Quick Recognition [Contrasting features of black spruce in brackets.] Needles shiny yellowish-green [dull grayish-green], curved [straight], 10–16 mm [8–15 mm]; lines of white dots faint [prominent]. Buds pointed, shiny chestnut-brown [blunt, dull brown-gray]. Twigs hairy, shiny yellow or yellowish-orange [dull yellowish-brown]; leaf-cushions rounded [flat]; grooves V-shaped [closed]. Cones rich chocolate-brown [purplish-brown], almost stalkless [with a short curved scale-covered stalk], falling within a year [persistent for 20–30 years]; open cones broadly ovoid [spherical]. Scale margins firm [thin and brittle], almost smooth [ragged]. Newly exposed bark light yellow or reddish-brown [olive-green or yellowish-green]. Lower principal branches long [short], sloping slightly [well below the horizontal], tips abruptly [gradually] upturned. Crown conical, broad [cylindrical, narrow; top often deformed by red squirrels].

Black Spruce
Bog spruce, swamp spruce

Picea mariana (Mill.) BSP

Épinette noire

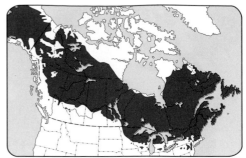

A transcontinental species, found southward to the United States. Usually a slow-growing wetland tree, but occurs frequently on upland sites where its growth is comparable to that of white spruce.

Leaves Straight, 8–15 mm long, blunt-pointed, dull grayish-green, lines of white dots more prominent on the undersurface; needles densely set along the twig, side needles at right angles, upper needles pointing forward.

Buds Conical, blunt-tipped, 3–5 mm long; outer scales dull brownish-gray, hairy, with long slender points projecting beyond the tip of the bud; inner scales darker brown, broader.

Twigs Dark orange-brown or yellowish-brown, dull, with a dark purplish stain at the base of the leaf-peg. Leaf-cushions flat, grooves closed, with many short brownish hairs, which may be crooked and/or tipped with a gland. Twigs of very young trees often hairless.

Seed Cones Broadly ovoid, 2–3 cm long, deep red to purple when young, changing to dark purplish-brown, blunt-pointed, gradually tapered to a curved, short, scale-covered stalk. Scales brittle, tight-fitting; margin irregularly toothed. Cones mature in September; remain on the tree up to 30 years with viable seeds; seeds are released gradually (quickly after a forest fire). Open cones almost spherical; scale margin thin, brittle, toothed. Some cones are produced nearly every year; good crops on young trees every 2 years, on older trees at longer intervals; cones often massed at the top of the tree.

Seeds Dark, about 2 mm long; wing 2–4 mm.

Vegetative Reproduction Layering, the rooting of attached branches, is common. An extending clump of trees all derived from one seedling can result. Often layering is the primary means of regeneration on organic soils, where conditions are favorable for layering and unfavorable for seed germination and survival, as in the northern part of the range.

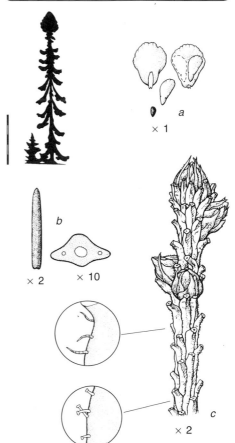

a. Cone scale; outer surface (left), inner surface (right), winged seed and seed (below). b. Needle and needle cross section. c. Twig terminal with details of twig hairs, some crooked (upper) and some gland-tipped (lower).

Bark Reddish- or grayish-brown, thin, scaly or shredded when young, becoming darker with larger scales. Newly exposed bark olive-green or yellowish-green.

Size and Form On poorly drained sites, trees small, up to 20 m high, 30 cm in diameter, and 200 years old; slow-growing; crown

5

Seed cones; scales thin, margins
irregularly toothed.

Young bark reddish- or
grayish-brown.

Mature bark has large thin scales.

narrow, spire-like. On well-drained upland sites, trees medium-sized, up to 30 m high, 60 cm in diameter. Principal branches short compared with other spruces, lower ones greatly drooping, tips upturned. Upper part of the crown often very dense, oddly shaped, with many cones. Root system very shallow, especially on organic soils with a high water table.

Habitat Grows on a variety of sites. Generally confined to wet poorly drained sites in the southern part of its range, in pure stands or with tamarack. Northward, usually grows on moist organic soils in extensive pure stands, or mixed with jack pine, white spruce, balsam fir, white birch, trembling aspen, and lodgepole pine. Moderately shade-tolerant.

Notes Important for the production of wood pulp and lumber. Vulnerable to flooding, windthrow, and fire damage. Reproduces well after wildfire, which opens the cones, releasing the seeds. Easily grown in tree nurseries, and set out in plantations. Where

the ranges of red and black spruce overlap, intermediate forms are common. Red squirrels (*Tamiasciurus hudsonicus*) chew off the tips of cone-bearing branches, resulting in an upper crown with a dense mass of small branches bearing many cones and a bare stretch of trunk just below it. Buds of black spruce open a week or two later than those of nearby white spruce, thus sustaining less damage from late spring frosts. The provincial tree of Newfoundland and Labrador.

Quick Recognition [See red spruce for contrasting features.] Needles 8–15 mm, blunt-pointed, straight, dull grayish-green. Terminal bud blunt, dull brownish-gray. Outer bud scales long, narrow, pointed. Twigs dull yellowish-brown, glandular hairy; leaf-cushions flat with closed grooves. Cones small, purplish-brown, with a short, curved, scale-covered stalk; spherical when open, persistent for 20–30 years; scale margin ragged, thin, brittle. Newly exposed bark olive-green or yellowish-green. Crown cylindrical, often deformed near the top by red squirrels.

Norway Spruce*

Picea abies (L.) Karst.
[syn. *P. excelsa* Link]

Épicéa commun

Large trees, up to 40 m high and 130 cm in diameter; native to Europe and Asia, where it occurs on upland sites. The major introduced spruce used in reforestation in eastern Canada and adjacent parts of the United States; also planted as ornamentals and for windbreaks. Possesses great natural variation; intensive breeding and selection have produced over 100 cultivated varieties. Hardy as far north as Zones C2, NA2.

    Needles straight, stiff, 12–24 mm long, sharp-pointed, dark green on all sides; bent away from the lower side of the twig, directed forward on the upper side. **Buds** conical, reddish to light brown, blunt-pointed, non-resinous; scales tight-fitting; outer scales sometimes with spreading tips. **Twigs** creamy-green, becoming light orange-brown with age, shiny, mostly hairless. **Seed cones** cylindrical, large, 10–18 cm long, light brown becoming reddish-brown or grayish-brown with age, pendulous, with tapered tip; scales thin, stiff, broadly tapered to a flat, slightly toothed tip. **Bark** reddish brown, wrinkled to smooth, or in small papery shreds; becoming, dark purplish-brown with small, hard, rounded scales with age. Drooping secondary **branches** distinctive. Grows on a variety of soils. Susceptible to drought and frost. Moderately shade-tolerant.

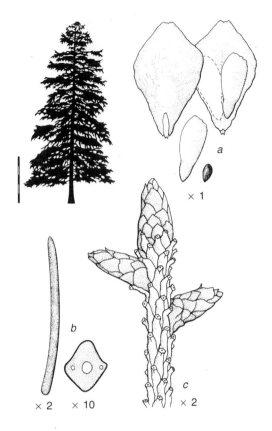

×1

×2　　×10　　×2

a. Cone scale; outer surface with bract (left), inner surface with winged seed (right), winged seed and seed (below). *b.* Needle and needle cross section. *c.* Twig terminal.

Serbian Spruce*

Picea omorika (Pančić) Purk.

Épicéa de Serbie

A medium-sized tree, native to the Balkans, where it grows on limestone slopes (800–1600 m) along the River Drina. Valued as an ornamental for its graceful slender form, drooping lower branches with upswept ends, and attractive foliage. Hybridizes easily with most other spruces; hybrid vigor shown by hybrids with Sitka and black spruces. Hardy as far north as Zones C3, NA4.

    Needles flattened, flexible, 10–20 mm long, blunt-pointed, 2 lines of white dots on upper surface; spreading widely on both sides of the twig. **Twigs** hairy; leafy branches dark green from above and whitish-green from below. **Seed cones** ovoid, 4–6 cm long, purplish-black, tapered gradually to a short, scaly, curved stalk; scales rounded, tight-fitting, margin weakly toothed.

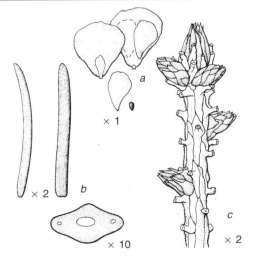

×1

×2　　*b*

×10

×2

a. Cone scale; outer surface with bract (left), inner surface with winged seed (right), winged seed and seed (below). *b.* Needles and needle cross section. *c.* Twig terminal.

Seed cones cylindrical, much larger than those of native species.

Norway Spruce

Mature bark has purplish-brown scales. Young bark (inset) wrinkled to smooth, reddish-brown.

Colorado Spruce*
Blue spruce, silver spruce

Picea pungens Engelm.

Épinette du Colorado

Seed cones; scales thin, flexible, loose-fitting.

Medium-sized trees, up to 30 m high, 90 cm in diameter, and 600 years old; native to the Rocky Mountains region of the United States. Many individuals have been selected with a striking bluish color; the best of these have been propagated by grafting and are frequently planted for ornamental purposes in Canada, the United States, and Europe. Also used as windbreaks. Hardy as far north as Zones C2, NA2.

Needles stiff, 15–30 mm long, very sharp-pointed, bluish-green; spreading out all around the twig, somewhat upswept and curved forward. **Buds** rounded to blunt-pointed, 10 mm long; scales papery, reflexed. **Twigs** very stout, shiny yellowish-brown, hairless. **Seed cones** cylindrical with tapered tip, 5–12 cm long; scales, shiny chestnut-brown, thin, flexible, loose-fitting, tapering slightly to a jagged tip, margin wavy. **Bark** purplish-gray to brown, flaky, becoming red-brown and furrowed with age. Hardy, drought-tolerant, and able to grow on a wide variety of soils.

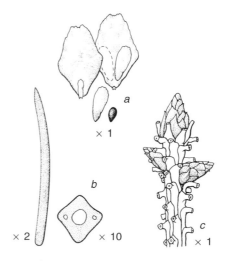

a. Cone scale; outer surface with bract (left), inner surface with winged seed (right), winged seed and seed (below). b. Needle and needle cross section. c. Twig terminal.

Douglas-Fir

Coast Douglas-fir, common Douglas-fir, green Douglas-fir, Douglas

Pseudotsuga menziesii (Mirb.) Franco var. *menziesii*
[syn. *P. taxifolia* (Lamb.) Britt.; *P. douglasii* (Carrière)]
Pinaceae: Pine Family

Douglas vert

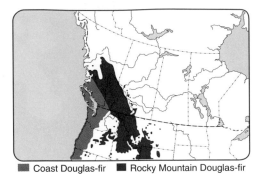

■ Coast Douglas-fir   ■ Rocky Mountain Douglas-fir

Worldwide the *Pseudotsuga* genus comprises 8 species; 2 are native to western North America and 6 to eastern Asia. The species native to Canada has 2 varieties — a coastal form, Douglas-fir, and an interior form, Rocky Mountain Douglas-fir. The coastal variety is described first; it occurs on the Pacific coast from central British Columbia to California.

Leaves Evergreen, remaining on the tree for 5–8 years; flat, flexible, 2–3 cm long, grooved above; often sharp-pointed; narrowed at the base into a slender short stalk, set on a leaf-cushion; upper surface bright yellowish-green, lines of white dots beneath. Needles spirally arranged along the twig; on horizontal branches, spreading out from the sides in 2 ranks; or spreading out from 3 sides and moderately parted on the upper side. Midvein prominent. Resin ducts (2) visible in cross section. Slight apple odor when crushed.

Buds Terminal bud narrowly conical, up to 10 mm long, sharp-pointed, with many shiny reddish-brown overlapping scales; a small cluster of lateral buds occurs just below and separated from the terminal bud by a few needles; other lateral buds scattered along the twig; pollen-cone buds are paler than the shoot buds; seed-cone buds are larger.

Twigs Moderately stout and flexible, greenish-brown becoming grayish-brown, hairy; leaf-cushion a low ridge of bark darkened at the forward end. New shoot and its needles preformed in the bud; the longest new side shoots develop just back of the annual node; those farther back are progressively shorter, mostly 2-ranked. Preformed shoot growth ceases in early summer; the leading shoot on vigorous young trees may continue to elongate by neoformed growth. Precocious shoot development from new buds may occur on vigorous shoots during summers with high rainfall. After the needles have fallen, the branchlets show slightly raised oval scars on the leaf-cushions.

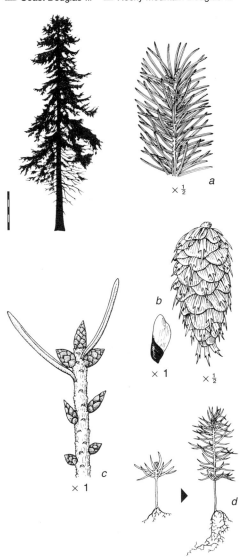

a. Twig and needles. *b.* Winged seed (left); cone (right). *c.* Twig terminal. *d.* Seedling development; cotyledon stage (left), end of 1st season (right).

Principal branches in irregular whorls, with lesser branches in between.

Pollen cones (left) pendulous; seed conelets (right) erect, at end of twig.

Seed cones; bracts 3-pronged, longer than the cone scales.

Young bark smooth, resin-blistered.

Trunk branch-free, cylindrical; mature bark deeply furrowed.

Pollen Cones Cylindrical to conical, fleshy, catkin-like, yellow to orange-red, 10–20 mm long, pendulous; borne in the upper-middle, middle, and lower parts of the crown, in leaf axils of previous year's twigs. Pollination takes place in spring as new leafy shoots start to elongate; cones then wither and fall away.

Seed Cones Conelets borne in the upper part of the crown, in leaf axils near the forward end of previous year's twig. At pollination, oblong, about 30 mm long, green to purple to red, erect, short-stalked, with distinctive 3-pronged bracts extending beyond the scales and partially obscuring them. After pollination, the cone stalk bends downward. Mature in late summer of 1st season.

Mature cones narrowly ovoid, 6–9 cm long, yellowish-brown to purplish-brown, pendulous on stout stalks. Scales numerous, broad, rounded, leathery; bracts prominent, 3-pronged, longer than the scales. Seeds shed throughout fall, winter, and spring; cones drop intact after seed dispersal. Seed production begins at 10–30 years and continues for several hundred years, with abundant seed crops at intervals of 2–11 years.

Seeds Somewhat triangular, 5–7 mm long, shiny reddish brown; wing 15–18 mm long. Viable for 10–20 years when kept dry and cool in airtight containers. Germination after storage is improved by moist prechilling.

Seedlings Newly germinated seedlings about 4 cm tall, with a stalk surmounted by a whorl of 5–10 upcurved, green cotyledons, about 2 cm long with smooth edges. Needles form above the cotyledons, clustered at first, later spreading along the new shoot, which can be up to 9 cm long. Seedlings require some shade in the 1st year, but then thrive in full sunlight.

Bark Gray, smooth, thin, resin-blistered when young; becoming deeply furrowed with irregular, broad, dark reddish-brown ridges; up to 30 cm thick.

Wood Moderately heavy and hard, exceptionally strong; heartwood reddish-brown, sapwood yellowish-white; marked contrast between earlywood and latewood; resin ducts present.

Size and Form One of the largest of Canada's native conifers; commonly up to 60 m high, 200 cm in diameter, and 500 years old. The tallest measured tree was more than 115 m high. Ring counts show some ages over 1000 years. After comparatively slow growth for the 1st decade, annual height growth up to 1 m may continue for up to 100 years or more.

Old trees characterized by long, branch-free, cylindrical trunks and short, columnar, flat-topped crowns; young trees by narrowly conical crowns that often extend to the ground. Principal branches in irregular whorls at the annual nodes, lesser branches in between. Dead branches may remain on the trunk for years. Root system strong, wide-spreading.

Habitat Found on a variety of soils, but grows best on deep, well-drained, sandy loams and where moisture in the soil and atmosphere is plentiful; does not thrive on compacted, poorly drained or limestone soils. Commonly a pioneer species that regenerates after forest fires, logging, and other disturbances. Less shade-tolerant than its associates, western hemlock, amabilis fir, western redcedar, and grand fir; in the course of succession, these species replace Douglas-fir. Under natural conditions, forest fires are necessary to continue the presence of Douglas-fir; old trees are particularly resistant to fire damage.

Notes The genus name *Pseudotsuga*, meaning "false hemlock," was created in 1867. Before then, the genus, first described by Archibald Menzies in 1791, had been classified as pine, hemlock, and fir.

A source of wood pulp and lumber; used for structural purposes, in shipbuilding, and in the production of such items as laminated beams, interior and exterior finishings, boxes, railway ties, and when impregnated with a preservative, in piling and decking for marine structures.

A source of food and shelter for wildlife. Frequently used for landscape and forestry purposes in North America (Zones C7, NA6), western Europe, New Zealand, and other cool moist temperate regions. Popular as Christmas trees.

Quick Recognition Large pendulous seed cones with protruding 3-pronged bracts separate Douglas-fir from all other conifers. Flat flexible needles with a prominent midvein distinguish it from spruce; stouter twigs from hemlock; reddish-brown sharp-pointed non-resinous buds from the true firs.

Rocky Mountain Douglas-Fir

Blue Douglas-fir, interior Douglas-fir

Pseudotsuga menziesii var. *glauca* (Beissn.) Franco

Douglas bleu

Occurs in the mountains of southern British Columbia, southward into the United States and Mexico. Differs from the coastal variety as follows: **Needles** distinctly bluish-green, with a strong odor when crushed, often not parted on the upper side. **Seed cones** less than 8 cm long, bracts usually bent back. **Trunk** tapered, long, limby **crown**; principal **branches** more ascending. Stocky, smaller, slower growing, less important commercially (up to 40 m high, 100 cm in diameter, and 300 years old). Occurs in pure stands or as a dominant component of mixed stands with ponderosa pine, lodgepole pine, western redcedar, western larch, and western white pine. It can thrive under colder and drier conditions than the coastal variety. Typical forms of Douglas-fir and Rocky Mountain Douglas-fir are quite distinct, but many intermediate forms occur.

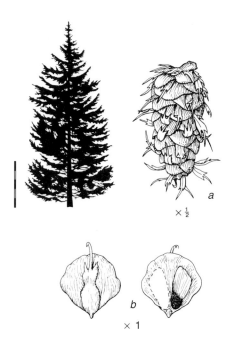

a. Seed cone with reflexed bracts. *b.* Cone scale outer surface with bract (left), inner surface with winged seed (right).

Hemlocks

Genus *Tsuga*
Pinaceae: Pine Family

Les pruches

Hemlocks occur in the temperate parts of North America and eastern Asia. Worldwide about 10 species are recognized; of the 4 species native to North America, 3 are found in Canada: western hemlock and mountain hemlock in British Columbia, eastern hemlock from Lake Superior to Cape Breton Island.

Leaves Evergreen, remaining on the tree for 3–10 years; flexible, up to 22 mm long; tip blunt, rounded or notched; narrowing abruptly at the base to a thread-like stalk pressed against the twig and attached to the forward end of a leaf-cushion. Needles spirally arranged, but on the 2 common species stalks are curved and twisted creating 2 ranks. Midvein prominent. A single resin duct evident in cross section just below vein.

Buds Terminal bud small, ovoid, about 2 mm long, non-resinous; lateral buds spaced along the twig. On more vigorous twigs, buds are found only near the base and tip, with short side branches intervening.

Twigs Slender, flexible, hairy; leaf-cushions prominent as ridges of bark extending about 3 mm below each leaf, separated by incised grooves. After the needles fall, the raised leaf scars face obliquely forward. Spring flush of shoot growth is preformed in the bud; subsequent shoot growth is neoformed progressively during the growing season. The more vigorous shoots bear neoformed side branches in the midposition. Side branches on previous year's twig decrease in length back from the annual node.

Pollen Cones Globular, catkin-like, stalks short and scaly, borne in the lower part of the crown, in the leaf axils of the previous year's twig. Pollination takes place in spring; cones wither and fall away.

Seed Cones Conelets, erect, solitary, borne at the tips of shorter twigs of previous year, mostly in the upper crown; bracts longer than the scales. Ready for pollination in spring just before the new shoots develop; mature in 1 season.
    Mature cones ovoid, pendulous from the branch tips, opening when mature but

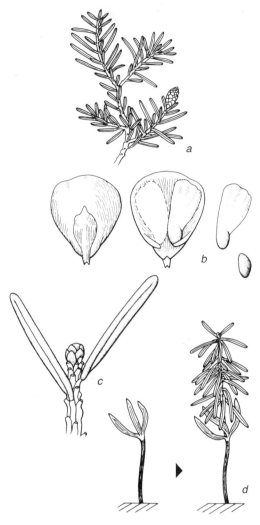

a. Twig with needles and seed conelet. *b.* Cone scale; outer surface with bract (left), inner surface with winged seed (center), winged seed and seed (right). c. Twig showing terminal bud, leaf-pegs, and leaf attachment. *d.* Seedling development; cotyledon stage (left), leaf stage at end of 1st growing season (right).

remaining on the tree until the following summer. Bracts usually shorter than the scales, separate from them.

Seeds Small, with relatively large wings, dotted with minute resin blisters.

Seedlings Newly germinated seedlings about 2 cm tall, with a stalk surmounted by a whorl of 3 upcurved, green cotyledons, about 15 mm long. Needles form above the cotyledons, clustered at first, later spreading along the new shoot.

eastern hemlock
Leading shoot oblique; principal branches irregularly spaced.

western hemlock
Seed conelets solitary, at tips of short twigs.

western hemlock
Pollen cones globular, on short scaly stalks.

Bark Furrowed, scaly, showing purplish layers when freshly cut.

Wood Hard compared with most conifers; sapwood not sharply distinct from heartwood, latewood distinct from earlywood; no resin ducts; fragrant when freshly sawn.

Size and Form Medium-sized to very large trees. Leading shoot oblique, slender, flexible, curving away from the prevailing wind. In young trees: crown dense, conical or columnar; principal branches horizontal or slightly drooping. In older trees: crown uneven; principal branches irregularly spaced, wide-spreading, with numerous shorter branches in between. Dead branches remain on the trunk for years. Root system shallow, wide-spreading.

Habitat Usually found in regions of high rainfall, and locally in moist locations such as northern slopes, the borders of streams and lakes, and near swampy areas; occurs in

pure stands or mixed with other species. Among the most shade-tolerant of trees; grows under large trees.

Notes Planted as ornamentals because of graceful branches and fine foliage of younger trees; good hedge trees, amenable to shaping and trimming. Important to wildlife as a source of browse, seeds, and shelter. The bark is rich in tannin.

Unrelated botanically to the herbaceous species of hemlock (*Conium* sp.) used by Socrates to poison himself. Also unrelated to ground-hemlock, or Canada yew (*Taxus canadensis*).

Quick Recognition Needles stalked; twigs slender, flexible, ridged, grooved, with forward-pointing leaf-pegs; cones solitary, pendulous at the tips of shorter branches; leading shoot oblique.

Western Hemlock
Pacific hemlock
Tsuga heterophylla (Raf.) Sarg.
Pruche de l'Ouest

Occurs on the Pacific coast, adjacent islands, and in the mountains up to 1500 m.

Leaves Sides parallel, flat, finely toothed, length variable, 5–20 mm, blunt-tipped; upper surface shiny dark green and grooved; whitened beneath with ill-defined lines of white dots on either side of the midvein; needles clearly arranged in 2 ranks, with a few shorter ones on the upper side pressed against the twig.

Buds Globular, 2–3 mm long.

Twigs Slender, brownish-gray, hairy.

Seed Cones Ovoid, 20–25 mm long, blunt-tipped, short-stalked, golden brown; scales rectangular, tip rounded, margin smooth or faintly toothed. Cones open in autumn; seeds shed gradually; cones may stay on the tree for 1 or 2 years.

Bark Smooth when young, reddish-brown; becoming darker, deeply furrowed with flat-topped scaly ridges.

Wood Moderately light, fairly hard and strong; whitish to dull light brown; little contrast between sapwood and heartwood.

Size and Form Very large trees, up to 50 m high, 120 cm in diameter, and 500 years old; occasionally larger. Lower trunk long, branch-free. Crown open, conical, becoming irregular with age; leading shoot oblique, bending away from prevailing wind. Principal branches coarse, spreading horizontally from the trunk; drooping sprays of branchlets giving a graceful appearance. Root system shallow, wide-spreading.

Habitat Occurs on a variety of soils; often the dominant species in a forest; grows in pure stands, but usually mixed with western redcedar, Douglas-fir, grand fir, black cottonwood, and red alder. Very shade-tolerant; regenerates well under a closed canopy; seedlings commonly found on rotten logs, or in partially decomposed forest litter. Thrives in full light and in shade. Plentiful moisture in soil and atmosphere required for regeneration and good growth.

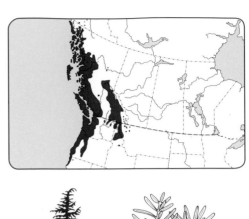

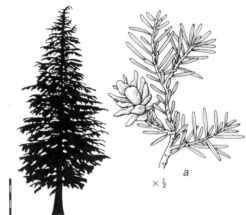

a. Twig with needles and open seed cone. *b.* Cone scale; outer surface with bract (left), inner surface with winged seed (center), winged seed inner surface and seed (right). c. Seed cone. *d.* Twig with lateral buds (left); needle cross section (right).

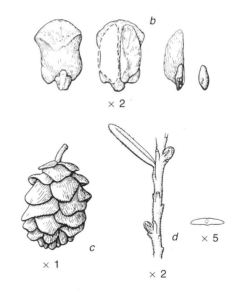

5

Seed cones ovoid; scales rectangular.

Young bark smooth, reddish-brown.

Mature bark deeply furrowed, with flat-topped scaly ridges.

Notes An important source of wood pulp and lumber; used in general construction, for siding, flooring, plywood, railway ties, boxes, and crates.

Susceptible to damage by forest fires and windthrow; regenerates well, grows rapidly. Used in forestry plantations in moister parts of western Europe.

Quick Recognition [See mountain hemlock for contrasting features.] Needles 2-ranked, variable in length, flat in cross section, lines of white dots on the lower surface; cones small with few scales; leading shoot oblique.

Mountain Hemlock
Black hemlock

Tsuga mertensiana (Bong.)
Carrière

Pruche subalpine

Found in the wetter areas of the subalpine forest at elevations of 750–1800 m; and at about sea level in the coastal and interior forests of northern British Columbia and Alaska.

Leaves Rounded in cross-section, 20–30 mm long, blunt-tipped; both surfaces dark bluish-green with faint lines of white dots; needles crowded, spreading all around the twig, especially on the upper side.

Buds Ovoid, 5 mm long, reddish-brown; outer scales with narrow, pointed tips.

Twigs Slender to stoutish, reddish-brown, hairy; side branches of unequal length, forming a tufted spray.

Seed Cones Oblong cylindrical, 30–80 mm long, purplish-brown; scales thickened, broad, fan-shaped, margins slightly roughened or toothed. Cones open in autumn, spreading very widely, becoming bent back towards the cone base during winter after the seeds have been shed; cones fall off during the spring or early summer.

Seedlings Seeds germinate in spring, even on snow; seedlings grow best in partial shade.

Vegetative Reproduction Trees growing on muskeg or near the alpine tree line may reproduce by layering (rooting of attached branches).

Bark Dark reddish-brown, scaly, divided into hard, narrow, flat-topped ridges.

Wood Moderately light, fine-grained, relatively hard and strong; heartwood light reddish-brown, not sharply distinct from the sapwood; similar to western hemlock.

Size and Form Small to large trees, usually up to 15 m high, 50 cm in diameter, and several hundred years old; occasionally 45 m high; low-spreading shrub on exposed ridges at high elevations. At lower elevations, trunk strongly tapered, bearing slender branches

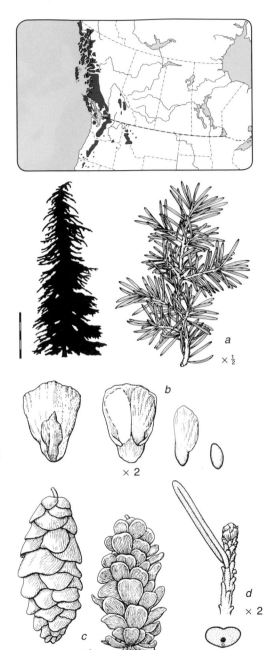

a. Twig and needles. *b.* Cone scales; outer surface with bract (left), inner surface and winged seed (center), winged seed and seed (right). *c.* Seed cones, closed (left) and open (right). *d.* Twig terminal with needle (above); needle cross section (below).

Seed cones cylindrical; scales fan-shaped.

Young bark smooth,
becoming scaly.

Mature bark has narrow, flat,-topped ridges.

with upturned tips, almost to the ground;
crown narrowly conical, becoming irregular
and bent or twisted on old trees; leading
shoot oblique. Root system shallow, wide-
spreading.

Habitat Grows best on deep moist soils on
north slopes; found in pure stands or mixed
with subalpine fir, amabilis fir, Engelmann
spruce, subalpine larch, whitebark pine, and
lodgepole pine.

Notes Useful in protecting steep slopes
against erosion and holding back snow; an
attractive part of mountain scenery.

Quick Recognition [Contrasting features
of western hemlock in brackets]. A tree of the
subalpine [coastal] forests; needles radially arranged [2-ranked], rounded in cross section [flat],
lines of white dots on both surfaces [only on
lower surface]; cones larger with more scales;
scales fan-shaped [rectangular]; branchlets form
a tufted [flat] spray.

Eastern Hemlock

Tsuga canadensis (L.) Carrière

Pruche du Canada

The only hemlock native to eastern Canada; a prominent tree in the forests of the Maritime provinces.

Leaves Slightly tapered, flat, finely toothed, 10–20 mm long, tip blunt, rounded or notched; upper surface shiny green and grooved; whitened beneath with well-defined lines of white dots on either side of the mid-vein, with a clear green margin; needles clearly arranged in 2 ranks, with a few shorter ones on the upper side.

Buds Ovoid, about 2 mm long, brownish, hairy.

Twigs Slender, yellowish-brown, hairy; arranged in flat sprays.

Seed Cones Ovoid, 12–20 mm long, purplish-brown, pointed; stalks 2–3 mm long, slender, hairy; scales few, thin, roundish, margins smooth or faintly toothed. Cones open in autumn; seeds shed in late autumn and winter; cones stay on the tree into the following season.

Bark Scaly when young; becoming deeply furrowed with dark brown, broad, flat-topped ridges. Outer bark with dull reddish-purple layers; inner bark bright reddish-purple.

Wood Weak; heartwood light orange-yellow.

Size and Form Medium-sized trees, up to 30 m high, 100 cm in diameter, and about 600 years old. Trunk strongly tapered; leading shoot oblique. Crown dense, conical, becoming ragged and irregular with age; branches slender and flexible, spreading horizontally from the trunk, drooping at the end. Dead branches persist. Root system shallow, wide-spreading.

Habitat Grows on various types of soil, but requires a cool moist site; found in pure stands or mixed with yellow birch, eastern white pine, red spruce, white spruce, sugar maple, and American beech. Very shade-tolerant; small trees persist in closed stands for many decades.

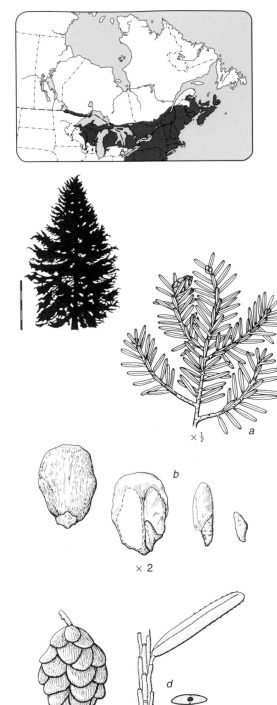

a. Twig with needles and seed conelet. b. Cone scale; outer surface with bract (left), inner surface with winged seed (center), winged seed inner surface and seed (right). c. Seed cone. d. Twig showing leaf-pegs and needle attachment (left); needle cross section (right).

Seed cones ovoid; scales rounded.

Young bark scaly.

Mature bark dark brown, deeply furrowed.

Notes Used for coarse lumber. Separation between the annual rings ("ring shake") and along radial lines ("star shake") often occurs in living trees, resulting in lumber that is subject to splitting and brittleness. Knots in the lumber are hard enough to dull a saw or deflect a nail. Not suitable for camp fires because the burning wood throws off sparks.

The bark was formerly used as a commercial source of tannin, and many bare logs were left in the woods.

Quick Recognition Distinguished from all eastern native conifers by its slender twigs and oblique leading shoot.

Group 6.

Needles evergreen, single, flat; twigs and buds green; single seed in a fleshy cup

Group 6 consists of one genus, yew (*Taxus*). Yews are distinguished by green twigs and buds and a single seed in a red fleshy cup-shaped aril. Their needles are similar to those of species in Group 5; however, the tip is drawn out to an abrupt point, the base narrows to a definite stalk, the edges are somewhat rolled under, and the lines of white dots are lacking.

Yews

Genus *Taxus*
Taxaceae: Yew Family

Les ifs

A small genus of 8–10 similar species, all shrubs or small trees; 3 species are native to North America, 2 to Canada: western yew and Canada yew, an eastern species often called ground hemlock. Introduced species and their cultivars are frequently used for landscaping. Species, especially cultivars, are difficult to identify.

Leaves   Evergreen, remaining on the tree for about 8 years, needle-like, 15–30 mm long, flattened, somewhat curved, midvein prominent, upper surface dark green, paler green beneath; edges rolled under; tip sharp, abruptly pointed; base tapered to a short slender stalk attached to the forward end of a flat leaf-cushion; spirally arranged, often appearing to be 2-ranked or spreading from 3 sides. No resin ducts.

Buds   Terminal bud green, small, rounded, with thin, closely overlapping scales; lateral buds similar but smaller.

Twigs   Green, becoming greenish-brown or reddish-brown, slender, flexible, hairless; leaf-cushions evident as flat ridges extending a few millimetres below each leaf, separated by incised grooves.
   Spring flush of shoot growth is preformed in the bud; subsequent shoot growth is neo-formed during the growing season. Vigorous new shoots often bear side shoots. Buds may develop precociously during the growing season.

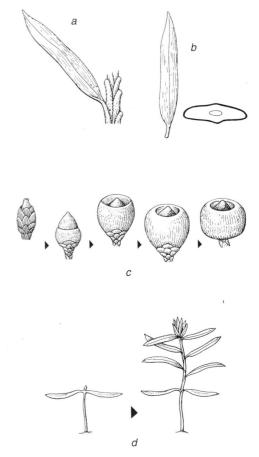

a. Needle on twig showing leaf-cushions. *b.* Needle and needle cross section. *c.* Aril development. *d.* Seedling development; cotyledon stage (left), leaf stage (right).

western yew
Pollen cones globular, on underside of twigs.

Japanese yew
Ovules minute, greenish, on separate trees
from pollen cones.

Pollen Cones Globular, small, yellow,
6–12 scales; on short stalks from winter
buds on the underside of previous year's
twigs. Pollen cones and seed-bearing
structure on separate trees.

Seed-bearing Structure A single greenish
ovule set on the end of a short scaly stalk
coming from a lateral winter bud on previous
year's twig. Wind-pollinated. Ripens the
same autumn.

Seeds Hard, dark-blue, about 8 mm long,
erect, set in a reddish, cup-shaped, fleshy
aril. Germination requires moist prechilling.
Dispersed by seed-eating birds.

Seedlings With 2, short, flat, sharp-pointed
cotyledons.

Vegetative Reproduction By stump
sprouts and layering (rooting of attached
branches); can be propagated by rooted
cuttings.

Bark Thin, with large scales, dark reddish-
brown to purplish-brown; newly exposed bark
dark reddish-purple.

Wood Heavy, hard, strong, resilient, even-
grained, fine-textured, decay-resistant, can
be polished; reddish to dark orange with yellow sapwood; annual rings distinct, rays not
visible; no resin ducts.

Size and Form Shrubs or small to
medium-sized trees up to 20 m high, 40 cm

in diameter, and several hundred years old.
Trunk often twisted and fluted; crown irregular, often with multiple tops; branches may
be as long as the tree is tall.

Habitat Usually a scattered tree of the
understory of conifer or northern broadleaf
forests. Very shade-tolerant; but thrives in full
sunlight.

Notes Yew wood is used for specialty items
such as canoe paddles, tool handles, and
carved ornaments; during the Middle Ages, it
was prized for archery bows.

Yews are planted for many landscape
purposes; they can withstand trimming and
shaping; are easily transplanted; and tolerate
urban pollution (but not road salt). Cultivars
vary widely in form from columnar to
spreading.

The needles of yew are poisonous to
horses and cattle, but yew browse is a common food for moose, deer, and elk; the seeds
are also poisonous, but the fleshy covering is
edible, at least for birds.

Taxol, a natural product of the bark,
needles, and twigs of western yew, is a
potent anti-cancer drug.

Quick Recognition On seed-bearing yews,
the single dark seed in a reddish fleshy cup.
Leaves plain green, flat, abruptly pointed, with
definite stalks; twigs green, with flat ridges
below each leaf. Buds green.

Western Yew

Pacific yew, yew

Taxus brevifolia Nutt.

If de l'Ouest

A small western tree, sometimes a shrub; up
to 20 m high; crown conical, with spreading
irregular branches, flat or slightly drooping
branchlets; needles 13–25 mm, needles
dark yellowish-green above, pale below; bud
scales keeled acute, loose.

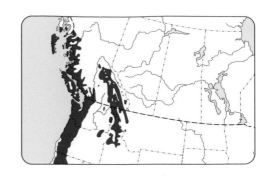

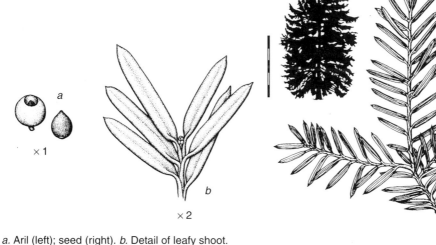

× 1

× 2

a. Aril (left); seed (right). *b.* Detail of leafy shoot.
c. Leafy shoot.

× ½

English Yew*

Taxus baccata L.

If commun

A medium-sized tree native to Europe, up to
25 m high, 100 cm in diameter; trunk often
massive, short; needles 10–30 mm, 2-
ranked; throws heavy shade; many cultivars.
Hardy as far north as Zones C7, NA6.

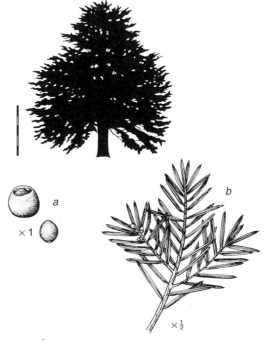

× 1

a. Aril (above); seed (below). *b.* Leafy shoot with
ovules.

× ½

6

Seed borne in center of fleshy aril.

Western Yew

Young bark smooth.

Trunk often fluted and twisted. Mature bark reddish-to purplish-brown.

Japanese Yew*

Taxus cuspidata Siebold & Zucc.

If du Japon

A small tree native to Japan; needles keeled, sickle-shaped, tend to radiate, pale green above, 2 yellowish-green bands below; bud scales triangular; many cultivars. Hardy as far north as Zones C4, NA4.

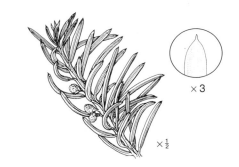

× 3

× ½

Leafy shoot with pollen cones; detail of undersurface of leaf at tip.

Canada Yew

Ground-hemlock

Taxus canadensis Marsh.

If du Canada

A straggling multistemmed shrub, common in eastern Canada; 2 m high; no trunk; stems trailing in the humus layers, with upturned tips; needles pale green; grows in colonies.

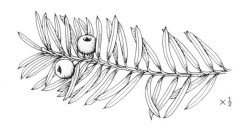

× ½

Leafy shoot with arils.

Group 7

Leaves fan-shaped, thin, notched; veins parallel, no midvein; seeds plum-like

The only species in Group 7 is ginkgo. Although its leaves are broad and deciduous, ginkgo is classified as a member of the division Pinophyta because its ovules are borne naked on stalks, not within the pistil of a flower.

Ginkgo*

Maidenhair-tree

Ginkgo biloba L.
Ginkgoaceae: Ginkgo Family

Ginkgo bilobé

Ginkgo is native to China, where it grows naturally only in a few locations and is mostly planted in temple gardens. It has become greatly favored for city planting in temperate regions throughout the world because of its attractive form and interesting leaves, freedom from insect pests and diseases, and tolerance to pollution. Hardy as far north as Zones C4, NA3.

The ginkgo family is represented by this one genus with one species; in previous geologic eras, there were several genera and species in various parts of the world.

Leaves Deciduous, fan-shaped, broad, about 7 cm wide, with shallow irregular teeth and 1 or more notches; green, turning yellow in autumn; veins appearing straight and parallel, but forked dichotomously, no midvein; stalks slender, about 5 cm long; in clusters of 3–5 on dwarf shoots, alternate and spiral on long shoots. Leaves neoformed during the growing season are more likely than preformed leaves to have more than 1 notch.

Buds Terminal bud broadly conical, 3 mm long, smooth, reddish-brown; lateral buds diverging from the twig.

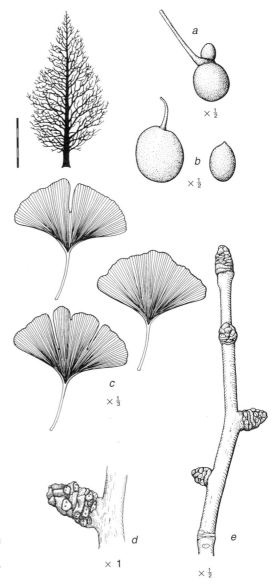

a. Ovules and stalk. *b.* Mature seed (left); seed with fleshy coat removed (right). *c.* Typical leaf shapes. *d.* Dwarf shoot showing leaf scars and annual rings. *e.* Twig and dwarf shoots.

Pollen cones catkin-like.

Ovules in pairs on stalks; on separate trees from pollen cones.

Seeds plum-like when ripe.

Young bark (inset) shallowly fissured. Mature bark rough, deeply fissured.

Twigs Stout, smooth, grayish-brown, leaf scars with 2 vein scars; dwarf shoots prominent, covered with circles of leaf scars, each marking a year's growth. A dwarf shoot may occur at the tip of a long twig.

Pollen Cones Catkin-like, 3–6 cm long, short-stalked, in small clusters, limp and pendulous from dwarf shoots, deciduous after pollen is shed. Pollen cones and seed-bearing structure on separate trees.

Seed-bearing Structure Develops with the leaves on dwarf shoots; consists of a long stalk with 2 ovules, each resembling a small acorn; one usually aborts, the other matures in the autumn. Mature structure resembles a plum, about 3 cm long; kernel (embryo) edible, enclosed in a hard shell covered by a thick fleshy coat. Seeds shed in autumn; kernel ripens in the succeeding weeks.

Seedlings Slender stalk surmounted by cotyledons.

Vegetative Reproduction By stem cuttings.

Bark Ash-gray, rough, becoming fissured.

Size and Form Medium-sized trees, up to 25 m high, 80 cm in diameter, and several hundred years old. Trunk tapering, distinct to the top. Crown slender with a few larger branches, mostly ascending, lower ones sometimes horizontal.

Notes Seed kernel an article of commerce in the Orient. Male trees are preferred for landscape purposes because they lack the offensive odor produced by the fleshy coating of the seed.

> **Quick Recognition** Leaves fan-shaped, with parallel veins; terminal bud broadly conical; twigs stout; dwarf shoots prominent.

The Broadleaf Trees

The second part of the book (Groups 8–12) deals with the broadleaf trees — in botanical terms, trees in the class Magnoliopsida within the division Magnoliophyta of the plant kingdom. These trees have true flowers. The essential characteristic of a flower is the presence of a pistil with an ovary containing ovules; fertilized ovules become seeds. Flowers vary from large and showy to small and inconspicuous; some are arranged in catkins, some in clusters, others singly. However, the seed-producing parts of the flower are always the same: the pistil with its ovary and ovules; and the stamens, which produce pollen. Pistils and stamens may occur in the same flower, or in separate flowers that are borne on the same tree or on separate trees. The "normal" flower has a **corolla** made up of petals, often colored; below the corolla is a **calyx** made up of leaf-like sepals, usually green; one or both may be absent.

The seeds of broadleaf trees are borne in fruits and contain two cotyledons (seed leaves). The wood (xylem) is distinctive; its strength derives from fibers — long (a few millimeters), narrow, thick-walled cells; its ability to conduct water comes from vessels — wide-diameter (up to 1 mm), open-ended cells arranged end to end to form tubes (seen as pores in cross section).

The broadleaf trees described in this book fall into 33 families and 65 genera.

Other terms used for broadleaf trees are angiosperms, dicotyledons, flowering trees, deciduous trees, and hardwoods.

Families and Genera
Groups 8 to 12

The Broadleaf Trees (Division Magnoliophyta, Class Magnoliopsida)

Synopsis of the families and genera in the broadleaf trees section of this book, based on Gleason and Cronquist (1991), and Harlow et al. (1991).

Family	Genus
Magnoliaceae, magnolia	*Liriodendron,* tulip-tree; *Magnolia,* magnolia
Annonaceae, custard-apple	*Asimina,* pawpaw
Lauraceae, laurel	*Sassafras,* sassafras
Cercidiphyllaceae, katsura	*Cercidiphyllum,* katsura-tree
Hamamelidaceae, witch-hazel	*Hamamelis,* witch-hazel; *Liquidambar,* sweetgum
Platanaceae, sycamore	*Platanus,* sycamore
Ulmaceae, elm	*Ulmus,* elm; *Celtis,* hackberry; *Zelkova,* zelkova
Moraceae, Mulberry	*Morus,* mulberry; *Maclura,* osage-orange
Myricaceae, wax-myrtle	*Myrica,* bayberry
Juglandaceae, walnut	*Juglans,* walnut; *Carya,* hickory
Fagaceae, beech	*Castanea,* chestnut; *Fagus,* beech; *Quercus,* oak
Betulaceae, birch	*Alnus,* alder; *Betula,* birch; *Carpinus,* blue-beech *Corylus,* hazel; *Ostrya,* hop-hornbeam
Tiliaceae, linden	*Tilia,* basswood
Salicaceae, willow	*Populus,* poplar; *Salix,* willow
Ericacaeae, heath	*Arbutus,* arbutus; *Rhododendron,* rhododendron
Rosaceae, rose	*Amelanchier,* serviceberry; *Crataegus,* hawthorn; *Malus,* apple; *Prunus,* cherry; *Sorbus,* mountain-ash
Caesalpiniaceae, cassia (formerly a subfamily of Leguminosae)	*Cercis,* redbud; *Gleditsia,* honey-locust; *Gymnocladus,* coffeetree
Fabaceae, bean (formerly a subfamily of Leguminosae)	*Caragana,* pea-tree; *Cladrastis,* yellow-wood; *Laburnum,* golden-chain; *Robinia,* black locust
Elaeagnaceae, oleaster	*Elaeagnus,* oleaster; *Hippophae,* sea-buckthorn; *Shepherdia,* buffalo-berry
Cornaceae, dogwood	*Cornus,* dogwood; *Nyssa,* tupelo
Celastraceae, staff-tree	*Euonymus,* euonymus
Aquifoliaceae, holly	*Ilex,* holly; *Nemopanthus,* mountain-holly
Rhamnaceae, buckthorn	*Rhamnus,* buckthorn
Hippocastanaceae, horsechestnut	*Aesculus,* horsechestnut
Aceraceae, maple	*Acer,* maple
Anacardiaceae, cashew	*Cotinus,* smoke-tree; *Rhus,* sumac; *Toxicodendron,* poison-sumac
Simaroubaceae, quassia	*Ailanthus,* ailanthus
Rutaceae, rue or citrus	*Phellodendron,* cork-tree; *Ptelea,* hoptree; *Zanthoxylum,* prickly-ash
Araliaceae, aralia	*Aralia,* aralia
Oleaceae, olive	*Fraxinus,* ash; *Syringa,* lilac
Bignoniaceae, trumpet-creeper	*Catalpa,* catalpa
Rubiaceae, madder	*Cephalanthus,* button-bush
Caprifoliaceae, honey-suckle	*Sambucus,* elder; *Viburnum,* viburnum

Group 8.

**Leaves in opposite pairs
(or subopposite or whorled);
blade simple or compound;
edges lobed, toothed, or smooth**

The opposite arrangement of leaves and buds is one of the best criteria for distinguishing different kinds of trees. It is unmistakable and holds true during any season, any day of the year, and can be recognized even when all branches are out of reach.

In this arrangement, two leaves occur at one position on opposite sides of the stem; adjacent opposite pairs are usually at right angles to each other. Trees with opposite leaves have buds in opposite pairs, and (except for horsechestnut) side branches in opposite pairs. The contrasting arrangement is alternate, only one leaf at any position.

In the subopposite arrangement, one leaf is slightly lower than the other. In the whorled arrangement, there are three or more leaves at one position.

A leaf is simple if the blade is in one piece, not divided into leaflets. It is compound if the blade is divided into two or more leaflets, each attached to a central leaf-stalk.

The edge of the blade may be lobed, with large projections and/or notches; toothed, with smaller projections, large enough to be seen with the naked eye; or smooth, with no projections.

All trees in Group 8 are deciduous; winter-creeper euonymus is evergreen.

Alternate-leaf dogwood is included in this group so that it will be with the other dogwoods, all of which have opposite leaves. European buckthorn and purple-osier willow, both with subopposite buds and leaves, key out to Group 8, but are described in Group 11, with other species of their respective genera. Dawn redwood, with opposite needles and buds, keys out to this group, but is described in Group 2, with other conifers.

Maples

Box-elders

Genus *Acer*

Aceraceae: Maple Family

Les érables

Worldwide there are over 100 species of maple, distributed throughout the North Temperate Zone. About two-thirds of them are in eastern Asia; 13 are native to North America; 10 to Canada. All species are woody; they vary in size from shrubs to large trees.

Many maples, both native and introduced, are planted for their colorful leaves. The brilliant colors of maple forests in autumn are among the most splendid natural spectacles of eastern North America.

The following is a general description of the species in the genus; features that differ for various species are given in brackets.

Leaves Deciduous, in opposite pairs, simple, long-stalked; 3–9 prominent veins radiating from the stalk at the base of the leaf; usually palmately lobed with the number of lobes corresponding to the number of prominent veins; lobes toothed. [Manitoba maple: leaves composed of 3–9 leaflets pinnately arranged along a central stalk.]

Buds Terminal bud usually present, with 1–8 pairs of scales; lateral buds smaller, in opposite pairs. [Vine and Japanese maple: terminal bud may be absent.] Leaf scars crescent-shaped, with 3 vein scars. [Bigleaf maple: with up to 9 vein scars.]

Twigs Stiff, straight. Spring flush of growth is preformed in the bud; subsequent shoot growth is neoformed. Shoot growth ends with the formation of a terminal bud or flower cluster. [Vine and Japanese maple: twigs often terminate in a withered stub between a pair of lateral buds.]

Flowers Small, with 5 sepals and 5 petals [petals absent in some species]; in clusters. Pollen flowers and seed flowers may be in the same cluster, in separate clusters, or on separate trees; or flowers may be perfect, producing both pollen and seeds; a tree may have 1, 2, or 3 types of flowers. Appear before or with the leaves. Pollinated by insects or wind.

Fruits Winged; in joined pairs (rarely in 3's) on a single stalk; often separating when shed. Each fruit (often called a key) consists

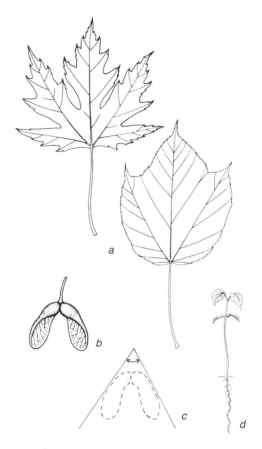

a. Leaf forms; deeply lobed (left); shallowly lobed (right). *b*. Fruit; paired keys on single stalk. *c*. Angle between outside edges of fruit wings is useful in species identification. *d*. Seedling with leaf-like cotyledons and primary leaves.

of a 1-seeded case and a long 1-sided wing. The angle between the wings in a pair is a useful feature in identifying species; the angle is measured with reference to the outer edge of the 2 wings. Wind-dispersed.

Seeds Remain within the fruit. Seeds ripening in autumn can be stored under cool moist conditions until the following spring; those ripening in spring or early summer are difficult to store.

Seedlings Newly germinated seedlings bear leaf-like cotyledons raised above the surface. [Silver maple: cotyledons green but may be retained within the seedcase.]

Vegetative Reproduction Often by stump sprouts; some species by layering (rooting of attached branches where they touch the ground).

8

Colorful maple foliage in autumn.

Wood Light-colored, straight-grained, uniform in texture; varies from species to species in hardness, toughness, and other properties; with a curly grain or bird's-eye figure in some trees. Diffuse-porous; rays small, but often visible without a hand lens.

Size and Form Shrubs to large trees. Leading shoot upright. Shoot growth from the pair of buds just below a terminal flower cluster frequently results in a forked stem.

Habitat Some species prefer wet sites; others grow mainly on uplands. Maples are a major component of many north temperate forests.

Notes Wood from some species is used for flooring, furniture, interior woodwork, plywood, veneer, and small woodenware.

Fruits, buds, and twigs are an important source of food for many species of birds and mammals.

Maple syrup and maple sugar can be derived from the sap of most species, most notably sugar maple; the sap flows in spring before the leaves appear.

The response of maple leaves to light is evident; leaves attached to the underside of a horizontal branch have longer stalks and larger blades than those on the upper side; leaf stalks are bent or twisted so that all leaves face up and are arranged so that all are well-exposed to light.

Quick Recognition Fruits paired, in clusters. Leaves opposite, palmately veined, and lobed [pinnately compound in Manitoba maple]. Terminal bud present in most species; bud scales opposite. Leaf scars crescent-shaped.

Sugar Maple
Hard maple, rock maple

Acer saccharum Marsh.

Érable à sucre

A characteristic tree of broadleaf forests in the Maritime provinces and in the southern parts of Ontario and Quebec. Frequently planted as a shade tree and for its spectacular autumn color (Zones C4, NA3).

Leaves 8–20 cm long, somewhat wider; 5 (occasionally 3) lobes with long blunt-pointed tips and a few irregular wavy teeth; central lobe almost square, separated from lateral lobes by wide rounded notches; upper surface deep yellowish-green, paler and hairless beneath; yellow to brilliant orange and bright red in autumn. Stalk 4–8 cm long.

Buds Terminal bud narrowly cone-shaped, 6–12 mm long, sharp-pointed, medium to dark brown, with 6–8 pairs of faintly hairy scales.

Twigs Shiny reddish-brown to green, hairless.

Flowers Small, with 5 greenish-yellow sepals but no petals, in drooping tassel-like lateral clusters (sometimes terminal) on slender stalks 30–70 mm long. Most flowers have both stamens and pistils, but usually only one of the organs is functional; both kinds of flowers on the same tree and often in the same cluster, where they tend to develop at different rates. Fully formed before the leaves appear.

Fruits Wings 30–35 mm long, slightly divergent. Seedcase plump. Keys in drooping clusters on slender stalks usually longer than the wings. Paired keys often shed as a unit; usually only one of the pair contains a viable seed. Seeds produced most years, with a good crop at intervals up to 7 years; often germinate and have fully expanded cotyledons in the spring while there is still snow on the ground.

Vegetative Reproduction Dormant buds at the base of most trees sprout vigorously if tree is cut down or damaged. Buried stems produce adventitious roots.

Bark Smooth, gray; becoming dark gray, divided into long, vertical, firm, irregular ridges that usually curl outward along one side, occasionally somewhat scaly.

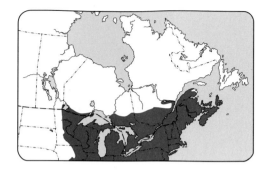

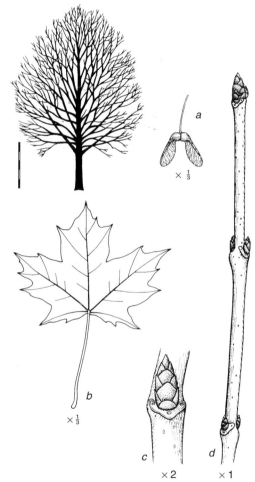

a. Fruit. *b*. Leaf. *c*. Lateral bud and leaf scar. *d*. Winter twig.

Wood Heavy, hard, strong, light yellowish-brown; often with curly grain, called bird's-eye. Diffuse-porous; rays easily visible. Turns green when a solution of ferric salts is applied (compare with red maple).

These flowers have functional pistils.

Seedcase plump; wings slightly divergent.

Young bark (inset) smooth, gray. Mature bark has vertical ridges curled outward along one side.

Size and Form Medium-sized to large trees, up to 35 m high, 90 cm in diameter, and 200 years old; occasionally larger. Trunk straight, often branch-free for two-thirds or more of its height. Crown narrow, round-topped, with short sturdy branches. Root system deep, wide-spreading.

Habitat Grows best on deep, fertile, moist, well-drained soils, with some lime content; also does well in deep soils on the Canadian Shield although they are low in lime. In pure stands, but usually mixed with other broad-leaf species, as well as with eastern white pine and eastern hemlock. Tolerates heavy shade and browsing for many years and then grows normally when released by an opening in the canopy. Decomposing leaves tend to enrich the soil by reducing the acidity and increasing the mineral content.

Notes Wood is used for furniture, flooring, toys, cabinetwork, veneer, plywood, turned woodenware, and cutting blocks.

Canada's national tree; a stylized version of its leaf is the central feature of the Canadian flag.

The sap of the sugar maple is the principal source for maple syrup and sugar. It requires about 40 litres of sap to make 1 litre of syrup.

Sugar maples occasionally exhibit a progressive dieback usually starting at the top. Some suspected causes are air pollution, soil conditions, drought and other weather conditions, and damage by diseases and insects.

Black maple is closely related and similar; intermediate forms are common.

Quick Recognition Leaves with 5 taper-pointed lobes; teeth few, irregular. Buds long-pointed, with 6–8 pairs of scales. Paired keys with stalks longer than the wings; wings slightly divergent. See black maple and Norway maple (leaves) for comparison.

Black Maple

Black sugar maple

Acer nigrum Michx.

Érable noir

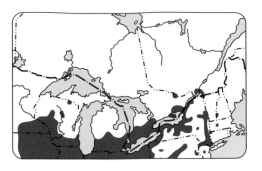

Similar to sugar maple; range extends farther west, but not as far north and northeast; features differing from those of sugar maple are noted below.

Leaves Usually 3 lobes, with a few, indistinct teeth; sometimes indistinctly 5-lobed; central lobe somewhat tapered, separated from lateral lobes by open, shallow notches; upper surface dark green; yellowish-green beneath, with dense, brownish, velvety hairs; yellow to brownish-yellow in autumn, seldom red like sugar maple. Stalk 6–10 cm long, hairy. Leaves have a characteristic wilted appearance.

Buds Scales blunt, dark brown, hairy.

Twigs Dull.

Flowers Stalks hairy, 18–50 mm long.

Fruits Wings parallel or converging slightly. Keys on hairy stalks about the same length as the wings; paired keys may separate when shed, leaving the stalk on the tree.

Bark Dark gray, with long, narrow, vertical, firm irregular ridges, deeply furrowed, often scaly.

Habitat Occurs most frequently on moist fertile floodplains and bottomlands — moister sites than those where sugar maple grows best.

Notes May be variety of sugar maple; trees with intermediate features frequently occur.

8

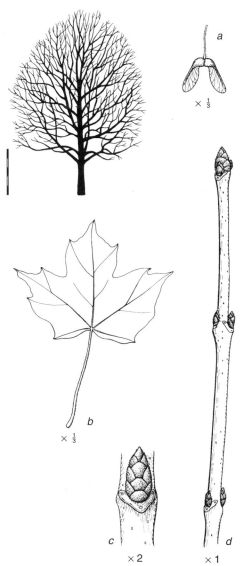

a. Fruit. b. Leaf. c. Lateral bud and leaf scar.
d. Winter twig.

These flowers have functional stamens.

Fruit stalks about the same length as the wings.

Mature bark deeply furrowed, often scaly.

Vertical ridges beginning to show in young bark.

Quick Recognition: Leaves with dense, brownish, velvety hairs beneath, appear wilted. Keys on hairy stalks, about same length as seed wings. Bark often scaly.

Bigleaf Maple
Broadleaf maple, Oregon maple
Acer macrophyllum Pursh
Érable à grandes feuilles

The only tree-size native maple on the Pacific coast. Its large leaves make bigleaf maple a distinctive species.

Leaves Very large, 15–30 (sometimes 60) cm wide, almost as long, deeply notched; 5 lobes with a few, irregular, blunt, wavy teeth; central lobe narrowing toward its base, sometimes overlapping with lateral lobes, but usually separated by narrow U-shaped notches; upper surface shiny dark green, paler and hairless beneath; bright orange or yellow in autumn. Leaf stalk exudes a milky sap when cut.

Buds Terminal bud 6–9 mm long, blunt, greenish to reddish, with 3–4 pairs of overlapping scales. Leaf scars with 5–9 vein scars.

Twigs Stout, reddish-brown, hairless.

Flowers Small, about 10 mm across, greenish-yellow, fragrant, in many-flowered drooping clusters, 10–15 cm long. Pollen flowers and seed flowers in the same cluster. Appear before the leaves.

Fruits Wings 30–40 mm long, slightly divergent. Seedcase swollen, hairy. Keys in elongated drooping clusters. Mature in early autumn.

Bark Grayish-brown, shallowly furrowed into narrow scaly ridges.

Wood Moderately hard, weak, light brown.

Size and Form Medium-sized trees, up to 30 m high, 100 cm in diameter, and 250 years old. In the forest, trunk may be branch-free for one-half or more of its length; crown narrow. In the open, trunk soon divides into a few large spreading and ascending limbs; crown broad, rounded. Root system shallow, wide-spreading.

Habitat Generally occurs on coarse, gravelly, moist soils; mixed with red alder, black cottonwood, Douglas-fir, western redcedar and western hemlock. Occasionally occupies newly disturbed sites; usually succeeded by conifers. Moderately shade-tolerant.

8

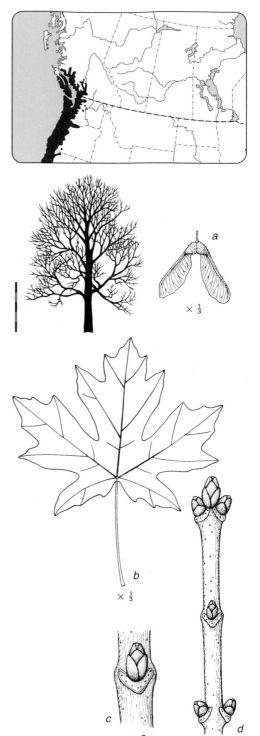

a

$\times \frac{1}{3}$

b

$\times \frac{1}{3}$

c

$\times 2$

d

$\times 1$

a. Fruit. *b.* Leaf. *c.* Lateral bud and leaf scar. *d.* Winter twig.

Seedcase densely hairy.

Flowers in drooping clusters; appear before the leaves.

Young bark smooth, grayish-brown.

Mature bark has narrow scaly ridges.

Notes Wood is used for furniture, flooring, musical instruments, and interior paneling. Because the bark retains moisture, the trunk and larger branches are often covered with mosses, liverworts, and ferns.

Quick Recognition Leaves very large, with 5 large lobes separated by narrow notches. Seedcase hairy, swollen; wings of paired keys near-parallel.

Red Maple
Swamp maple, soft maple

Acer rubrum L.

Érable rouge

A common species of the forests of eastern North America, ranging as far south as Florida. Widely planted as a shade tree (Zones C3, NA3). The twigs, buds, flowers, immature fruits, leaf stalks, and autumn leaves are usually bright red.

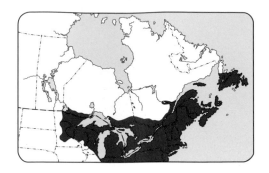

Leaves 5–15 cm long, about as wide; 3–5 lobes with sharp irregular teeth; central lobe with its 2 sides almost parallel to the midvein, separated from lateral lobes by wide, shallow, angular notches; upper surface light green, whitened beneath; bright red in autumn. Stalk 5–10 cm long.

Buds Terminal bud 3–4 mm long, blunt, almost twice as long as wide, shiny, reddish, smooth, usually with 4 pairs of scales. Flower buds stout; become larger during the winter.

Twigs Shiny red to grayish-brown, hairless. Dwarf shoots occur on most branchlets, often bearing clusters of flower buds; flower buds also occur side by side at some leaf scars.

Flowers Noticeably red, with 5 very small petals and sepals, in tassel-like clusters, on slender stalks. Pollen flowers and seed flowers usually on different branches of the same tree. Young trees may bear only one type of flower. Appear in late winter, long before the leaves; one of the first maple species to flower in the spring.

Fruits Wings 12–25 mm long, angle between them about 60°. Seedcase swollen. Keys mature and are shed individually in early summer.

Vegetative Reproduction Dormant buds at the base of most trees sprout vigorously if the tree is cut down or damaged.

Bark Smooth, light gray when young; becoming dark grayish-brown, with scaly ridges fastened at the center and loose at the ends.

Wood Moderately heavy, hard, and strong; light brown. Rays scarcely visible on a tangential face. Turns blue when a solution of ferric salts is applied (compare with sugar maple).

8

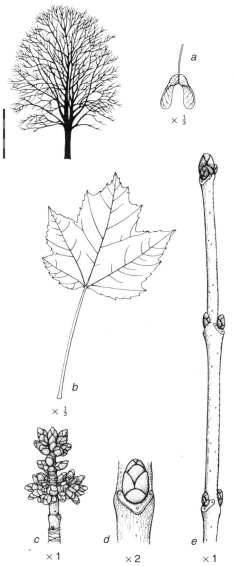

a. Fruit. *b.* Leaf. *c.* Clustered flower buds on dwarf shoots. *d.* Lateral bud and leaf scar. *e.* Winter twig.

Pollen flowers (left) and seed flowers (right) usually on different branches of the same tree.

Angle between wings about 60°.

Young bark (inset) light gray. On forest-grown trees, trunk straight and branch-free.

Size and Form Medium-sized trees, up to 25 m high, 60 cm in diameter, and 100 years old. In the forest, trunk usually branch-free for half its length; crown short, narrow. In the open, trunk divides near the ground into a few ascending limbs that bear widely diverging and ascending branches; crown rather long, dense. Root system shallow, wide-spreading.

Habitat Often occurs in swamps and on other moist soils, but thrives on a great variety of soils and sites. Moderately shade-tolerant.

Notes Highly variable, with several named varieties. Hybridizes readily with silver maple; intermediate forms often occur.

A source of browse for wildlife.

Quick Recognition [See silver maple for contrasting features.] Leaves with 3 or 5 shallow lobes with parallel sides or tapered toward the tip, irregularly toothed. Flowers red; appear before the leaves. Wings small, angle between about 60°.

Silver Maple

Soft maple

Acer saccharinum L.

Érable argenté

Resembles red maple, especially when leaf-less; hybrids common where the two species occur together; range does not extend as far north or south as red maple; widely planted as a shade tree. Features differing from those of red maple are noted below.

Leaves About 15–20 cm long; 5–7 lobes, widest above the base, with coarse, sharp, irregular teeth; central lobe narrowing to the center of the leaf, separated from lateral lobes by deep, narrow notches; upper surface light green, silvery-white beneath; pale yellow or brownish (rarely red) in autumn.

Twigs Unpleasant odor when bruised.

Flowers Small, greenish-yellow, on short stalks. Appear in late winter at least a week before red maple and long before the leaves; the earliest species of maple to flower.

Fruits Wings 40–70 mm long, angle between them about 90°. Seedcase ribbed. Often seed develops in only one of the paired keys. Keys mature and are shed individually in late spring about the time the leaves are fully developed.

Vegetative Reproduction Vigorously from dormant buds in a cut stump; by stem cuttings, both leafy and dormant.

Bark Smooth, gray when young; becoming dark reddish-brown, with long, thin, narrow flakes fastened at the center and free at both ends, giving the tree a shaggy appearance.

Size and Form Medium-sized to large trees, up to 35 m high, 100 cm in diameter, and 130 years old. Fast-growing. In the forest, trunk long, with ascending branches; crown high, open. In the open, trunk short, dividing near the ground into a few sharply ascending branches; crown broad, rounded at the top. Larger branches from the trunk arch outward and downward before turning upward at the ends.

Habitat Grows best on rich, moist bottomlands bordering streams, swamps, and lakeshores. Less shade-tolerant than red maple.

8

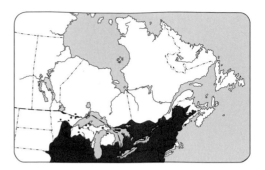

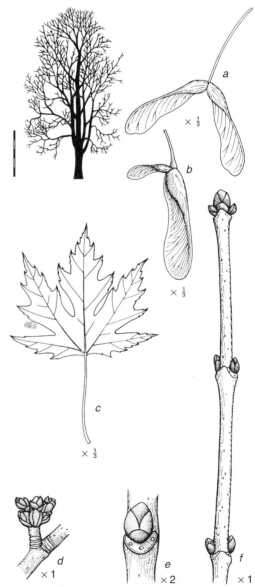

a. Fruit. *b.* Fruit with one key undeveloped. *c.* Leaf. *d.* Clustered flower buds on dwarf shoot. *e.* Lateral bud and leaf scar. *f.* Winter twig.

Keys large, wing angle about 90°.

Pollen flowers (above); seed flowers (below).

Young bark slightly darker than that of red maple.

Trunk appears shaggy; bark in long, thin, narrow flakes.

Notes Trunks of silver maple are often hollow; cavities provide dens for squirrels, raccoons, and other mammals, and nesting space for wood ducks and other birds.

Drawbacks as a street tree include heavy seed and leaf fall, brittle limbs likely to break during ice storms, large size, dull autumn colors, and aggressive roots that often clog sewer pipes. Cultivars with deeply lobed leaves are frequently planted.

Quick Recognition Leaves deeply lobed; lobes 5–7, irregularly toothed, broadest above the base. Keys very large, wing angle about 90°. Bark shaggy.

Manitoba Maple
Box-elder, ashleaf maple

Acer negundo L.

Érable à feuilles composées

Unique among native maples in that the leaves are compound, resembling ash. Frequently planted (Zones C2, NA2) and naturalized beyond its natural range.

Leaves Pinnately compound; composed of 3–9 leaflets on a central stock. Leaflets 5–12 cm long, shallowly and irregularly coarsely toothed or lobed, often asymmetrical; upper surface light green, grayish-green beneath, usually hairless; yellow in the autumn. Preformed leaves have only 3 leaflets; neoformed leaves have more leaflets and a more irregular margin.

Buds Terminal bud ovoid, 3–8 mm long, blunt, with 2 or 3 pairs of brownish-green or purple scales, coated with fine white hairs. Lateral buds almost as large, pressed against the twig; located within the base of the leaf stalk, hence not visible until the leaf falls off. Leaf scars V-shaped, meeting each other around the twig.

Twigs Moderately stout, hairless, shiny greenish-purple or brown; often covered with a waxy powder that is easily rubbed off.

Flowers Small, with pale green sepals, no petals. Pollen flowers and seed flowers on separate trees. Pollen flowers on slender single stalks in lateral bundles. Seed flowers in loose drooping clusters with a central stem; borne toward the base of new shoots.

Fruits Wings 30–50 mm long, incurved, angle between them usually less than 45°. Seedcase elongated, wrinkled. Keys in drooping clusters. Mature in autumn; often remaining on the tree over winter.

Vegetative Reproduction Sprouts readily from stumps and roots.

Bark Smooth, light grayish-brown; with age, becoming darker, furrowed into narrow firm ridges.

8

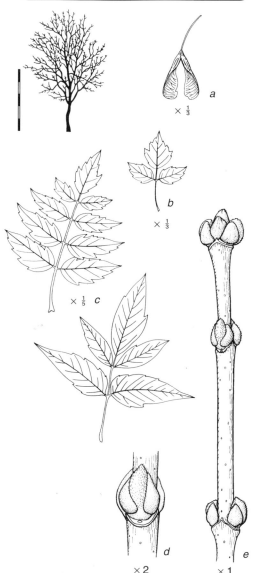

a. Fruit. *b.* Preformed leaf. *c.* Neoformed leaves.
d. Lateral bud and leaf scar. *e.* Winter twig.

Pollen flowers (left) and seed flowers (right) on separate trees.

Hanging fruit clusters will persist on tree in winter.

Young bark (inset) light grayish-brown. Mature bark with narrow firm ridges.

Size and Form Small trees, up to 20 m high, 75 cm in diameter, and 60 years old. Fast-growing. Trunk dividing near the ground into a few long, spreading, rather crooked limbs that branch irregularly; crown broad, uneven; in forest-grown trees the undivided portion of the trunk sometimes long, straight. Root system shallow, fibrous; sometimes with a taproot in deep soils.

Wood Moderately light, soft, weak.

Habitat Grows on lakeshores and stream banks and on sites that are seasonally flooded; colonizes disturbed sites. Intolerant of shade.

Notes Seeds are an important source of winter food for birds and small mammals.

A variable species; one common variety regularly has only 3 leaflets with hairy leaf stalks and twigs.

Easy to grow, survives on most soils, and resists frost and drought; however, branches are weak, liable to break during storms. Hardy in the Prairie provinces, where it has been planted for shade and in shelterbelts.

Quick Recognition [Contrasting features of ash in brackets.] Leaves pinnately compound; leaflets 3–9 [5–9], irregularly coarsely toothed or lobed [regularly toothed or with smooth margin]. Buds hairy. Leaf scars meeting each other [well-separated]. Twigs with a waxy powder. Fruits abundant, in drooping clusters; angle between incurved wings of the paired keys is narrow [fruits single, wing symmetrical]; seed-case wrinkled [smooth].

Mountain Maple

Acer spicatum Lamb.

Érable à épis

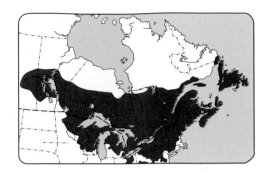

A characteristic understory tree of the forests of eastern Canada.

Leaves 6–12 cm long, almost as wide; 3–5 lobes, coarsely and irregularly toothed, with edges of teeth usually curved outward; central lobe triangular, separated from lateral lobes by wedge-shaped notches; neoformed leaves more likely to be 5-lobed; upper surface yellowish-green, white and hairy beneath; red, yellow, or brown in autumn. Stalk slender, reddish; usually longer than the blade.

Buds Terminal bud slender, stalked, 2–3 times as long as wide, with 1 pair of visible scales that meet along their edges; covered with gray hairs.

Twigs Slender, yellowish-green to reddish-brown or pink, coated with very short gray hairs giving a somewhat dull, velvety look.

Flowers 10 mm across, with 5 petals and sepals, pale yellowish-green to creamy white, on slender stalks (often branched); arranged along a central stem in dense erect terminal clusters. Pollen flowers, seed flowers, and flowers producing both pollen and seeds may occur on the same tree. Appear after the leaves are fully grown.

Fruits Wings about 20 mm long, incurved, angle between them less than 90°; often bright red, later turning yellow or pinkish-brown. Seedcase indented on one side. Keys on short, sometimes branched, stalks arranged around a central stem; in drooping clusters; mature in late summer, often remaining on the tree into winter.

Vegetative Reproduction Often by layering, producing thickets.

Bark Thin, dull, reddish to grayish-brown, smooth or slightly grooved, often with light-colored blotches.

Size and Form Large shrubs or very small trees, up to 5 m high and 15 cm in diameter. Trunk short, crooked, irregularly divided into a few ascending, slender, rather straight limbs; crown unevenly rounded, open. Root system very shallow.

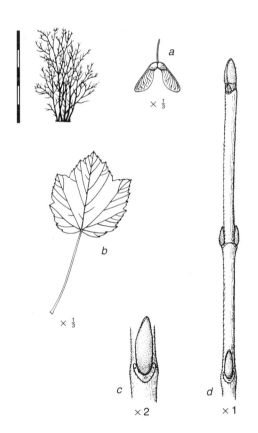

a. Fruit. *b.* Leaf. *c.* Lateral bud and leaf scar. *d.* Winter twig.

Habitat Grows on well-drained moist soils along streams, in ravines, and on moist rocky hillsides. Common on recently cut-over northern forest land, often forming thickets. Shade-tolerant, seldom thriving in the open.

8

Flowers on slender stalks arranged along a central stem; close-up right.

Seedcase indented; wing angle less than 90°.

Bark grayish-brown, smooth, slightly grooved.

Notes When leafless, mountain maple resembles some of the dogwoods. The keys and fine hairs on the twigs identify it as a maple.

Quick Recognition Shrubby trees. Leaves 3-lobed, single-toothed. Buds stalked, with 2 hairy bud scales. Twigs slender, hairy, reddish and/or green. Flowers in erect terminal clusters. Keys in drooping clusters on a central stalk; seedcase indented on one side.

Striped Maple
Moosewood, moose maple

Acer pensylvanicum L.

Érable de Pennsylvanie

A shrubby understory tree distinctive because of its striped bark; occurs in northern forests from Ontario east into the Maritime provinces; also planted for ornamental purposes (Zones C2, NA3).

Leaves Large, 10–16 cm long, often wider than long; 3 shallow lobes on upper portion with long, fine tips pointing forward; uniformly double-toothed; central lobe broadly triangular; neoformed leaves often without lobes; both surfaces pale yellowish-green, hairless; yellow in autumn. Stalk 3–8 cm long.

Buds Terminal bud stalked, about 10 mm long, almost twice as long as wide, with 1 pair of visible scales that meet along their edges, hairless; lateral buds much smaller.

Twigs Rather stout for a maple, smooth, shiny reddish-brown or greenish, hairless.

Flowers Large, 6 mm across, bell-shaped, with 5 greenish-yellow petals and sepals; arranged along a central stem in drooping terminal clusters. Pollen flowers and seed flowers usually on separate trees; may be in separate clusters on the same tree. A tree may bear only seed flowers one year and only pollen flowers the next. Appear after the leaves are fully grown.

Fruits Wings 25–30 mm long, somewhat divergent, angle between them 90°. Seedcase indented on one side. Keys on short stalks, in pendulous terminal clusters. Mature in autumn.

Bark Smooth; green or brownish-green when young; conspicuously marked after 1 or 2 years by long, vertical, whitish stripes; becoming greenish-brown with darkened stripes.

Size and Form Large shrubs or very small trees, up to 10 m high, 25 cm in diameter, and 100 years old. Trunk short, dividing into a few irregular, ascending and arching branches. Crown broad, uneven, flat-topped to rounded. Upright branches often evenly forked. Root system shallow, wide-spreading.

8

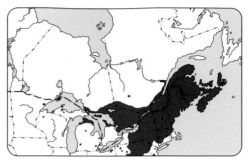

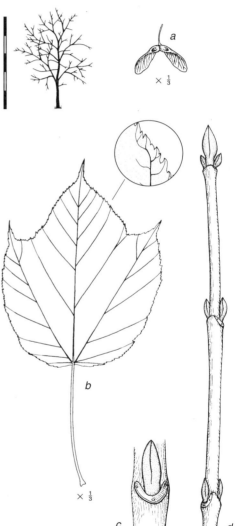

a. Fruit. b. Leaf with detail of double-toothed margin. c. Lateral bud and leaf scar. d. Winter twig.

Pollen flowers (left) and seed flowers (right) usually on separate trees.

Seedcase indented; wing angle greater than 90°.

Young bark (inset) smooth, greenish with conspicuous vertical white stripes. Bark darkened and rough on older trees.

Habitat Grows best on well-drained cool moist soils in deep valleys and on northern slopes. Shade-tolerant.

Notes Leaves and young shoots are a favorite food of moose and deer, and thus the tree is sometimes called "moosewood" or "moose maple." In winter, birds feed on the buds; beavers and porcupines eat the bark. Tolerates heavy browsing.

Quick Recognition   Shrubby trees. Bark green with vertical white stripes. Leaves large, with 3 forward-pointing lobes, evenly double-toothed. Buds stalked; 2 scales.

Douglas Maple
Rocky Mountain maple

Acer glabrum Torr. var. *douglasii*
(Hook.) Dippel

Érable nain

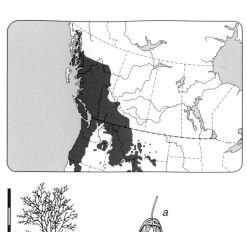

A shrubby western tree. Planted as ornamentals in small gardens because of its size and attractive autumn coloring (Zones C5, NA4).

Leaves 7–14 cm long, about as wide, thin; 3–5 lobes, coarsely double-toothed with the outer edges of the teeth often curved outward; central lobe narrowing toward its base; notches usually shallow and narrowing to sharp slits, occasionally dividing the blade into 3 leaflets; upper surface dark green, grayish-green beneath, hairless; dull red in autumn. Stalks 3–12 cm long, slender, often reddish.

Buds Terminal bud 2–5 mm long, blunt, smooth, bright red or sometimes yellow, with 1 visible pair of scales.

Twigs Slender, smooth, light green to reddish-brown or purplish; often faintly many-sided in cross section.

Flowers About 5 mm across, yellowish-green, with 5 petals, in loose drooping clusters at the end of new shoots and along the sides of the branchlets. Pollen flowers and seed flowers usually on separate trees. Appear with the unfolding leaves.

Fruits Wings 18–22 mm long, nearly parallel, angle between them less than 45°; often rose-colored, turning light brown in autumn, hairless. Seedcase strongly wrinkled, indented. Keys in drooping clusters. Mature in midsummer.

Bark Thin, smooth, dark reddish-brown; rough with age.

Size and Form Large shrubs or very small trees, up to 10 m high and 25 cm in diameter. Trunk short, with a few slender and sharply ascending limbs dividing into many small branches. Crown irregular, uneven-topped. Root system shallow, wide-spreading.

Habitat Occurs along streams and on other moist sites.

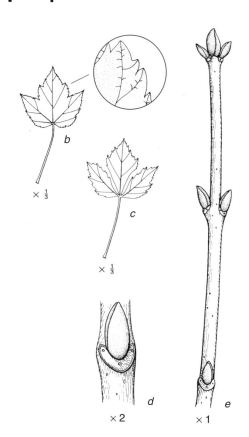

a. Fruit. b. Leaf with detail of coarsely double-toothed margin. c. Deeply notched leaf form. d. Lateral bud and leaf scar. e. Winter twig.

8

Seed flowers (above) and pollen flowers usually on separate trees; appear with unfolding leaves.

Seedcase strongly wrinkled; wing angle less than 45°.

Young bark thin, smooth.

Tree is often multistemmed and shrubby in form.

Notes The typical variety of this species, *A. glabrum* Torr. var. *glabrum*, a tree of the Rocky Mountains in the United States, can be found in a few localities in Alberta.

Quick Recognition A shrubby tree. Leaves coarsely double-toothed. Fruit wings nearly parallel.

Vine Maple
Acer circinatum Pursh
Érable circiné

A shrubby tree of southern British Columbia; often a constituent of the forest understory.

Leaves Almost circular, 3–10 cm across; usually 7–9 radiating lobes and a corresponding number of veins; single- or double-toothed; lobes narrowly triangular, separated by narrow V-shaped notches; upper surface bright yellowish-green; pale green beneath and downy in spring becoming hairless; red or yellow in autumn.

Buds Terminal bud lacking (except where a flower bud occupies the terminal position). Lateral buds small, with 1–2 pairs of red hairless scales; inner scales hairy.

Twigs Slender, hairless, green, often becoming red in autumn.

Flowers About 12 mm across, with 5 purple or red sepals and 5 smaller white petals; arranged on slender branched stalks, in loose, drooping terminal clusters. Appear when the leaves are half grown.

Fruits Wings 25–38 mm long, angle between them almost 180°; red or rose-colored during the summer. Seedcase swollen, ridged, hairless. Mature in late autumn.

Bark Thin, greenish becoming bright reddish-brown, smooth, or sometimes marked by shallow crevices.

Size and Form Large shrubs or very small trees, up to 10 m high and 15 cm in diameter. Trunk short, crooked, often prostrate, with a few twisted, spreading limbs. Crown low, broad, irregular.

Habitat Commonly occurs along stream banks; also found in forest openings and on recently logged areas. Shade-tolerant. Sometimes grows in clumps or patches.

Notes Used by wildlife for browse and habitat.

8

Quick Recognition Shrubby trees. Terminal bud usually lacking. Leaves almost round; lobes 7–9, long and narrow. Wings spreading widely.

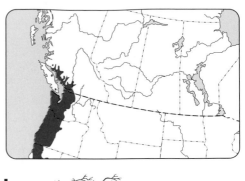

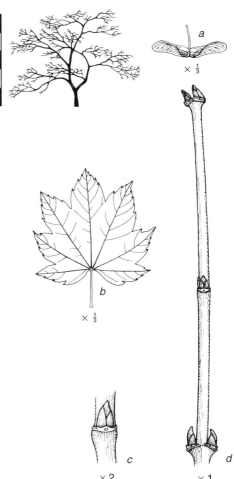

a. Fruit. *b.* Leaf. *c.* Lateral bud and leaf scar. *d.* Winter twig.

Flowers have red or purple sepals and white petals.

Vine Maple

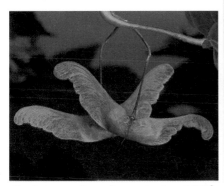

Seedcase swollen; wing angle almost 180°.

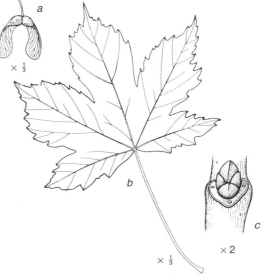

Young bark (inset) tinged green. Mature bark remains smooth.

Sycamore Maple*

Sycamore

Acer pseudoplatanus L.

Érable sycomore

Introduced from Europe; a large tree commonly planted as a shade tree in southern Canada, the United States, and Europe. Hardy as far north as Zones C5, NA4. **Leaves** 6–14 cm long, about as wide, 5-lobed, single-toothed, thick, wrinkled, dark green on the upper surface, whitish- or purplish-green beneath. **Buds** ovoid, pointed; scales green with dark margin. **Flowers** 15 mm across, greenish-yellow, in drooping branched clusters; appear after the leaves are grown. **Fruits** with wings 30–50 mm long, hairless; angle between them less than 30°.

a
× ⅓

b
× ⅓

c
× 2

a. Fruit. *b*. Leaf. *c*. Lateral bud and leaf scar.

Norway Maple**
Acer platanoides L.
Érable de Norvège

Medium-sized trees introduced from Europe; frequently planted as a street tree in eastern North America. Naturalized in many places. Hardy as far north as Zones C5, NA3. Leaves similar to sugar maple; features of that species shown in brackets.

   Leaves 5−7 [3−5] lobes; lobes and teeth bristle-tipped [blunt-pointed]; upper surface dark green [yellowish-green], undersurface lustrous green; green or yellow in autumn [usually red or orange]. A milky juice exudes from cut leaf stalks, bud scales, and twigs [no milky juice]. Terminal **bud** large, plump [slender], blunt [pointed], with 3−4 [4−8] pairs of fleshy [dry] scales, purplish-green or reddish-purple [brown]. Leaf scars tend to meet around the twig. **Twigs** stout [not so stout], smooth, hairless, colored like the buds; branchlets greenish-brown, with prominent lenticels. Branch tips often forked; the result of a terminal flower cluster. **Flowers** large [smaller], 10 [5] mm across with 5 [0] petals and 5 sepals, greenish-yellow, in erect [drooping] terminal [usually lateral] clusters; appear with [before] the leaves. **Fruits** with wings spread very wide [almost parallel]; 35−50 [30−35] mm long. Seedcase flat [swollen]. Seeds produced abundantly almost every year [every few years]; germinate readily during the following spring [while there is still snow on the ground]. **Bark** very dark [dark] gray, with firm, low, intersecting ridges [with vertical strips curled outward along one side]; pattern very regular [irregular]; not scaly [may be scaly].

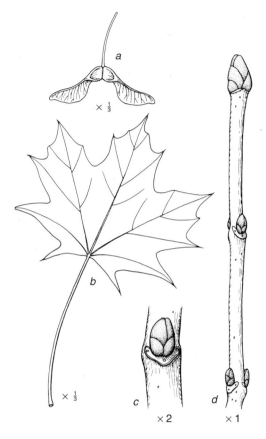

a. Fruit. b. Leaf. c. Lateral bud and leaf scar.
d. Winter twig.

   Many cultivars have been developed; Schwedler maple with purplish-red foliage in spring has been planted in North America for over a century; Crimson King has leaves of a rich maroon color during the whole growing season.

Amur Maple*
Acer ginnala Maxim.
Érable ginnala

Shrubby trees, native to eastern Asia; often with brilliant red foliage in autumn. Hardy as far north as Zones C2, NA2. **Leaves** 8−10 cm long, narrowly triangular, with shallow notches just above 2 short basal lobes; coarsely toothed. **Flowers** and **fruits** in drooping terminal clusters; fruit wings nearly parallel.

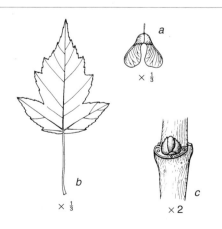

a. Fruit. b. Leaf. c. Lateral bud and leaf scar.

8

Flowers erect in terminal clusters; appear with the leaves.

Norway Maple

Seedcase flat; wings widely spread.

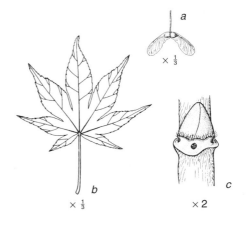

Young bark (inset); ridges beginning to form. Mature bark dark, with regular intersecting ridges.

Japanese Maple*

Acer palmatum Thunb.

Érable palmé

A shrubby tree with a crown often wider than high. Japanese maple and a number of similar species native to eastern Asia are frequently used for landscape planting; hundreds of horticultural forms have been developed. Hardy as far north as Zones C6, NA5. **Leaves** with 5–9 lobes, double-toothed, taper-pointed, separated by deep, narrow, V-shaped notches; yellow, bronze, or red in autumn. **Buds** in pairs at shoot tip; small, green, or red, surrounded by a raised leaf scar. **Twigs** slender, green in summer, red in autumn, hairless, often covered with a waxy powder. **Fruits** with wings 12–15 mm long, much incurved, angle between them 100°; turning red in autumn.

a. Fruit. *b.* Leaf. *c.* Lateral bud and leaf scar.

Horsechestnuts

Buckeyes, chestnuts

Genus *Aesculus*

Hippocastanaceae: Horsechestnut
Family

Les marronniers

The horsechestnut genus comprises
13 species of trees and shrubs found in the
North Temperate Zone; 5 species are native
to North America; one of these, Ohio
buckeye, has recently been found in southwestern Ontario.

Leaves Deciduous, in opposite pairs, palmately compound; composed of 5–7 stalkless leaflets on a long central stalk.

Buds Terminal bud very large, with several
pairs of overlapping scales; lateral buds
small, in opposite pairs. Leaf scars large,
shield-shaped, with 3 or more vein scars.

Twigs Very stout; unpleasant odor when
bruised.

Flowers Each with 5 large, colored, irregular
petals. In cone-shaped erect terminal clusters; upper flowers only bearing pollen.
Appear after the leaves.

Fruits Large, leathery, green or brown
capsules in small clusters; splitting into
3 parts when ripe, containing 1–3 seeds.

Seeds Very large, brown, mostly covered
with a smooth shiny seed coat except for a
large pale rough spot on the lower end. Not
edible.

Seedlings Cotyledons white, swollen,
fleshy, retained within the seed coat. Newly
germinated seedlings produce a shoot with
scale-like leaves; subsequent leaves increasingly resemble adult ones.

Bark Gray or brown, smooth, becoming
scaly with age; unpleasant odor when
bruised.

Size and Form Small to medium-sized
trees, with short trunks and spreading
crowns.

Wood Light, weak, uniform in texture, close-
grained, pale; "ripple-marks" may show on a
radial surface.

Notes In Europe, the resemblance of the
inedible seed of *Aesculus* to the edible chestnuts of the genus *Castanea* gave rise to the
name "horsechestnut". In America, the name
"buckeye" was applied because of the pale
spot on the brown seed.

Wood is used for carving and for the
manufacture of small woodenware.

Quick Recognition Leaves opposite, palmately compound. Large terminal bud on stout
twigs. Fruit a large leathery green capsule;
seeds 1–3 per capsule, large shiny brown with
a pale spot.

8

Ohio Buckeye
Fetid buckeye

Aesculus glabra Willd.

Marronnier glabre

A common species in the midwestern United States as far north as Michigan; occasionally planted in milder parts of Canada; recently a population of large trees was found in a natural forest in southwestern Ontario on Walpole Island at the north end of Lake St. Clair. Prefers moist sites such as river bottoms; occurs mixed with other broadleaf trees.

Leaves 5−7 leaflets on a central stalk about as long as a leaflet. Leaflets 6−15 cm long, taper-pointed, unevenly saw-toothed, yellowish-green on the upper surface, paler and hairy beneath; orange or yellow in autumn; almost stalkless.

Buds Terminal bud 15−18 mm long, widest at the middle, pointed; with a non-sticky powdery coating; scales reddish, triangular, with a central ridge. Leaf scars horseshoe-shaped, with 3 vein scars.

Twigs Stout, reddish-brown, becoming hairless; lenticels orange.

Flowers 15−35 mm long, narrowly bell-shaped, with 4 yellowish-green petals; unpleasant odor.

Fruits 25−50 mm long, light brownish-green, covered with blunt spines; 1−3 seeds, 20−35 mm wide.

Bark Gray, becoming dark brown, rough and furrowed, with scaly plates.

Size and Form Small trees up to 15 m high, 50 cm in diameter, and 80 years old.

Notes An unpleasant odor is produced by most parts of the tree when bruised; hence the name fetid buckeye.

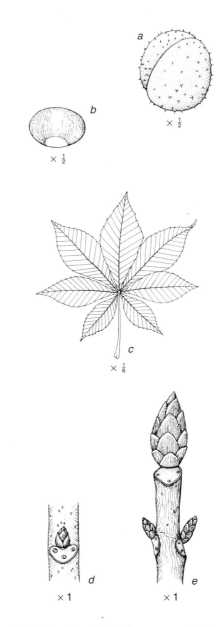

a. Fruit. b. Seed. c. Palmately compound leaf. d. Lateral bud and leaf scar. e. Winter twig.

Horsechestnut**

Common horsechestnut, chestnut

Aesculus hippocastanum L.

Marronnier d'Inde

Native to the Balkans in southeastern Europe; frequently planted in Canada for its spectacular spring flowers and large spreading crown. Tolerates urban conditions. Naturalized in various places. Hardy as far north as Zones C5, NA3

Leaves 5–9 (usually 7) leaflets; terminal leaflet the largest. Leaflets 10–25 cm long, broadest above the middle, abruptly pointed, teeth of 2 sizes; stalkless.

Buds Terminal bud 20–40 mm long, dark brown, sticky, with several pairs of scales, no external hairs; in cross section, the bud is packed with many white hairs.

Twigs Stout, light brown.

Flowers Very showy, 20–30 mm long, bell-shaped, white or cream-colored, with red and yellow spots.

Fruits Globular, 50–60 mm across, green turning brown, with pointed spines; borne in drooping lateral clusters (prominent round scars mark location of cluster after it has been shed). Good seed crops occur most years. Seeds up to 5 cm wide, 1 or 2 per capsule; viable until the following spring under moist cool conditions.

Bark Dark gray to dark brown; becoming fissured and scaly; inner bark orange-brown.

Size and Form Medium-sized trees, up to 25 m high, 50 cm in diameter, and 100 years old. Crown spreading, rounded. Principal branches ascend near the trunk, curve down in the middle portion, and ascend again making the tips almost upright.

Notes Esculin is extracted from the leaves and bark for use as a skin protectant. Despite the esculin in the seeds, squirrels eat the embryo stalk.

8

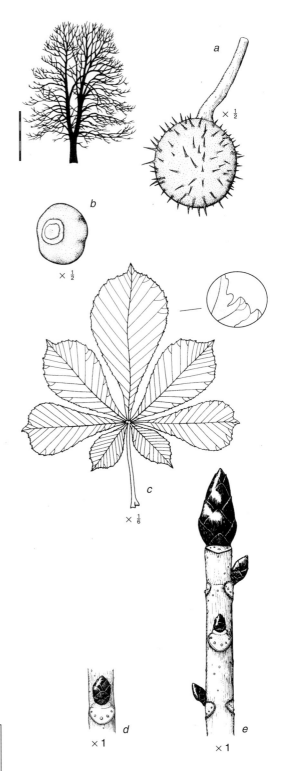

$\times \frac{1}{2}$

$\times \frac{1}{2}$

$\times \frac{1}{6}$

$\times 1$

$\times 1$

Quick Recognition Leaves opposite, palmately compound. Terminal bud large, sticky. Flowers in large showy erect clusters at the branch tips in spring. Fruits large, greenish, globular, spiny. Seeds large, shiny brown, with a large pale spot.

a. Fruit. *b.* Seed. *c.* Leaf with detail of double-toothed margin. *d.* Lateral bud and leaf scar. *e.* Winter twig.

Flowers showy; arranged in cone-shaped terminal clusters.

Horsechestnut

Fruit a spiny capsule containing 1 or 2 large seeds.

Young bark (inset) dark gray, smooth. Mature bark fissured and very scaly.

Red Horsechestnut*

Aesculus ×*carnea* Hayne
(*A. hippocastanum* × *A. pavia*)

Marronnier rouge

Resembles common horsechestnut. Occasionally planted for its red flowers. Hardy as far north as Zones C5, NA4. **Leaves** composed of 5 leaflets. **Buds** somewhat sticky.

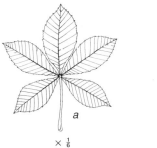

$\times \frac{1}{6}$

a

b

×1

a. Leaf. *b*. Twig with terminal bud, lateral bud, and leaf scars.

Ashes

Genus *Fraxinus*
Oleaceae: Olive Family

Les frênes

The ash genus comprises about 60 species of trees and shrubs; 16 are native to North America, 4 to Canada. Native and introduced species are frequently planted for landscape purposes.

Leaves Deciduous, in opposite pairs, pinnately compound; composed of 5–11 leaflets; all about the same size, or terminal one larger.

Buds Terminal bud broadly pyramidal; 1–3 pairs of scales, with a soft, granular surface. Leaf scars oval to semicircular or crescent-shaped, with many small vein scars.

Twigs Stout, straight, often flattened beside the leaf scars. In mature trees, most shoot growth preformed in a bud; in vigorous saplings, summer shoot growth often neoformed.

Flowers Small, dark, borne in compact, many-flowered clusters, arranged laterally on twig, rarely at tip. Pollen flowers and seed flowers on the same or separate trees. Flowers of native species lack petals; the sepals are minute or missing. Appear in early spring before the leaves. Wind-pollinated.

Fruits Winged, dry, 1-seeded; in most species, base encircled by remnants of the flower; borne in large drooping lateral clusters. Mature in autumn; remain on the tree during the winter [blue ash sheds sooner]. Dispersed by wind and water.

Seeds Remain within the fruit; can be stored dry in sealed containers for several years; require exposure to cool moist conditions for some months before germination.

Seedlings Cotyledons green, leaf-like, strap-shaped, 2–3 cm long; raised above the surface. First true leaves are simple; later leaves are compound with an increasing number of leaflets.

Vegetative Reproduction By stump sprouts.

Bark Varies from finely furrowed with firm ridges to scaly and so soft that it rubs off easily with hand pressure.

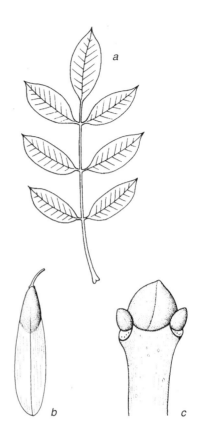

a. Pinnately compound leaf. *b.* Fruit 1-seeded, single-winged. *c.* Twig tip with terminal bud and uppermost pair of lateral buds.

Wood Medium weight, strength, and hardness; tough, shock-resistant, straight-grained, attractive tangential grain-pattern, grayish-brown. Ring-porous; rays inconspicuous.

Size and Form Small to large trees; fast-growing. Trunk straight, often distinct almost to the top of the crown. Crown formed of ascending and spreading straight branches with few side branches. Leading shoot upright.

Habitat Occupy a variety of sites, but grow best on rich moist soils. Some species occur in swamps or along streams; others on poor, dry upland soils. Shade-tolerant when young; tolerance decreases with age.

8

red ash

Crown formed of ascending straight branches with few side branches.

red ash

Fruit in drooping lateral clusters.

Notes Wood is flexible; used to make such items as tool handles, sports implements, snowshoe frames, chair backs, and church pews. Seeds are an important food for many birds and small mammals.

A gall mite can cause male flowers to develop abnormally into galls, which remain on the tree for several months.

Quick Recognition Leaves opposite, pinnately compound; leaflets 5–11. Terminal bud pyramidal. Fruits with long terminal wing; in large drooping lateral clusters.

White Ash
Fraxinus americana L.

Frêne blanc

An eastern species; the most common native ash. Individual trees stand out in autumn because of their bronze-purple leaves.

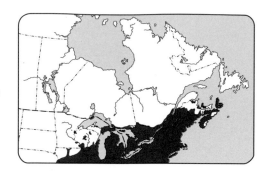

Leaves 5–9 (usually 7) leaflets on a central stalk 15–25 cm long. Leaflets 6–15 cm long (basal leaflets the shortest), oval to lance-shaped, gradually tapering to each end; margin smooth or with infrequent rounded teeth; upper surface dark green; very pale beneath, hairless except along the veins; bronze-purple in autumn; stalks 5–15 mm long. Leaflets tend to fall one at a time in autumn.

Buds Terminal bud 5–14 mm long, wider than long, 4-sided, blunt, reddish-brown. Uppermost pair of lateral buds very close to the terminal bud, with no bark showing in-between. Leaf scars semicircular with a notch in their upper surface.

Twigs Stout, shiny, purplish, or with a gray-ish skin, hairless.

Flowers Purple. Pollen flowers and seed flowers on separate trees.

Fruits 2.5–5 cm long. Wing encloses only the tip of the seedcase. Good seed crops occur about every 3 years.

Bark Light gray, finely furrowed into thin firm intersecting ridges in a regular diamond pattern.

Wood Heavy, straight-grained, hard, strong, tough, light brown.

Size and Form Medium-sized trees, up to 30 m high, 150 cm in diameter, and 200 years old. An erect graceful tree. Trunk long, straight, often branch-free almost to the top. Crown narrow, pyramidal. Root system deep where soil conditions permit.

Habitat Usually occurs on deep, well-drained, upland soils, mixed with other broadleaf trees and occasional conifers. Moderately shade-tolerant.

8

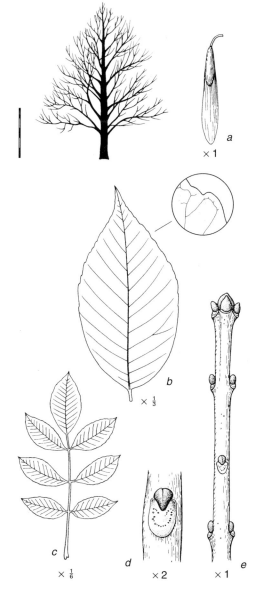

a. Fruit. *b*. Leaflet with detail of infrequent, rounded teeth along margin. *c*. Leaf. *d*. Lateral bud and leaf scar. *e*. Winter twig.

Pollen flowers (left) and seed flowers (right) on separate trees.

Mature bark has intersecting ridges in a regular diamond pattern.

Fruit clusters; seed wing encloses only the tip of the seedcase.

Young bark light gray, smooth.

Notes White ash populations sometimes undergo declines; the trees exhibit a progressive dieback called "ash yellows"; these declines are probably triggered by climate-induced stress, which predisposes the trees to other damaging agents.

The main source of commercial ash wood.

Quick Recognition Usually 7 stalked leaflets, with no teeth or a few rounded ones. Bronze-purple in autumn. Uppermost pair of lateral buds close to the terminal bud. Fruit wing enclosing only the tip of the seedcase. Twigs hairless.

Red Ash

Fraxinus pennsylvanica Marsh.

Frêne rouge

The most widely distributed ash, ranging from Saskatchewan to Nova Scotia. Aggressively invades urban spaces; frequently planted for landscape purpose (Zones C2, NA3).

Leaves 5−9 (usually 7) leaflets on a hairy central stalk 15−20 cm long. Leaflets 10−15 cm long, oval, taper-pointed; toothed above the middle; upper surface yellowish-green, paler and densely hairy beneath; yellowish-brown in autumn; stalks hairy, with 2 narrow green wings. Leaflets tend to fall one at a time in autumn.

Buds Terminal bud 3−8 mm long, reddish-brown, hairy. Uppermost pair of lateral buds very close to the terminal bud. Leaf scars semicircular, sometimes with a slight notch in their upper surface.

Twigs Moderately stout, grayish-brown; densely hairy, less so in the winter.

Flowers Pollen and seed flowers on separate trees; stalks densely hairy.

Fruits 3−6 cm long. Wing enclosing one-half or more of the seedcase, often notched at the tip. Abundant seed crops almost every year; seeds sometimes lie dormant in the litter for several years before germinating.

Bark Grayish-brown, often tinged with red on young branchlets, becoming broken into firm, narrow, irregular, slightly raised ridges that intersect and form a diamond-shaped pattern.

Size and Form Small to medium-sized trees, up to 25 m high, 60 cm in diameter, and 100 years old. Fast-growing. Variable in form; a small, shrubby tree with a leaning or twisted trunk in one place, a slender tree with a straight trunk and rounded crown in another.

Habitat Commonly on bottomlands; confined to river valleys in the Prairie provinces; often mixed with willow, silver maple, and eastern cottonwood. Moderately shade-tolerant. Can withstand many weeks of flooding during the dormant season.

8

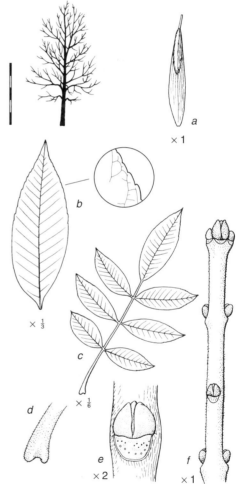

a. Fruit. *b*. Leaflet with detail of toothed margin. *c*. Leaf. *d*. Detail of hairy leaf stalk. e. Lateral bud and leaf scar. *f*. Winter twig.

Quick Recognition Twigs and leaf undersurface densely hairy. Leaflets usually 7, toothed above the middle; with winged stalks. Uppermost pair of lateral buds close to the terminal bud. Fruit wing enclosing one-half or more of the seedcase.

Pollen flowers (left);
seed flowers (right).

Red Ash

Fruit (left); seed wing encloses about one-half of the seedcase.
Young bark (center) smooth, often tinged with red. Diamond
pattern of mature bark (right) less regular than that of white ash.

Green Ash

Fraxinus pennsylvanica
var. *subintegerrima* (Vahl) Fern.

Frêne vert

Differs from the typical red ash in that its leaf
stalks, leaves, twigs, flower stalks, and fruit stalks
are almost hairless. More prevalent toward the
west; the common ash in the Prairie provinces.

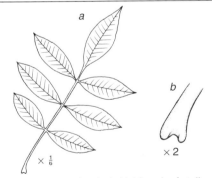

$\times \frac{1}{6}$

$\times 2$

a. Leaf. *b*. Hairless leaf stalk.

Northern Red Ash

Fraxinus pennsylvanica
var. *austini* Fern.

Frêne d'Austin

Differs from the typical red ash in having
prominently toothed leaflets, shorter fruits and
broader, spatula-shaped fruit wings. More common
in eastern Canada than red ash.

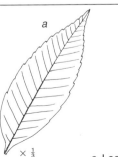

$\times \frac{1}{3}$

$\times 1$

a. Leaflet. *b*. Fruit.

Black Ash

Swamp ash, hoop ash

Fraxinus nigra Marsh.

Frêne noir

Commonly occurs in northern swampy woodlands from eastern Manitoba into the Atlantic provinces. The only ash native to Newfoundland, where it is a small, rare tree.

Leaves 7–11 leaflets on a central stalk 15–30 cm long. Leaflets 10–14 cm long, elongated oval, base unequally rounded or broadly wedge-shaped, tapered to a long, slender tip; stalkless; finely and sharply toothed; both surfaces dark green and hairless, except for dense tufts of reddish-brown hairs where leaflet attaches to the central leaf-stalk; reddish-brown in autumn. Leaves, rather than leaflets, tend to fall in autumn.

Buds Terminal bud 4–10 mm long, broad, pointed, dark brown. Uppermost pair of lateral buds set distinctly below the terminal bud, with the bark clearly visible above them. Leaf scars large, rounded.

Twigs Very stout, bright green with dark purple, raised lenticels when young, becoming dull gray.

Flowers Pollen and seeds usually produced in the same flower (perfect); occasionally in separate flowers on separate trees; or sometimes pollen flowers, seed flowers, and perfect flowers on the same tree. Flower clusters more likely to occur at the twig tip than in other ash species.

Fruits 2.5–4 cm long. Wing broad, often twisted, sometimes slightly notched at the tip, extending to the base of the flattened seedcase. No flower remnants at the base of the fruit. Good seed crops occur at irregular intervals; up to 7 years. Germination does not take place until the 2nd season after seedfall.

Bark Light gray, soft, with corky ridges that are easily rubbed off by hand or indented with a fingernail; becoming scaly.

Wood Straight-grained, flexible.

Size and Form A small tree, up to 20 m high, 50 cm in diameter; occasionally larger. Trunk slender, sometimes bent or leaning. Crown narrow, open, with coarse, ascending branches. Root system shallow, widespreading.

8

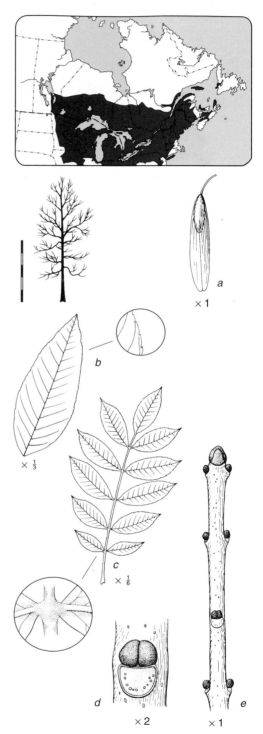

a. Fruit. *b.* Leaflet with detail of sharp teeth.
c. Leaf with detail of hairs at junction of leaflet and leaf stalk. *d.* Lateral bud and leaf scar. e. Winter twig.

Fruit clusters; seedcase entirely enclosed by a broad, twisted wing.

Pollen flowers (above) and seed flowers (below).

Young bark corky and soft to touch.

Mature bark scaly; root system wide-spreading.

Habitat Can tolerate standing water for many weeks. Sometimes in pure stands, but usually grows mixed with black spruce, balsam fir, eastern white-cedar, speckled alder, red maple, and silver maple. Intolerant of shade.

Notes Wood can be permanently bent and is favored for making snowshoe frames, barrel hoops, canoe ribs, and woven basketware, bolts of wet wood can be made to separate along the annual rings into thin layers by repeated pounding.

> **Quick Recognition** Leaflets 7–11, toothed, stalkless. Uppermost pair of lateral buds set down from the terminal bud. Fruit wing extends along the seedcase. Bark in corky ridges, soft to the touch. Found in wet places.

Blue Ash
Fraxinus quadrangulata Michx.

Frêne anguleux

Occurs in a few locations in southwestern Ontario.

Leaves Composed of 5–11 leaflets on a central stalk 13–25 cm long; leaflets 8–14 cm long, coarsely toothed, short-stalked, with an asymmetrical base.

Twigs With 4 conspicuous ridges or wings, making them 4-sided.

Flowers Perfect, producing both pollen and seeds.

Fruits Wing broad, twisted, often notched at the tip, enclosing a flattened seedcase.

Bark With age, broken into irregular fissures and loose scaly plates, appearing shaggy; inner bark turns blue when exposed.

Wood Hard, heavy, strong, coarse-grained.

Size and Form Small trees up to 20 m high and 25 cm in diameter; trunk slightly tapered with a narrow, often irregular crown of spreading branches.

Habitat Occurs as a scattered tree, mixed with white ash, black ash, chinquapin oak, black walnut and other southern broadleaf trees; found on floodplains and on limestone outcrops. The most drought-resistant of the native ashes.

Notes Blue ash is an increasingly rare Canadian tree and has been designated a threatened species by the Committee on the Status of Endangered Wildlife in Canada.

8

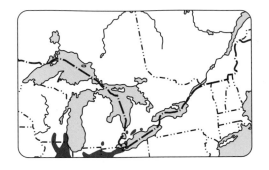

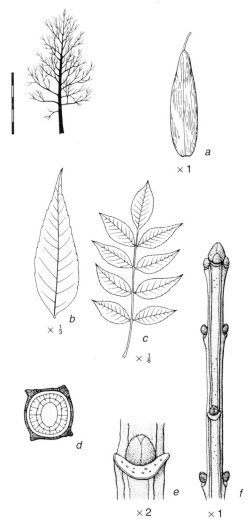

a. Fruit. *b*. Leaflet. *c*. Leaf. *d*. Twig cross section showing conspicuous ridges. *e*. Lateral bud and leaf scar. *f*. Winter twig.

Flowers perfect, producing pollen and seeds.

Blue Ash

Fruit clusters; seedcase flattened, entirely enclosed by seed wing.

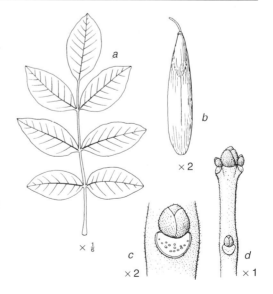

Young bark (inset) smooth but soon fissured and scaly. Mature bark broken into loose scaly plates.

Oregon Ash

Fraxinus latifolia Benth.

Frêne de l'Orégon

Medium-sized to large trees of the Pacific coast region of North America; found in southern British Columbia, but natural occurrence in Canada uncertain; may be naturalized. **Leaves** with 5–9 (usually 7) leaflets; each 4–10 cm long, with a tapered base and broad pointed tip, usually smooth-margined. Terminal **bud** 3–8 mm long, broad, pointed, with 4 pairs of hairy scales; leaf scars large, crescent-shaped. **Twigs** densely hairy in spring, becoming reddish-brown to gray. Pollen **flowers** and seed flowers in separate clusters. **Fruits** flattened; wing uniformly broad. Mature **bark** deeply fissured with broad, flat ridges, reddish-brown to gray.

a. Leaf. *b*. Fruit. *c*. Lateral bud and leaf scar. *d*. Winter twig.

Pumpkin Ash
Fraxinus profunda (Bush) Bush

Frêne pubescent

Medium-sized trees up to 30 m; native to the
eastern United States and only recently identified
in Canada in a few locations at the west end of
Lake Erie. Usually occurs in wet places. Leaves
and fruit larger than other ashes. **Leaves** 45 cm
long with leaflets up to 25 cm. **Fruits** up to 8 cm
long. Pollen **flowers** and seed flowers on separate
trees.

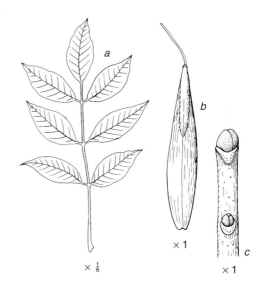

a. Leaf. *b*. Fruit. *c*. Winter twig.

European Ash*
Fraxinus excelsior L.

Frêne commun

8

Medium-sized to large trees introduced from
Europe; frequently planted for landscape purposes. Hardy as far north as Zones C5, NA5.
Leaves with 9–15 leaflets; each 5–8 cm long,
finely and sharply toothed, dark green, stalkless,
hairy on the midvein but not at the base. **Buds** inky
black, hairy. **Twigs** grayish-brown, hairless. **Fruits**
with a wide wing enclosing the seedcase. **Bark**
firm, ridged.

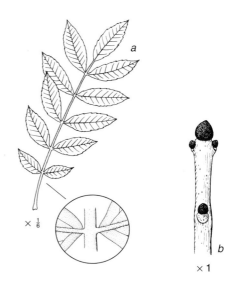

a. Leaf with detail of stalkless leaflets and hairless
leaf stalk. *b*. Winter twig.

Amur Corktree*

Phellodendron amurense Rupr.
Rutaceae: Rue or Citrus Family

Arbre liège de Chine

Small trees, native to eastern Asia. Hardy as far
north as Zones C3, NA3.

    Leaves deciduous, in opposite pairs, pinnately
compound; composed of 5–13 leaflets on a central
stalk 10–15 cm long; each leaflet 6–10 cm long,
narrowly ovate, taper-pointed, smooth-margined,
shiny dark green on the upper surface, aromatic,
with translucent dots. **Buds** small, naked, with
2 hairy immature leaves; leaf scars U-shaped,
almost encircling the bud; vein scars in 3 groups.
Twigs yellowish, coarse, with prominent lenticels.
Flowers small, yellowish-green, in terminal clusters; pollen flowers and seed flowers on separate
trees. **Fruit** a small berry, 10 mm across, dark
blue, aromatic, with 5 hard seeds. **Bark** yellowish-
gray, becoming furrowed with corky ridges; inner
bark bright yellow.

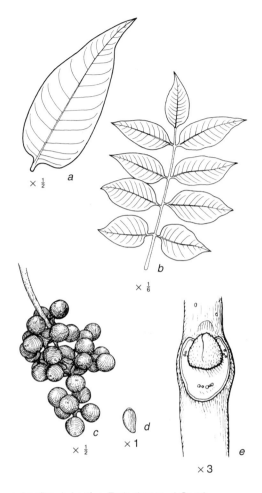

a. Leaflet. b. Leaf. c. Fruit cluster. d. Seed.
e. Lateral bud and leaf scar.

Elders
Elderberries

Genus *Sambucus*
Caprifoliaceae: Honeysuckle Family

Les sureaux

The elder genus comprises about 30 species of small trees and shrubs occurring in the Northern Hemisphere; 7 tree-size species are native to North America, 3 to Canada. Several native shrub species are briefly described because they are common and resemble the tree-size species.

Leaves Deciduous, in opposite pairs, pinnately compound; composed of 5–11 leaflets, toothed, base often uneven.

Buds Terminal bud usually absent; lateral buds with several pairs of scales; leaf scars large, with 5–7 vein scars.

Twigs Stout; woody part thin, enclosing a wide pith. Lenticels often prominent.

Flowers Small with a 5-lobed corolla; borne in large many-branched terminal clusters. Insect-pollinated.

Fruits Berry-like, round, with 3–5 hard seeds, dispersed by fruit-eating birds and mammals.

Vegetative Reproduction By sprouts from roots and damaged stumps.

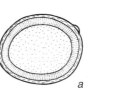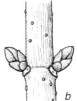

a. Twig cross section showing wide pith. *b.* Twig with lateral buds, leaf scars, and lenticels.

Bark Rough, brownish.

Size and Form Coarse shrubs or very small trees up to 10 m high.

Notes Some species produce berries that are poisonous to humans. Twigs and bark are important as browse for wildlife.

> **Quick Recognition** Leaves opposite, pinnately compound; leaflets 5–11, toothed. Flowers in large many-branched terminal clusters. Fruits small, berry-like, with 3–5 hard seeds. Stems stout; pith very thick relative to the size of the stem and the thickness of the wood cylinder.

Black-Berry Elder
Blackbead elderberry

Sambucus melanocarpa A. Gray
[syn. *S. racemosa* var. *melanocarpa*]

Sureau arborescent

Shrubs of the Rocky Mountains; found at elevations above 1000 m. **Twig** pith brownish. **Flowers** and **fruits** in loose globular clusters about 7 cm across; berries shiny black.

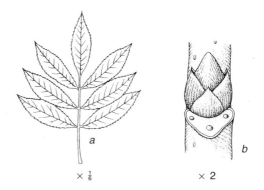

× ⅙   × 2

a. Leaf *b.* Lateral bud and leaf scar.

American Elder

White elder

Sambucus canadensis L.

Sureau blanc

Very small trees or large shrubs of eastern Canada. **Leaves** with 5–11 leaflets; terminal leaflet often largest; lower pair sometimes sub-divided. Leaflets 5–15 cm long, sharply toothed, short-stalked. **Twig** pith white. **Flowers** white, in broad flat clusters; appear in midsummer after the leaves have developed. **Fruits** purplish-black; ripen in autumn; edible.

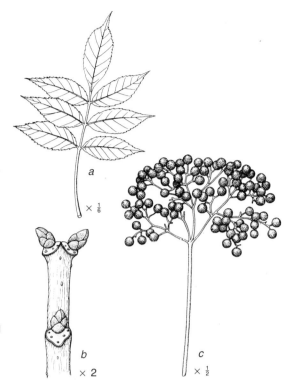

a. Leaf. *b.* Lateral bud and leaf scar. *c.* Flat-topped fruit cluster.

Eastern Red Elderberry

Eastern red-berry elder

Sambucus pubens Mich.

Sureau rouge

Shrubs of eastern Canada. **Leaves** with 5–7 leaflets; with taper-pointed tips. **Buds** plump, reddish, with 1 pair of scales. **Twig** pith orange to brown. **Flowers** in pyramidal clusters; appear early with the unfolding leaves. **Fruits** red; ripen in mid-summer; not considered edible.

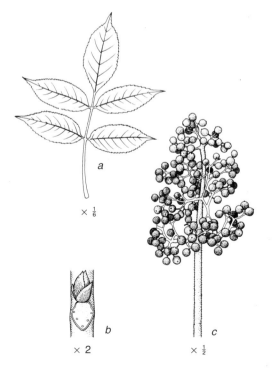

a. Leaf. *b.* Lateral bud and leaf scar. *c.* Round-topped fruit cluster.

Blue-Berry Elder

Blue elderberry, blue elder

Sambucus cerulea Raf.
[syn. *S. glauca* Nutt.]

Sureau bleu

Tall, irregularly shaped shrubs or very small trees, up to 10 m high and 20 cm in diameter. The largest of the Canadian elders; occurs in southern British Columbia in forest clearings, usually on dry open sites with gravelly or stony soils.

Leaves 5–9 leaflets on a grooved central stalk 6–12 cm long. Leaflets 5–15 cm long, lance-shaped, with a long, slender, sharp tip; coarsely and sharply toothed; occasionally subdivided; bluish-green above, paler beneath, hairless; unpleasant odor when crushed; stalks 6–12 mm long.

Buds Large, 10–20 mm long, with 2 or 3 pairs of greenish or brownish scales; leaf scars large, triangular, with 5–7 vein scars.

Twigs Very stout, often ridged, light green to brownish; pith white.

Flowers Creamy white, in flat-topped, 5 branched clusters, 7–15 cm across; appear in summer after the leaves are fully grown.

Fruits Bluish-black, 3–5 mm across, thin-fleshed, juicy, sweet, with 3 seeds; ripen in early autumn.

Bark Thin, light brown, often reddish, with narrow intersecting scaly ridges and warty lenticels.

Notes Occasionally planted for ornamental purposes and as a source of food for birds (Zones C3, NA3).

Quick Recognition Leaves opposite, pinnately compound; leaflets 5–9, unpleasant odor. Twigs stout, with warty lenticels; pith white. Flowers and fruits in large flat-topped 5-branched clusters. Berries bluish. Prefers dry sites.

8

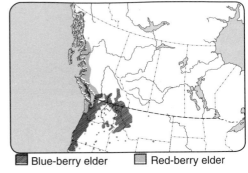

☒ Blue-berry elder   ☐ Red-berry elder

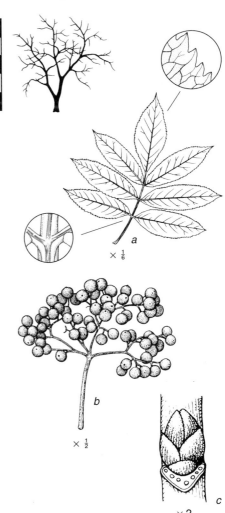

$\times \frac{1}{6}$ *a*

$\times \frac{1}{2}$ *b*

c $\times 2$

a. Leaf with details of sharp teeth (above) and grooved stalk (below). *b.* Fruit cluster. *c.* Lateral bud and leaf scar.

Fruit bluish-black when ripe.

▾ **Red-berry elder** ▸

Flowers in rounded clusters.

Fruit ripens in midsummer.

◂ **Blue-berry elder**

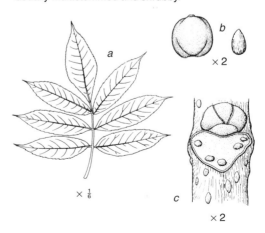

Young bark (inset) with warty lenticels. Tree usually multistemmed and shrubby.

Red-Berry Elder

Pacific coast red elder

Sambucus callicarpa Greene
[syn. *S. racemosa* var. *arborescens*]

Sureau rouge du Pacifique

Shrubs or very small trees of the Pacific coast. **Leaves** with 5–7 leaflets; each 6–16 cm long, lance-shaped, sharply toothed, hairy along the veins. **Flowers** yellowish-white, in rounded clusters about 10 cm across; appear in spring. **Fruits** red, 4–6 mm across; ripen in midsummer; **POISONOUS.** Prefers moist sites.

a

b
× 2

× ⅙

c
× 2

a. Leaf. *b*. Fruit (left); seed (right). *c*. Lateral bud and leaf scar.

Viburnums

Cranberry-bushes, highbush-cranberries

Genus *Viburnum*

Caprifoliaceae: Honeysuckle Family

Les viornes

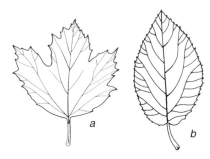

a. 3-lobed viburnum leaf. *b.* Ovate viburnum leaf.

The viburnum genus comprises about 150 species of shrubs and trees found in temperate and subtropical regions; 15 are native to North America, 9 to Canada. Several native and introduced species are planted for their showy clusters of flowers and as a source of food for birds.

    Leaves deciduous, in opposite pairs, simple, toothed or 3-lobed, often with warty glands on the leaf stalk just below the blade.

Flowers with 5 small petals; arranged in branched, flat-topped or rounded, terminal clusters. **Fruits** berry-like, with a single flat seed; seeds dispersed by fruit-eating birds.

    In the absence of fruits viburnums with 3-lobed leaves may be mistaken for maples.

Sweet Viburnum

Nannyberry, sheepberry

Viburnum lentago L.

Viorne flexible

A flowering shrubby tree; often planted as an ornamental (Zones C2, NA2).

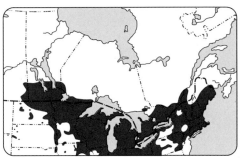

 Leaves 5–10 cm long, ovate to narrowly oval; tip slender, sharp, short-tapered; base rounded; finely and sharply toothed; upper surface yellowish-green, tiny dark brown dots beneath; hairless; stalk 1–2 cm, grooved, winged along each side by narrow extensions of the leaf blade.

Buds Terminal bud 20–30 mm long, slender, naked, with one pair of brownish-gray granular immature leaves; terminal flower bud about twice as long with a bulbous base. Lateral buds smaller.

Twigs Slender, smooth, light brown; unpleasant odor when bruised.

Flowers Creamy white; borne in round-topped clusters, 5–10 cm wide, with several branches radiating from shoot tip.

Fruits Bluish-black, 8–12 mm long, with thin sweet edible flesh.

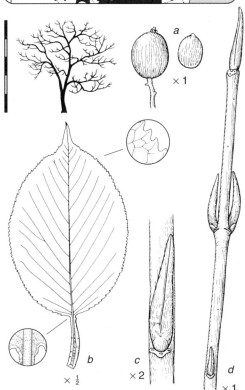

a. Fruit (left); seed (right). *b.* Leaf with details of finely toothed margin (above) and winged leaf stalk (below). *c.* Lateral bud and leaf scar. *d.* Winter twig.

8

Flowers in round-topped clusters
without a central stock.

Mature bark rough, scaly.

Fruit hard and green in summer; when
ripe, in autumn, bluish-black, edible.

Sweet Viburnum

Young bark smooth, brownish.

Vegetative Reproduction By root sprouts.

Bark Grayish-brown, with very small irregular scales.

Wood Hard, heavy, close-grained, dark brown.

Size and Form Large shrubs or very small trees, up to 10 m high and 15 cm in diameter. Trunk slender, crooked. Crown open, irregular, with a few arching branches.

Habitat Occurs along riverbanks, lakeshores, forest edges, and roadsides.

Quick Recognition Leaves opposite, ovate-oval, finely toothed, with a sharp tip. Terminal bud long, narrow, naked, often bulbous at the base, gray. Flowers and fruit in clusters without a central stalk.

European Cranberry Viburnum**

European highbush-cranberry, European cranberry-bush, guelder-rose

Viburnum opulus L.

Viorne obier

Introduced from Europe; similar to the North American cranberry, but larger; frequently planted and naturalized in many urban locations. Hardy as far north as Zones C2, NA3. Saucer-like glands on the leaf stalk. Fruits bitter, remaining on the tree over winter. The 'Snowball' is a cultivar of this species with sterile showy flowers.

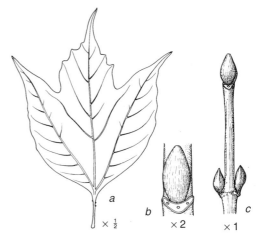

a. Leaf. *b*. Lateral bud and leaf scar. *c*. Winter twig.

Wayfaring Viburnum**

Wayfaring tree

Viburnum lantana L.

Viorne mancienne

Very small trees; introduced from Europe; frequently planted for ornamental purposes and as a source of food for birds. Hardy as far north as Zones C2, NA3. **Leaves** 5-12 cm long, ovate, with branched hairs on the lower side; margin toothed. **Buds** naked, 1 or 2 pairs of immature leaves visible; with yellowish-white hairs; flower buds large, squat, much wider than long. **Twigs** yellowish-gray, mealy. White **flowers** and black **fruits** in flat-topped clusters.

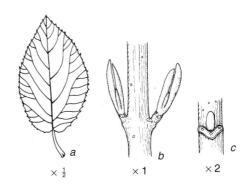

a. Leaf. *b*. Buds protected by immature leaves. *c*. Lateral bud and leaf scar.

8

Squashberry Viburnum

Virburnum edule (Michx.) Raf.

Viorne comestible

Shrubs occurring across Canada; an understory species in many forests. **Leaves** 3-lobed, crinkled on the upper surface, hairy beneath, long-stalked; red in autumn. **Flowers** white, in small clusters, 1–2.5 cm wide. **Fruits** red, bitter, in clusters of 2–5.

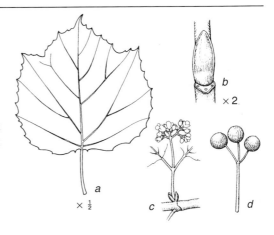

a. Leaf. *b*. Lateral bud and leaf scar. *c*. Flower cluster. *d*. Fruit cluster.

Fruit orange to red, often persistent, conspicuous in winter.

Cranberry Viburnum

Highbush-cranberry, cranberry-bush

Viburnum trilobum Marsh.
[syn. *V. opulus* (L.) var. *americanum* (Ait.)]

Viorne trilobée

Shrubs occurring across Canada; commonly planted (Zones C2, NA2). **Leaves** 5–11 cm long, almost as wide, 3-lobed, maple-like; margin smooth or sparsely toothed; stalks 2–4 cm long, grooved, with several small club-shaped glands where it meets the blade. **Buds** reddish, with a pair of fused scales; terminal bud usually lacking. **Flowers** in large flat-topped terminal clusters up to 10 cm across; outer flowers white, showy, sterile; inner flowers small, functional. **Fruits** orange to red, 8–10 mm across, edible.

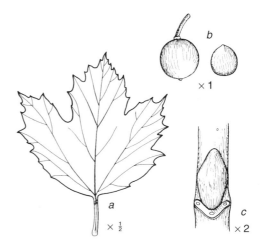

a. Leaf. *b.* Fruit (left); seed (right). *c.* Lateral bud and leaf scar.

Dogwoods
Cornels

Genus *Cornus*
Cornaceae: Dogwood Family

Les cornouillers

The dogwood genus comprises about
40 species of small trees, shrubs, and a few
herbs widely distributed throughout the North
Temperate Zone. About 12 species are
native to Canada, including 3 small trees,
several shrubs, and 2 herbaceous peren-
nials. Species with floral bracts around the
flower clusters are a conspicuous feature of
the understory in spring before most trees
are in leaf. Several species, both native and
introduced, are planted for ornamental
purposes.

Leaves Deciduous, in opposite pairs (except
for the alternate-leaf dogwood), simple;
smooth-margined; veins prominent, curving
forward and following the margin toward the
leaf tip (arcuate).

Buds Terminal bud often a large, swollen,
globular flower bud, conspicuously so in the
species with floral bracts; leaf buds slender,
pointed, with 2 pairs of scales in most
species. Leaf scars with 3 vein scars.

Twigs Slender, with prominent lenticels.
Pith round in cross section; white in some
species, brownish in others. In some species,
neoformed branches develop near the end of
the new shoots.

Flowers Small, white, greenish-white, or
yellow; arranged in compact terminal clusters
surrounded by large bracts, or in loose terminal clusters without bracts. Insect-pollinated.

Fruits Small, berry-like, with 1 or 2 hard
stones; borne in terminal clusters. Dispersed
by birds.

Seeds Germinate after some months of cool
moist storage.

Seedlings Cotyledons raised above the
surface.

Vegetative Reproduction Some species by
root sprouts or by layering (rooting of
attached branches where they touch the
ground).

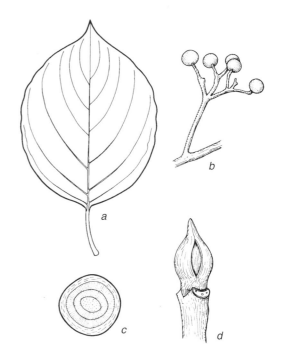

a. Leaf with arcuate venation. b. Fruit cluster.
c. Cross section of 2-year-old twig. d. Terminal bud
and elevated leaf scar.

Size and Form Herbs, shrubs, or small
trees. Branches of tree-sized species often
form horizontal layers.

Wood Heavy, hard, strong, fine-grained,
resistant to abrasion.

Habitat Found on moist fertile soils; along
fencerows and roadsides, at forest edges, in
openings and understories.

Notes Daggers were made from the hard
tough wood of these species and thus may
account for the name "dogwood" (dagwood
or daggerwood).
   Stems are browsed by rabbits and deer;
fruits are an important source of food for
birds and small mammals.

Quick Recognition Leaves with arcuate veins,
in opposite pairs (except alternate-leaf dogwood), smooth-margined. Fruits berry-like, in
terminal clusters. Species with floral bracts are
distinctive when in flower.

8

Inconspicuous flowers surrounded by 4 white floral bracts; appear before the leaves.

Eastern Flowering Dogwood

Cornus florida L.

Cornouiller fleuri

An understory tree of the deciduous forests of eastern North America; occurring north into southern Ontario; **similar to western flowering dogwood** (next page). Planted for landscape purposes (Zones C6, NA5).

　　Floral bracts broad, round-tipped or notched, usually 4; forming a "flower" 5–6 cm across; appearing before the leaves in early spring. **Fruit** clusters consist of a few separate red fruits. **Leaves** turn red in autumn. **Bark** rough, breaking into quadrangular plates. Very small trees, up to 10 m high and 20 cm in diameter.

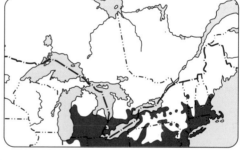

a. Central flower cluster and floral bracts.
b. Leaf. *c*. Fruit (left); fruit cluster (right).

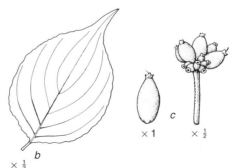

Western Flowering Dogwood

Pacific dogwood

Cornus nuttallii Audub.

Cornouiller de Nuttall

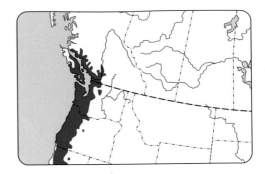

A Pacific coast species, spectacular because of its floral bracts; frequently planted as an ornamental (Zones C8, NA8). Ripe red fruits and bright flowers may appear together on the same tree.

Leaves Somewhat clustered at the shoot tips; 8–15 cm long, widest near the middle; tapered to both ends; tip short, sharp; base broad, wedge-shaped; margin wavy; upper surface deep green, grayish-green beneath; dull red in autumn; fine hairs on both surfaces.

Buds Terminal flower bud globular; terminal leaf bud long-tipped, 8–12 mm long, dull, grayish-green to brownish, with 1 pair of scales covered with fine white hairs. Lateral buds minute.

Twigs Light green, hairy when young; becoming purplish, smooth. Lenticels prominent.

Flowers Small, dull purple or green, inconspicuous, in compact clusters surrounded by 4–6 white (rarely pink) pointed floral bracts, 5–10 cm long, forming a "flower" about 10–20 cm across. Floral bracts originate as bud scales. Blooming from April to June and sometimes again in late summer.

Fruits Bright red, 10–12 mm long, in densely packed globular clusters of 30–40. Ripening in autumn.

Bark Thin, light gray to reddish-brown, smooth, breaking into small plates with age.

Size and Form Small trees, up to 18 m high and 30 cm in diameter. Forest-grown trees: trunk slightly tapered, straight, distinct to the top of a narrow short crown. Open-grown trees: trunk short; crown rounded, about as wide as high. Principal branches ascending when young, then spreading horizontally. Root system quite deep, occasionally with a taproot.

8

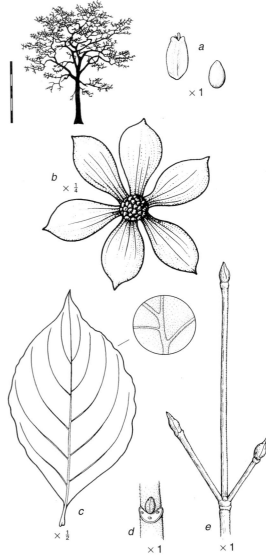

a. Fruit (left); seed (right). *b.* Central floral cluster and conspicuous floral bracts. *c.* Leaf with detail of fine hair on undersurface. *d.* Lateral bud and leaf scar. *e.* Winter twig showing terminal floral buds.

Flower cluster usually surrounded by 6 floral bracts.

Fruit flattened and angular; in densely packed clusters.

Young bark thin, smooth.

Mature bark broken into small rectangular plates.

Habitat Occurs on deep, coarse, well-drained sites, such as on water-formed terraces, along riverbanks, and near the bottom of valleys. Often a major component of the understory in open conifer stands.

Quick Recognition A small western tree with opposite leaves and buds. Leaves with arcuate veins. Terminal bud present; leaf buds slender; flower buds globular. Flowers in tight terminal clusters surrounded by 4–6 large white floral bracts. Fruits red, in globular terminal clusters.

Alternate-Leaf Dogwood

Pagoda dogwood

Cornus alternifolia L. f.

Cornouiller à feuilles alternes

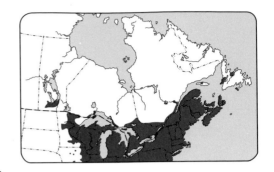

An understory species of the deciduous forests of eastern North America; planted as an ornamental and as a source of food for birds (Zones C3, NA3). Unique among native dogwoods because the leaves are alternate rather than opposite. Branches in distinct tiers on the main trunk; hence, the name "pagoda dogwood".

Leaves Alternate, may appear opposite or whorled on very short twigs, noticeably clustered at the ends of shoots; 4–13 cm long, ovate, widest near the middle; tip tapering, long, slender; base rounded or tapered; margin slightly wavy; upper surface dark green, whitish beneath; red in autumn; hairless.

Buds Terminal buds small, pointed, with 2 or 3 reddish-brown scales, not tight against the bud.

Twigs Shiny, polished, greenish-red to dark reddish-brown, or dark purplish-red. Many shoots ending in a flower bud have 1 or 2 slender neoformed side branches near the tip, longer than the stout central axis.

Flowers Small, white or cream-colored, on a jointed stalk, arranged in large open irregularly rounded terminal clusters flattened at the top. Opening after the leaves.

Fruits Berry-like, 8–10 mm across, dark blue or bluish-black, on red stalks; ripening in midsummer.

Bark Thin, reddish-brown, smooth, with age separating into shallow ridges.

Size and Form Very small trees or large straggly shrubs, up to 10 m high and 15 cm in diameter. Crown formed of several spreading, horizontal tiers, resulting in a flat, layered appearance.

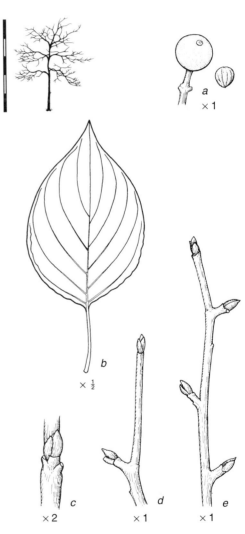

a. Fruit (left); seed (right). *b.* Leaf. *c.* Lateral bud. *d.* Neoformed side branch at shoot tip. *e.* Winter twig.

8

Flowers open after leaves have developed, lack conspicuous floral bracts.

Fruit berry-like; ripens in midsummer.

Young bark (inset) reddish-brown, smooth. Tree often shrubby but can be a single-stemmed.

Habitat Occurs on well-drained deep soils, in the understory of open woodlands, at forest edges, along streams, bordering swamps and near the bottom of steep slopes.

Quick Recognition Leaves alternate; veins arcuate. Terminal bud small, with loose scales. One or two slender neoformed side branches often longer than central shoot. Fruits dark blue on red stalks. Branches in tiers.

Cornelian-Cherry*

Cornel dogwood

Cornus mas L.

Cornouiller mâle

Very small branchy trees, native to Europe and Asia, cultivated there for centuries; occasionally planted in Canada and the United States. Hardy as far north as Zones C5, NA4. **Leaves** 5-10 cm long, ovate, with 4 or 5 pairs of veins; hairy on both surfaces; green into late autumn. **Buds** yellow; terminal bud with 1 pair of scales; flower buds globular, on short side shoots. **Twigs** hairy, red and green, with ridges; pith white. **Floral bracts** 4, pale yellow, boat-shaped, hairy, appear early in spring before the leaves, soon shed. **Fruits** single, cherry-like but 2-seeded, oblong, 15 mm long, edible, bright red in midsummer.

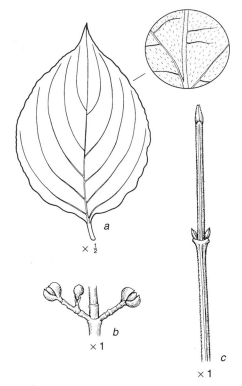

a. Leaf with detail of fine hair on surface. b. Flower buds on short side shoots. c. Winter twig.

Kousa Dogwood*

Cornus kousa Hance

Cornouiller de Kousa

Very small trees, native to eastern Asia, occasionally planted in Canada and the United States. Hardy as far north as Zones C6, NA5. **Leaves** 5–10 cm long, elliptic, with 4–6 pairs of veins; upper surface dark green, hairy below with tufts of hairs in the vein axils; reddish-purple in autumn. Terminal **bud** dark brown, with 2 scales; flower buds terminal on side shoots, globular, with 2 scales covered with silky hairs and forming a sharp point. **Twigs** slender, light brown with tinges of purple and green. **Flowers** in clusters on a stalk 5 cm long, at the tip of a short shoot; floral bracts creamy-white, 5 cm long, taper-pointed, appear with the leaves, lasting for 6 weeks. **Fruits** pinkish-red, in globular clusters. Outer layers of bark shed, leaving a mosaic of grays and browns.

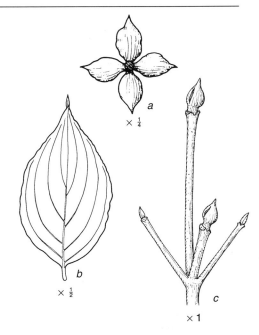

a. Flower cluster and bracts. b. Leaf. c. Winter twig with neoformed lateral shoots.

Flowers in cone-shaped clusters; appear in mid-May.

Common Lilac**

Syringa vulgaris L.
Oleaceae: Olive Family

Lilas commun

Shrubby trees, native to southeastern Europe. Probably the most commonly planted flowering tree in Canada. Frequently found near old homesteads, in clumps originating from root sprouts. Hardy as far north as Zones C2, NA3.

Leaves deciduous, simple, opposite, 5–12 cm long, heart-shaped, pointed, smooth-margined; stalk 2–3 cm long. **Buds** large, stout, ovoid, with 3–5 pairs of green and brown scales; usually no terminal bud. **Flowers** purple, pink, or white, fragrant, about 10 mm long, with 4 petals forming a tube, borne in large cone-shaped clusters, usually in pairs at the tip of previous year's twigs; appearing with the leaves. **Fruit** a flattened 2-seeded capsule, 10 mm long; persisting over winter.

Hundreds of cultivars of this genus have been developed; some differ from the common lilac in that the flowers are in terminal clusters on leafy shoots.

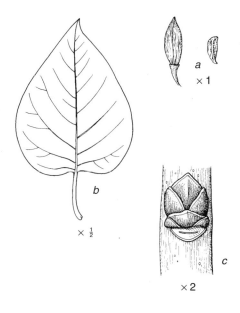

a. Fruit capsule (left) and seed (right). b. Leaf. c. Lateral bud and leaf scar.

Catalpas
Catawba
Genus *Catalpa*
Bignoniaceae: Trumpet-Creeper Family

Les catalpas

The catalpa genus comprises about 12 tree species, 2 of which are native to North America. Planted in Canada and United States beyond their natural range because of their showy flowers.
   Leaves deciduous, opposite (or whorled), large, simple; 10–30 cm long, heart-shaped, pointed at the tip, occasionally with 1–3 large teeth or shallow lobes, softly hairy beneath; leaf stalks 10–16 cm long.
   Buds 2–5 mm long, rounded, with 6 brown scales; no terminal bud. Leaf scars oval or rounded, depressed in the center, with many vein scars forming a ring. **Twigs** stout; tip tapered to a blunt withered point.
   Flowers large, showy, with 5 petals fused into a tube, white, with yellow and purple spots; in large terminal clusters; appearing in midsummer, when the leaves are fully grown.
Fruits long cylindrical capsules containing many 2-winged seeds; borne in clusters of 1–5; remaining on the tree into winter. **Bark** with irregular shallow fissures and large thick scales. Can be heavily trimmed to form a small globular crown that does not produce flowers.

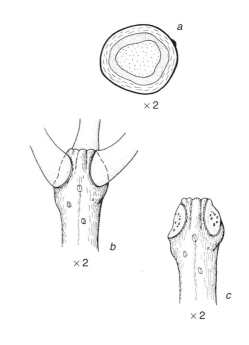

a. Cross section of 1-year-old twig. *b*. Twig tip and leaf stalks, showing whorled arrangement. *c*. Twig tip showing large depressed leaf scars.

Northern Catalpa*
Catawba, bean-tree

Catalpa speciosa Warder

Catalpa à feuilles cordées

Medium-sized trees, up to 30 m high, native to the central Mississippi Valley. Hardy as far north as Zones C5, NA4. **Leaves** long-pointed. **Flowers** very large, 6–7 cm across, petals white, lightly spotted with yellow and purple; calyx greenish-purple; borne in few-flowered clusters. **Fruit capsules** 25–60 cm long, in clusters of 1–3. **Bark** dark brown.

8

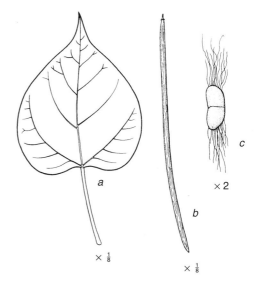

a. Leaf. *b*. Fruit capsule. *c*. 2-winged seed.

Flower clusters showy; flower petals fused
into a large tube.

Northern Catalpa

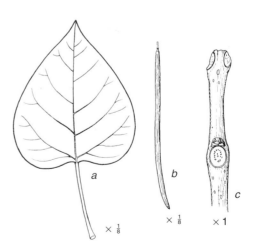

Fruit capsules conspicuous; often persist on
trees into winter.

Southern Catalpa*

Catalpa bignonioides Walt.

Catalpa commun

Similar in many features to the northern catalpa.
Small trees, up to 15 m high, native to the southeastern United States. Hardy as far north as
Zones C5, NA4. **Leaves** short-pointed. **Flowers**
smaller, 3–5 cm across, densely spotted with
purple and yellow; appearing about 2 weeks later;
more in a cluster. **Fruit capsules** smaller,
15–40 cm long, in clusters of 1–5; seeds with tufts
of white hairs at the tips of the wings. **Bark**
reddish-brown.

a $\times \frac{1}{8}$ *b* $\times \frac{1}{8}$ *c* $\times 1$

a. Leaf. *b*. Fruit capsule. *c*. Winter twig with lateral
bud and leaf scar.

Euonymus
Spindle trees, burning-bushes, wahoos
Genus *Euonymus*
Celastraceae: Staff-tree Family

Les fusains

The euonymus genus comprises about 170 species, found mostly in Asia; 4 shrubs and small trees are native to North America; 1 tree-size species is native to Canada. Several introduced shrubs and trees are used for landscape purposes.

   Leaves mostly deciduous, in opposite pairs, simple; 5–12 cm long, toothed, short-stalked. Terminal **bud** present. **Twigs** smooth, brown or green, 4-sided or with corky wings.

   Flowers small, 4-parted, on slender stalks; in few-flowered clusters in the axils of the leaves; appear in early summer when the leaves are fully grown. **Fruit** a capsule, usually with 4 prominent lobes, pink when ripe in autumn; splitting to reveal 4 seeds, each enclosed in a fleshy aril; remain on the tree after the leaves have fallen; dispersed by birds.

Burning-Bush Euonymus
Spindle tree, eastern wahoo
Euonymus atropurpureus Jacq.

Fusain pourpre

The only euonymus species native to Canada reaching tree size. Shrubs or shrubby trees up to 6 m high; occurring along streams, on floodplains, and in moist woodlands; occasionally in rocky woods.

   Leaves 5–12 cm long, oval, long-pointed, finely toothed, red in autumn; stalk about 1 cm long. Terminal **bud** 2–4 mm long, ovoid, pointed, green tinged with red, with 6 narrow scales; lateral buds pressed against the twig. **Twigs** smooth, greenish, somewhat 4-sided. **Flowers** small, purplish. **Fruit capsules** 10–14 mm across, with 4 prominent lobes; splitting to reveal 4 seeds, each enclosed in a bright orange-red fleshy aril; pink when ripe in autumn; remain on the tree after the leaves have fallen; dispersed by birds. **Bark** greenish-gray, streaked with reddish-brown.

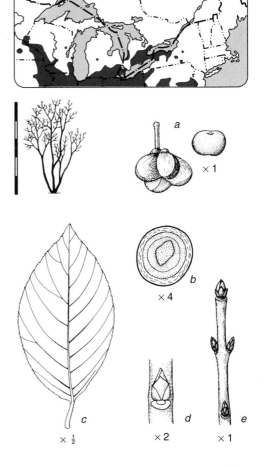

a. Fruit capsule (left); aril (right). *b*. Cross section of 2-year-old twig. *c*. Leaf. *d*. Lateral bud and leaf scar. *e*. Winter twig.

8

Flowers purplish, on slender stalks.

Burning-Bush Euonymus

Fruit capsule 4-angled; turns pink when ripe.

Young bark (inset) grayish, streaked. Tree often multistemmed.

European Euonymus*
European spindle tree

Euonymus europaeus L.

Fusain d'Europe

Very small trees, introduced from Europe, occasionally planted in Canada. Hardy as far north as Zones C4, NA3. Similar in many respects to the burning-bush euonymus. **Buds** plump, not pressed against the stem; scales diverging. **Flowers** greenish or yellowish-white. **Seeds** covered with an orange aril.

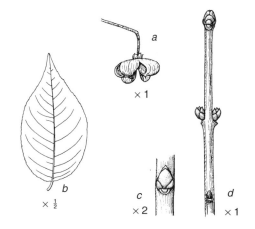

a. Fruit capsule. *b.* Leaf. *c.* Lateral bud and leaf scar. *d.* Winter twig.

Winter-Creeper Euonymus*

Euonymus fortunei (Turcz.)
Hand.-Mazz.

Fusain rampant

A creeping shrub introduced from China, often
putting out aerial roots and climbing to a height of
20 m. The tree-form of this species is artificially
made by grafting winter-creeper scions on an
upright stem of another species of euonymus; up
to 3 m high; interesting because it is one of the few
broadleaf evergreens with a tree-form in eastern
Canada. Hardy as far north as Zones C5, NA5.
   Leaves evergreen, shed during the 2nd
season, ovate, thick, shiny, bright green, marked
with white veins or white margins in some cultivars, short-stalked. **Buds** plump, green, sharp-
pointed, with 3 or 4 pairs of scales. **Twigs** stout,
green, somewhat 4-sided. **Flowers** white; **fruit
capsules** pink to red; **seeds** covered with an
orange-red aril.

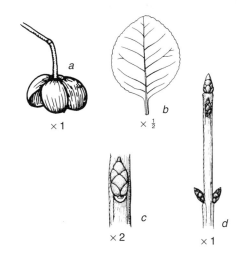

a. Fruit capsule. *b.* Leaf. *c.* Lateral bud and leaf scar.
d. Winter twig.

Winged Euonymus*

Euonymus alatus (Thunb.)
Siebold

Fusain ailé

8

Shrubs or shrubby trees, introduced from
northeastern Asia, occasionally planted in
Canada. Hardy as far north as Zones C3, NA3.
Leaves opposite or subopposite, 3–7 cm long,
often widest above the middle, tapered to the base,
finely toothed, short-stalked, dark green, bright red
in autumn. **Buds** greenish or brownish, with
6–8 pairs of scales, diverging from the stem.
Twigs 4-sided, with 4 prominent corky ridges,
interrupted by the buds. **Flowers** yellowish-green;
fruit capsules red; **seeds** covered with an
orange-red aril.

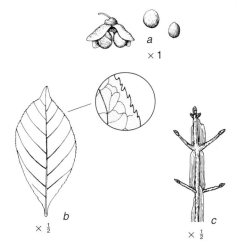

a. Fruit capsule (left); aril (center); seed (right).
b. Leaf with detail of finely toothed margin. *c.* Winter
twig with corky wings.

Button-Bush

Cephalanthus occidentalis L.
Rubiaceae: Madder Family

Céphalante occidental

A common shrub, occasionally a very small tree up
to 10 m high and 20 cm in diameter; occurs along
lakeshores and on other wet sites in southern
Ontario, east to Nova Scotia and south to the Gulf
of Mexico. **Leaves** deciduous, opposite or
whorled, simple; 6–18 cm long, lance-shaped,
taper-pointed, smooth-margined, thin, dark green
on the upper surface, slightly hairy beneath;
stipules sharp-pointed. Terminal **bud** infrequent;
twigs green. **Flowers** in many-flowered clusters,
2–5 cm across, at the tips of the shoots; **fruit
clusters** globular, about 2 cm across, on long
stalks. A source of food and shelter for wildlife.

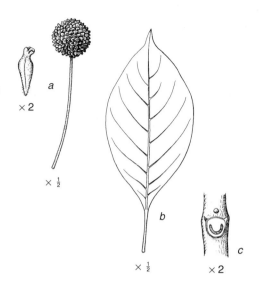

a. Fruit (left); fruit cluster (right). *b.* Leaf. *c.* Lateral
bud and leaf scar.

Katsura-Tree*

Cercidiphyllum japonicum Siebold
& Zucc.
Cercidiphyllaceae: Katsura Family

Cercidiphyllum du Japon

Medium-sized trees, native to China and Japan;
occasionally planted as shade trees in milder parts
of Canada. Hardy as far north as Zones C5, NA4.
A family with 1 genus and 1 species.
  Leaves deciduous; subopposite on new
shoots, singly on dwarf shoots along the branchlets, simple; 5–10 cm long, ovate to round, base
heart-shaped; weakly toothed, with 3–5 main
veins radiating from the leaf stalk; purplish in
spring, blue-green in summer, red and yellow in
autumn, with a spicy odor. **Buds** small, 2–4 mm
long, with 2 scales; no terminal bud. **Twigs**
slender, swollen below the buds, brownish.
Flowers small, on dwarf shoots; pollen flowers and
seed flowers on separate trees; appear in spring
before the leaves. **Fruit** a small pod, about 2 cm
long; seeds thin, winged.

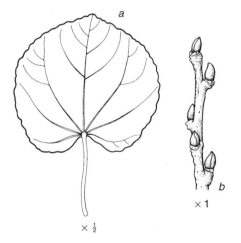

a. Leaf. *b.* Winter twig.

Silver Buffalo-Berry

Thorny buffalo-berry, buffalo shepherdia,
silver-berry

Shepherdia argentea Nutt.
Elaeagnaceae: Oleaster Family

Shépherdie argentée

Large shrubs or very small trees of central and
western Canada; the silvery brown scales on the
leaves, twigs, and other parts of the plant are
characteristic of the oleaster family. **Leaves** deciduous, opposite, simple; 2–6 cm long, oblong to
lance-shaped. **Buds** with 1 or 2 pairs of scales.
Twigs thorny. **Flowers** in small clusters on previous year's twigs; pollen flowers and seed flowers
on separate trees. **Fruits** berry-like, 4–6 mm
across, sour, edible, red. Occasionally planted as a
hedge.

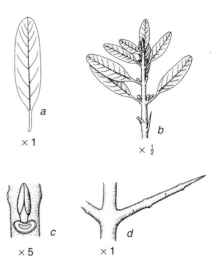

a. Leaf. *b.* Leafy twig. *c.* Lateral bud and leaf scar.
d. Thorn

8

Group 9.

Leaves alternate, compound (divided into 3 or more leaflets)

The alternate arrangement of leaves is by far the most common. Species with alternate leaves occupy the rest of the book. These leaves are usually arranged in a spiral, but for a few species they are in two ranks or rows on opposite sides of the stem.

Compound leaves are usually easy to recognize; the leaf blade is divided into two or more leaflets, each attached to a central stalk. Leaves can be distinguished from leaflets by the presence of a bud in the leaf axil, the place where the leaf is attached to the stem.

In Canada most trees with compound leaves have leaflets that are pinnately arranged along the central stalk. A few species such as horsechestnut (Group 8) have leaflets that are palmately arranged, radiating out from the end of the stalk. The palmate arrangement is more common in tropical trees. Compound leaves composed of three leaflets are considered to have a pinnate arrangement.

Leaves may be doubly compound; that is, the central stalk bears a number of side stalks, which in turn bear the leaflets.

All trees in Group 9 are deciduous. Unless otherwise indicated, the leaflets of species in Group 9 are arranged in opposite pairs.

Walnuts
Butternuts

Genus *Juglans*
Juglandaceae: Walnut Family

Les noyers

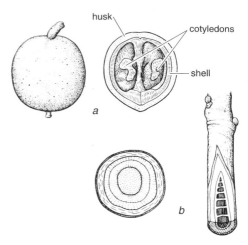

a. Fruit (left); fruit longitudinal section (right).
b. Cross section of 2-year-old twig (left); twig showing chambered pith (right).

Worldwide, there are about 20 species of walnut, distributed in North and South America, southeastern Europe, eastern Asia, and northern India. Of the 6 species native to North America, only black walnut and butternut have ranges extending north into eastern Canada. Walnuts are frequently cultivated for their nuts and high-quality wood.

Leaves Deciduous, alternate, pinnately compound, composed of 5–23 leaflets on a central stalk 20–60 cm long. Leaflets oblong, 5–11 cm long, tip pointed, base somewhat rounded, short-stalked, toothed, strongly aromatic.

Buds Terminal bud distinctive, ovoid, irregularly flattened, with several scales. Lateral buds smaller, often 2 or more in line above a leaf scar; immature pollen catkins with a cone-like appearance may be present. Leaf scars in 5 rows along the twig, shield-shaped to triangular, raised, with 3 conspicuous U-shaped vein scars, or 3 U-shaped clusters of vein scars, 1 in each corner of the leaf scar. Bud scales enlarge markedly during the spring flush of growth.

Twigs Stout. Pith with alternating cavities and partitions after the 1st season. Shoot growth ends in midsummer with the formation of a terminal bud.

Flowers Small, green, inconspicuous. Pollen flowers and seed flowers on the same tree. Pollen flowers with 8–40 stamens, borne in drooping catkins up to 15 cm long, on previous year's twigs or at the base of a new shoot, singly or in 2's. Seed flowers in small erect clusters at the tip of a new shoot. Appear in spring with the leaves. Wind-pollinated.

Fruits Large, hard-shelled nut enclosed in a firm fleshy green husk that does not split open; shell irregularly furrowed (in native species), or shallowly figured; kernel edible. Borne singly or in small clusters. Ripen in 1 season, shed in autumn. Dispersed mainly by squirrels. Good seed crops every few years.

Seeds Kernel (embryo) consists of 2 large irregularly lobed cotyledons and a small stem–root axis. Viable for 1 or 2 year(s) in moist conditions at temperatures just above freezing; some remain dormant until the 2nd spring after being buried in the surface litter.

Seedlings Newly germinated seedlings produce a shoot which at first bears simple toothed leaves; later leaves are compound with an increasing number of leaflets. Cotyledons white, retained within the shell.

Vegetative Reproduction Stump sprouts may form when young trees are cut.

Wood Moderately hard. Semi-ring-porous; rays visible with a hand lens.

Size and Form Medium-sized to large trees. Crown open, with coarse irregular ascending branches. Leading shoot upright; most side branches occur at an annual node.

Habitat Found mostly on well-drained, fertile soils, usually in mixed stands. Intolerant of shade and competition

Notes Squirrels store the fruits by burying them in the autumn; the germination of some of these is a major mode of reproduction. Wood used in cabinetmaking.

Quick Recognition Leaves large, pinnately compound, with an aromatic odor. Twigs end in a terminal bud. Leaf scars prominent, with 3 vein scars. Crown consisting of a few large irregular branches with stout twigs.

9

black walnut
Walnut crown is open, with large limbs and coarse ascending branches.

Butternut

White walnut

Juglans cinerea L.

Noyer cendré

A species of eastern North America; occurs in southern New Brunswick, Quebec, and Ontario.

Leaves 11–17 almost stalkless leaflets on a stout hairy central stalk, 30–60 cm long; yellowish-green and rough above, paler and densely hairy beneath; sticky when young. Terminal leaflet usually present and about the same size as the adjacent leaflets; lateral leaflets progressively smaller toward the base.

Buds Terminal bud elongated, somewhat flattened, 12–18 mm long, blunt-tipped, pale yellow, hairy, with lobed outer scales. Leaf scars flat on upper margin, edged with a pad of hairs.

Twigs Stout, orange-yellow, hairy. Pith reddish-brown.

Flowers Pollen flowers with 8–12 stamens, in catkins 6–14 cm long. Seed flowers in erect clusters of 4–7.

Fruits Elongated, pointed, 5–8 cm long, in drooping clusters of 1–5. Husk with dense sticky hairs. Surface of nut shell with irregular, jagged ridges. Kernel sweet, oily.

Bark Light gray, smooth when young; becoming ridged and grooved with narrow, shallow, dark fissures and wide, irregular, flat-topped, intersecting ridges.

9

Wood Light, soft, weak, coarse-grained, reddish-brown.

Size and Form Medium-sized trees, up to 25 m high, 75 cm in diameter, and 80 years old. Trunk often short, forked. Crown open, broad, irregular in outline and rounded on top, with a few large ascending branches. Smaller side branches tending to bend downward, then upward at the tips. Root system deep, wide-spreading, usually with a taproot.

Habitat Occurs on a variety of sites, including dry rocky soils (particularly those of limestone origin); grows best on well-drained, fertile soils in shallow valleys and on gradual slopes; singly or in small groups mixed with other species. Intolerant of shade.

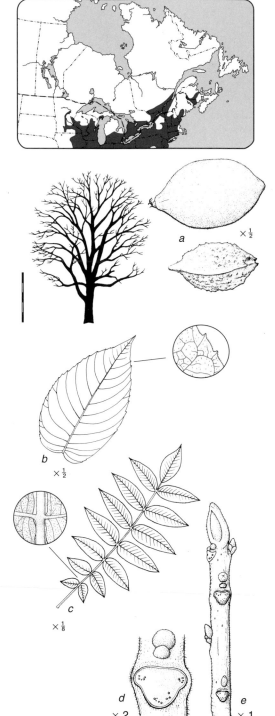

a. Fruit with husk (above); nut (below). *b.* Leaflet with detail of toothed margin. *c.* Leaf with detail of nearly stalkless leaflets. *d.* Lateral bud and leaf scar. *e.* Winter twig.

Seed flowers in erect clusters at the shoot tip.

Pollen flowers with 8–12 stamens;
catkins unbranched, drooping.

Fruit elongated, densely hairy, sticky.

Young bark (inset) at first smooth, becomes
ridged. Mature bark pale gray, fissured.

Notes Wood used in cabinetry and turnery.
Trees increasingly attacked by butternut can-
ker (*Sirococcus clavignenti-juglandacearum*),
a fatal disease.

Quick Recognition Leaves very hairy below;
terminal leaflet as large as adjacent leaflets.
Leaf scar flat on the upper margin with a pad
of hairs. Fruits elongated; shell surface with
jagged ridges. See black walnut text for contrasting features.

Black Walnut
American walnut

Juglans nigra L.

Noyer noir

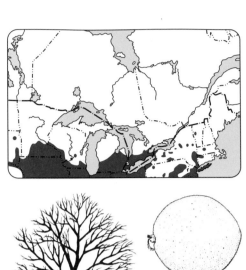

Uncommon in Canada, occurring naturally in southern Ontario; frequently planted for its fruit and as an ornamental within its range, as well as northward in eastern Canada and westward in southern British Columbia (Zones C3, NA4).

Leaves 14–22 short-stalked leaflets (15–23, with terminal leaflet) on a moderately stout central stalk, 20-60 cm long; yellowish-green and smooth above, faintly hairy beneath. Terminal leaflet missing, or much smaller than largest lateral leaflets; middle leaflets larger than those at either end of leaf.

Buds Terminal bud ovoid, slightly flattened, 8–10 mm long, blunt, pale gray, slightly hairy. Leaf scars deeply notched on upper margin, hairless.

Twigs Stout, orange-brown, faintly hairy. Pith orange-yellow.

Flowers Pollen flowers with 20–30 stamens, in catkins 5–10 cm long. Seed flowers in erect clusters of 1–4.

Fruits Globular, 4–6 cm in diameter, in drooping clusters of 1–3. Husk slightly hairy. Surface with deep grooves and smooth-topped ridges. Kernel strongly flavored, oily.

Bark Light brown, scaly when young; becoming darker, with rounded, almost black, intersecting ridges.

Wood Heavy, hard, strong, dark brown to black, sapwood nearly white, resistant to decay, straight-grained.

Size and Form Medium-sized trees, up to 30 m high, 120 cm in diameter, and 150 years old. Trunk straight. Crown open, rounded, with a few large ascending branches. Root system deep, wide-spreading, usually with a taproot.

Habitat Occurs on well-drained, fertile lowlands; found singly or mixed with other broadleaf trees. Intolerant of shade.

Notes Valued for its wood, which is easily worked and finished, not likely to shrink or warp, and has an attractive grain. Isolated stands in the Ottawa Valley may have been planted or be remnants of warmer climates.

9

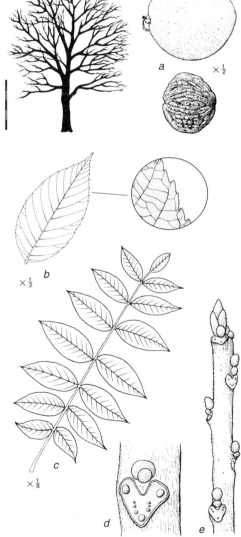

a. Fruit with husk (above); nut (below). *b.* Leaflet with detail of finely toothed margin. *c.* Leaf. *d.* Lateral bud and leaf scar. *e.* Winter twig.

Black Walnut

Pollen flowers (left) with 20–30 stamens.
Seed flowers (right) in clusters of 1–4.

Fruit globular; husks slightly
hairy.

Young bark (inset) light brown, scaly. Mature bark
dark, with intersecting ridges.

A toxic substance, juglone, leaches from
fallen leaves and is exuded by roots; it inhib-
its the growth of many broadleaf plants, including walnut seedlings.

Quick Recognition [Contrasting features of
butternut in brackets.] Leaves slightly [very]
hairy, terminal leaflet small or missing [normal
size]. Leaf scar notched [flat] on the upper
margin, not hairy [with a hairy pad]. Fruit globular [elongated]; shell surface with rounded
[jagged] ridges.

English Walnut*
Persian walnut, common walnut

Juglans regia L.

Noyer commun

Native to Europe and Asia. Hardy as far north as
Zones C6, NA5. **Leaves** with 5–9 leaflets;
terminal leaflet large, margins smooth. **Buds**
grayish-brown, hairy. **Twigs** smooth, greenish-
brown. **Fruits** large, globular, easily opened; shell
surface figured. The source of commercial walnuts.
Wood very valuable.

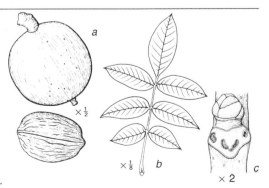

a. Fruit with husk (above); nut (below). *b.* Leaf.
c. Lateral bud and leaf scar.

Hickories
Pecans

Genus *Carya*
Juglandaceae: Walnut Family

Les caryers

Hickories occur naturally in 2 areas within the North Temperate Zone: eastern North America and eastern Asia. About 12 species are native to North America, 4 of them to Canada.

The genus is divided into 2 parts: the true hickories, which include shagbark, shellbark, and red hickory; and the pecan hickories, which include the well-known pecan, and one Canadian species, bitternut hickory. Features of the true hickories native to Canada are given below; bitternut features are noted in brackets.

Leaves Deciduous, alternate, pinnately compound, composed of 5–11 leaflets on a central stalk 12–25 cm long. Terminal leaflet large; lateral leaflets progressively smaller toward the base. Leaflets finely toothed, base asymmetrical, stalkless, or nearly so, highly aromatic.

Buds Terminal bud distinctive, ovoid, 6–20 mm long, with several overlapping scales [irregular, with 2–4 scales that meet edge to edge]. Lateral buds much smaller. Leaf scars in 5 rows along the twig, conspicuous, raised, 3-lobed, with many vein scars.

Twigs Stout to moderately slender. Pith solid, angled. Shoot growth ends in July with the formation of a terminal bud.

Flowers Small, green, inconspicuous. Pollen flowers and seed flowers on the same tree. Pollen flowers in 3-branched, long drooping catkins on the previous year's twigs or at the base of new shoots. Seed flowers in small erect clusters at the tip of new shoots. Appear in spring with the leaves. Wind-pollinated.

Fruits Hard-shelled nut enclosed in a firm greenish-brown husk that splits into 4 sections. Shell smooth, with 4 lines or ridges. Ripen in one season; shed in autumn. Dispersed by squirrels and in flowing water. Good seed crops occur every 1–3 years. [Bitternut husk is thin, with 4 ridges extending from the tip to the middle; shell can be cut with a knife.]

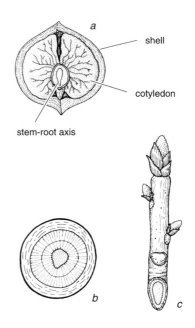

a. Longitudinal section through fruit. *b*. Cross section of 2-year-old twig. *c*. Twig showing continuous pith.

Seeds Kernel (embryo) consists of 2 large irregularly lobed cotyledons and a small stem-root axis. Viable for several years in moist conditions at temperatures just above freezing.

Seedlings Cotyledons white, retained within the shell. First true leaves simple; later leaves are compound with an increasing number of leaflets.

Vegetative Reproduction By stump sprouts.

Wood Hard, strong, tough. Ring-porous, pores easily visible; rays visible with a hand lens.

Size and Form Medium-sized trees with medium coarse branching. Leading shoot upright. Side branches usually occurring at an annual node. Root system deep, with a central taproot.

Habitat Found singly or in small groups mixed with other broadleaf trees. Moderately shade-tolerant. Susceptible to damage by fire because of the thin bark.

9

bitternut hickory
Hickory crown is less coarsely branched than crown of walnut.

Notes Wide variations in form and habit occur within each species of hickory, making identification difficult. Shagbark and bitternut are the most common hickories in Canada and are relatively easy to identify. Mockernut hickory (*Carya tomentosa* Nuttall) was once thought to be native to southern Ontario.

The tough, strong wood is unequalled for products such as sporting implements and tool handles. The nuts are an important source of food for squirrels, other mammals, and larger birds.

Quick Recognition Leaves pinnately compound, with 5–11 leaflets; terminal one as large or larger than the laterals; strong aromatic odor. Fruit husk splits open at maturity. Pith solid.

Shagbark Hickory
Upland hickory

Carya ovata (Mill.) K. Koch

Caryer ovale

Occurs from southern Ontario along the St. Lawrence River into Quebec.

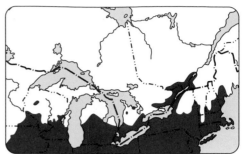

Leaves 5 (sometimes 7) leaflets on a central stalk 15–25 cm long. Leaflets widest at the middle, pointed at both ends, finely toothed, with 2 or 3 tufts of hairs per tooth; upper surface yellowish-green, paler and sparsely hairy beneath.

Buds Terminal bud ovoid, 12–18 mm long, blunt-pointed, greenish-brown; 4–6 overlapping scales, the outer ones loosely spreading and often broken, the inner ones densely hairy. Lateral buds diverge from the twig.

Twigs Stout, shiny, reddish-brown to grayish-brown.

Fruits Almost globular, 3–5 cm long, wider than long, solitary or in pairs. Husk thick, woody, splitting to the base when the fruit is ripe. Shell of nut thin, hard. Kernel sweet, edible.

Bark Dark gray; with age, separating into long plates, free at their lower ends or at both ends, giving the trunk a shaggy look.

Size and Form Medium-sized trees, up to 25 m high, 60 cm in diameter, and 200 years old. Trunk straight, slender, spreading at the base, often branch-free for three-quarters of its length. Crown composed of short, ascending, spreading branches, widening at the top to become almost flattened.

Habitat Occurs on rich moist soils, on hillsides and in valleys; mixed with other broadleaf trees.

Notes The main source of edible hickory nuts and an important food for squirrels. The best quality hickory wood.

    Isolated stands near Lake Huron and Georgian Bay may have been started by native people.

9

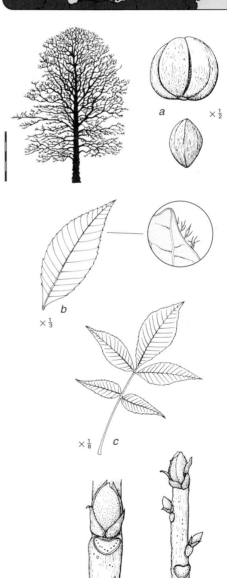

a. Fruit (above); nut (below). *b.* Leaflet with detail of tufts of hair on margin. *c.* Typical leaf. *d.* Lateral bud and leaf scar. *e.* Winter twig.

Pollen flowers in 3-branched catkins, at the base of new shoots.

Seed flowers in small clusters at tips of new shoots.

Fruit almost globular; husk thick and woody.

Trunk straight, slender, branch-free for much of its length.

Bark is noticeably loose and "shaggy".

Quick Recognition   Leaflets 5; edges toothed, with tufts of hairs. Buds large, with loose overlapping scales. Nuts large, thin-shelled, husks splitting to the base. Bark in long loose plates.

Shellbark Hickory

Big shagbark hickory, kingnut

Carya laciniosa Michx. f.

Caryer lacinié

Rare in Canada; thinly scattered in the area north of Lake Erie in southern Ontario. Similar to shagbark hickory, but with larger leaves, buds, twigs, and nuts.

Leaves 7 (sometimes 9) leaflets on a central stalk 25–30 cm long. Leaflets widest near the middle; margin finely toothed, hairy but not in tufts; upper surface dark yellowish-green, paler and hairy beneath. Central stalk often remains on twig after leaflets are shed.

Buds Terminal bud 20–25 mm long; 10–12 scales. Lateral buds diverge slightly from the twig.

Twigs Dull yellowish-brown to dark orange-cinnamon, slightly hairy.

Fruits Globular, 5–7 cm long, in small clusters. Husk 6–12 mm thick, woody, splitting along 4 lines to the base when the fruit is ripe. Shell of nut moderately thick, hard. Kernel sweet, edible.

Bark Dark gray; with age, separating into long, shaggy plates free at their lower ends or at both ends, giving the trunk a shaggy look.

Size and Form Medium-sized trees, up to 30 m high, 90 cm in diameter, and 200 years old. Trunk branch-free for more than one-half its length, often strongly tapered from the base upward. Crown narrow, open, with short, sturdy, ascending branches that spread out toward the top. Taproot deep, strong, except in swamps.

Habitat Occurs on moist to wet sites, in valleys and along stream banks; mixed with other broadleaf trees.

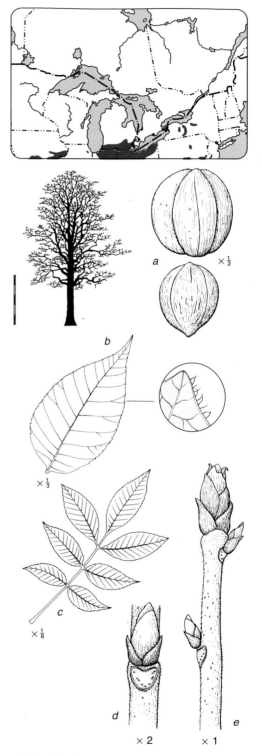

a. Fruit (above); nut (below). *b.* Leaflet with detail of single hairs along margin. *c.* Typical leaf. *d.* Lateral bud and leaf scar. *e.* Winter twig.

9

Fruit the largest of native hickories.

Pollen catkins in dense clusters.

Bark dark gray and shaggy.

Trunk often strongly tapered.

Quick Recognition Resembles shagbark hickory but is a larger tree, with larger leaves, often 2 more leaflets, larger buds, larger nuts with thicker shells. Hairs on the edges of the leaflets not in tufts; central stalk often remaining attached to the twig after leaflets are shed. Orange-cinnamon of twigs is distinctive, but not always present. Grows on moist to wet sites.

Red Hickory

Pignut hickory, false shagbark hickory

Carya glabra (Mill.) Sweet var.
odorata (Marsh.) Little

Caryer glabre

Rare in Canada; occurs in scattered locations in southern Ontario.

Leaves 5–7 leaflets on a central stalk 15–25 cm long. Leaflets dark yellowish-green above, paler and hairy on the main veins beneath; tip narrow-pointed, base wedge-shaped; margin finely toothed, usually hairless on mature leaves (in any case hairs not in tufts).

Buds Terminal bud variable in shape, 6–9 mm long, somewhat pointed; outer scales shed in early autumn leaving a stout, densely hairy bud. Lateral buds smaller, broad, blunt-tipped.

Twigs Slender, often with long ridges, shiny, gray to reddish-brown, hairless.

Fruits Somewhat pear-shaped, 25–50 mm long, in small clusters. Husk thin, smooth, glossy, friable, 4-ridged, splitting readily from the top to the base. Nut 20–25 mm long, slightly flattened. Shell moderately thick. Kernel bitter, not edible.

Bark Thin, gray, becoming scaly and shallowly fissured, resulting in narrow intersecting ridges.

Size and Form Small trees, up to 20 m high, 50 cm in diameter, and 200 years old. Trunk often branch-free and with little taper. Crown irregularly narrow, with short crooked branches. Branch tips and lower branchlets often bending downward; the longest, heaviest branches often near the top.

Habitat An upland species; occurs on well-drained sites; mixed with other broadleaf trees. Intolerant of shade.

9

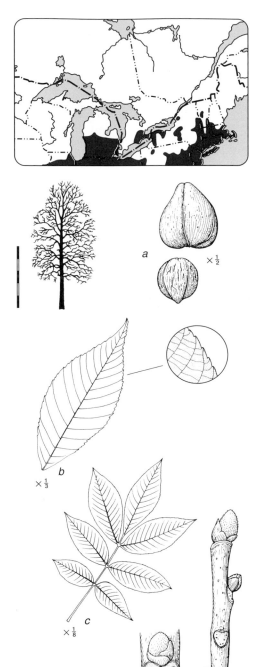

a. Fruit (above); nut (below). *b*. Leaflet with detail of hairless margin. *c*. Typical leaf. *d*. Lateral bud and leaf scar. *e*. Winter twig.

Pollen flowers just before pollen is released.

Seed flowers beginning to open.

Fruit somewhat pear-shaped.

Mature bark tight, fissured, with intersecting ridges.

Young bark smooth not shaggy.

Quick Recognition Leaflets hairless, 5−7.
Terminal bud small. Fruits pear-shaped; husk
splitting to the base; nuts flattened.

Bitternut Hickory
Swamp hickory

Carya cordiformis (Wangenh.)
K. Koch

Caryer cordiforme

Occurs in southern Ontario and southern Quebec. The most abundant and widespread of Canadian hickories.

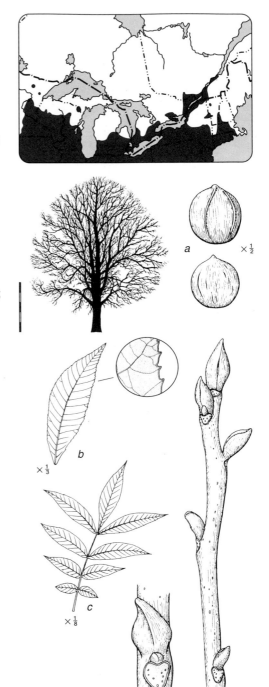

Leaves 7–11 leaflets on a central stalk 12–20 cm long. Terminal leaflet seldom larger than the uppermost lateral pair. Leaflets long-pointed, narrow, scythe-shaped; finely toothed; upper surface shiny dark green; paler, hairy, and dotted beneath.

Buds Terminal bud slender, tapering, flattened, 10–18 mm long, dark yellow, granular; 2–4 scales, with edges that meet but do not overlap. Lateral buds smaller, usually more than 1 above a leaf scar, angular, often stalked. Leaf scars small, oval to 3-lobed, slightly raised. Bud scales enlarge markedly as the new shoot emerges.

Twigs Slender, smooth, shiny, greenish- to grayish-brown, often slightly hairy.

Fruits Globular, broadest toward the tip, 20–35 mm long, solitary or in pairs. Husk thin, covered with yellowish matted hairs; 4 ridges extending below a short sharp tip. Nut broadest toward the base, flattened. Shell thin, can be cut with a knife. Kernel reddish-brown, very bitter.

Bark Greenish-gray, with grayish-yellow irregular vertical lines when young; remaining smooth for many years; separating into shallow narrow fissures and scaly ridges with age; never with loose plate-like scales.

Size and Form Medium-sized trees, up to 25 m high, 50 cm in diameter, and 150 years old. Trunk long, branch-free, with little taper. Crown short, rounded, with slender ascending branches that spread out toward the top. Side branchlets often curving downward.

Habitat Occurs on moist lowlands, also on rich soils on higher ground; mixed with other broadleaf trees. Moderately shade-tolerant.

a. Fruit (above); nut (below). *b.* Leaflet with detail of finely toothed margin. *c.* Typical leaf. *d.* Lateral buds and leaf scar. *e.* Winter twig.

Pollen flowers just before opening.

Seed flowers at time of pollination.

Fruit with 4 ridges extending below sharp tip.

Trunk with little taper; bark tight, with shallow fissures.

Young bark with irregular vertical lines.

Notes Wood used for producing smoke that gives hams and bacon a hickory-smoked flavor.

The only pecan hickory native to Canada, but with a non-edible nut kernel.

Quick Recognition Leaflets curved, 7–11. Terminal bud yellow, granular; scales not overlapping. Fruit husk with 4 ridges; shell thin; kernel bitter.

Kentucky Coffeetree

Gymnocladus dioicus (L.) K. Koch
Caesalpiniaceae: Cassia Family

Chicot févier

The leaf of Kentucky coffeetree is doubly compound and by far the largest of any native tree. A rare tree in Canada, occurring in southwestern Ontario; planted as a landscape tree well beyond its natural range (Zones C5, NA3); easily transplanted; tolerant of urban conditions.

The genus *Gymnocladus* has only 2 species; the other occurs in China.

Leaves Deciduous, alternate, doubly pinnately compound; composed of about 70 leaflets on 3–7 pairs of branches from a central stalk 30–90 cm long; central stalk easily mistaken for a stem. Leaflets ovate, about 5 cm long, smooth-margined, bluish-green, short-stalked; seldom opposite each other; no terminal leaflet. Tree is leafless more than half the year.

Buds No terminal bud; tip of the twig tapered to a blunt point. Lateral buds small, 6–9 mm long, blunt, with several scales, covered with dark silky hairs; 2 or 3 in a group above the leaf scar, the upper one larger. Leaf scars large, inversely heart-shaped, with many vein scars.

Twigs Very stout, grayish-brown, widely spaced. Pith large, deep orange-red.

Flowers Greenish-white, in large, open, many-branched terminal clusters. Pollen flowers and seed flowers on separate trees. Appear in spring with the leaves. Insect-pollinated.

Fruits Pods 12–20 cm long; husk hard, dark reddish-brown, leathery, usually with a powdered appearance; hanging on a stout stalk 2–3 cm long; remaining on the tree through the winter.

Seeds Large, about 2 cm long, rounded, slightly flattened, hard-shelled, dark brown; only a few in each pod, imbedded in a sweet, sticky pulp. Viable for several years in cool dry conditions. Seed coat must be rendered permeable to water for germination.

9

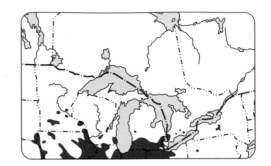

a. Seed (left); fruit pod (right). *b.* Leaflet. *c.* Doubly compound leaf. *d.* Lateral bud group and leaf scar. *e.* Winter twig.

Flowers in many-branched terminal clusters. Seed flower (inset) and pollen flowers on separate trees.

Thick-husked fruit pods persist through-out winter.

Young bark (inset) soon breaks into conspicuous flaky scales. Mature bark with scaly ridges curling outward.

Seedlings Cotyledons remain within the seed coat as germination proceeds.

Vegetative Reproduction Frequently by root sprouts forming colonies.

Bark Dark gray, hard, firm, with thin scaly ridges curling outward along their edges.

Wood Moderately heavy, hard, decay-resistant, reddish-brown. Ring-porous; rays scarcely visible.

Size and Form Medium-sized trees, up to 25 m high, 60 cm in diameter, and 75 years old. Principal branches ascending, forming a narrow crown.

Habitat Occurs mainly on deep rich soils, mixed with other broadleaf trees.

Notes Seeds and husks bitter, seldom eaten by wildlife. Roasted seeds are thought to have been used by early settlers as a sub-stitute for coffee beans; however, due to their toxic properties this practice is not recom-mended.

The name *Gymnocladus*, Greek for "naked branch", derives from the relatively late leaf-out in spring and early leaf-fall in autumn.

> **Quick Recognition** Leaves doubly compound; central stalk long, with 3–7 pairs of branches. No terminal bud; lateral buds in a group of 2 or 3, above a prominent leaf scar. Twigs very coarse, widely spaced. Fruit a pod, large, thick, persisting through the winter. Bark with out-curving scales.

Honey-Locust
Thorny-locust

Gleditsia triacanthos L.
Caesalpiniaceae: Cassia Family

Févier épineux

Rare in Canada; occurs in southwestern Ontario; commonly planted beyond its range because of its tolerance to droughty and alkaline soils (Zones C4, NA3); frequently becoming naturalized.

    The genus *Gleditsia* consists of about 12 species occurring in temperate and tropical regions; 2 species are native to North America, 1 to Canada.

Leaves Deciduous, alternate, singly or doubly pinnately compound. Those singly compound form early on dwarf shoots or toward the base of long shoots; bear 14−30 leaflets (no terminal leaflet) on a central stalk 15−20 cm long; preformed in buds. Those doubly compound bear 4−7 pairs of branches each resembling a singly compound leaf; neoformed during the growing season. Leaflets 25−40 mm long, widest near the base; tip rounded, often with a small point; sometimes minutely toothed.

Buds No terminal bud; twig ends in a withered stub. Lateral buds small, in vertical rows, mostly hidden beneath the bark. Leaf scars U-shaped, with 3 vein scars.

Twigs Long shoots zigzag, brownish; short shoots (scarcely projecting beyond the bark) bear leaves and flowers. Thorns smooth, sharp, reddish, 3-branched or more, occurring on the trunk and stems.

Flowers Greenish-white, regular, small, about 5 mm across. Pollen flowers and seed flowers on same tree; often on separate branches. Perfect flowers may also be present. Pollen flowers in long many-flowered clusters (racemes) 5−7 cm long. Seed flowers in few-flowered clusters 7−9 cm long. Appear in spring with the leaves. Insect-pollinated.

Fruits Pods 15−40 cm long, flat, curved, twisted, brownish; husk leathery; falling in winter without opening.

Seeds Bean-like; with a hard, impermeable seed coat. Viable for many years in cool dry conditions. Seed coat can be rendered permeable to water by cool, moist chilling, dipping in boiling water or strong acid.

9

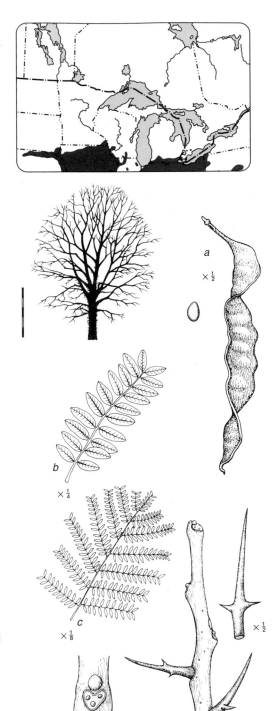

a. Seed (left); fruit pod (right). *b.* Singly compound preformed leaf. *c.* Doubly compound neoformed leaf. *d.* Lateral bud and leaf scar. *e.* Winter twig and detail of 3-branched thorn.

Pollen flowers (above) and seed flowers on same tree.

Fruit pods twisted, singly on stalks; persist after leaf-fall.

Young bark (inset) smooth, shiny. Mature bark has flat scaly ridges. Most cultivars lack thorns.

Seedlings Cotyledons fleshy; released from the seed coat and turning green during germination. First true leaves preformed in the seed.

Bark Smooth, brownish, with horizontal lenticels; with age becoming deeply furrowed with scaly ridges.

Wood Heavy, hard, strong, decay-resistant, reddish-brown. Ring-porous; rays conspicuous.

Size and Form Medium-sized trees, up to 30 m high, 90 cm in diameter, and 120 years old. Trunk typically short, often bearing long branched thorns and sprouts. Crown broad, open, flat-topped. Root system deep, wide-spreading; one of the few tree species that has root hairs.

Habitat Occurs on moist, rich bottomlands, as scattered individuals mixed with other broadleaf trees. Intolerant of shade.

Notes Seeds and pods are a source of food for mammals and birds. Scions or cuttings from branches with only pollen flowers will develop into trees with such flowers, but will not bear fruit. Most cultivars used in landscaping have neither thorns nor fruits; one common cultivar has yellow leaves. A superior lawn tree because it casts a light shade.

Quick Recognition Leaves singly and doubly compound. Thorns branched and sub-branched. Fruit a pod, large, curved, twisted.

Black Locust**
False acacia

Robinia pseudoacacia L.
Fabaceae: Bean Family

Robinier faux-acacia

Native to the eastern United States; quite
hardy, widely planted, and naturalized in
much of southern Canada. Hardy as far
north as Zones C4, NA3.
   The genus *Robinia* consists of about
15 species of trees and shrubs all native to
the United States.

Leaves Deciduous, alternate, pinnately
compound; composed of 7–19 leaflets
(terminal leaflet present) on a central stalk
20–30 cm long; 2 spines (modified stipules)
at the base of each leaf. Leaflets oval,
30–50 mm long, dull green, bristle-tipped;
smooth-margined.

Buds No terminal bud. Lateral buds tiny, in
small clusters, covered with overlapping
scales, formed under the base of the leaf,
embedded in the bark. Leaf scars triangular
to 3-lobed, with 3 vein scars.

Twigs Slender, brittle, smooth, reddish-
brown; 3 narrow ridges descend from each
leaf scar. Two spines beside each bud persist
for years; vary greatly in size, much larger on
young trees and on vigorous shoots of older
trees.

Flowers Showy, white, pea-like, fragrant; in
loose, drooping clusters about 14 cm long,
arising from leaf axils near the tip of a new
shoot. Appear in early summer about a
month after the leaves. Insect-pollinated.

Fruits Pods 7–10 cm long, flat; husk thin-
walled, smooth, dark reddish-brown; several
on a central stalk; remaining on the tree
during the winter. Some seeds are produced
every year, with abundant crops every
2 or 3 years.

Seeds Dark, bean-like, 3–5 mm long, with
a hard impermeable coat, 4–8 per pod.
Viable for many years in cool dry conditions.

Seedlings Cotyledons fleshy; released
from the seed coat, becoming green, and
raised above the surface during germination.
First true leaves simple; later ones with increasing numbers of leaflets.

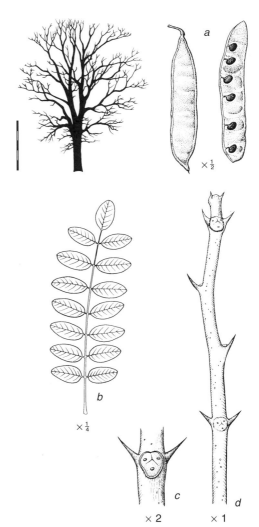

a. Fruit pod (left); interior of fruit showing seeds
(right). b. Leaf. c. Lateral bud, leaf scar, and paired
spines. d. Winter twig.

Vegetative Reproduction Frequently by
root sprouts forming colonies.

Bark Smooth, brown; with age, becoming
thick, deeply furrowed, scaly, dark brown.

Wood Heavy, hard, very strong, durable in
contact with soil. Ring-porous; rays barely
visible.

Size and Form Medium-sized trees, up to
25 m high, 60 cm in diameter, and 90 years
old. Trunk irregular; crown open, irregular;
branches short, brittle.

9

Showy pea-like flowers appear after leaves have developed.

Fruit pods flat, several on a central stalk; persist into winter.

Young bark (inset) has conspicuous lenticels and large sharp spines. Mature bark thick, with deep furrows.

Habitat Occurs along fencerows and roadsides, in pastures and moist woodlands; thrives on limestone soils. Intolerant of shade.

Notes Seeds eaten by birds and small mammals; twigs and bark are poisonous to domestic livestock.

Used to reforest mine spoils, gravel pits, and other wasteland areas where few trees will survive.

Frequently attacked by the locust borer (*Megacyllene robiniae*), which spoils the wood for most uses (except fence posts), and has reduced its popularity for landscaping.

Quick Recognition Leaves pinnately compound with a terminal leaflet; a pair of spines at the base. Flowers conspicuous, fragrant, white, drooping clusters in early summer. Fruit a pod, flat, about the size of a pea pod, with small, hard, dark seeds.

Siberian Pea-Tree*

Siberian pea-shrub, caragana

Caragana arborescens Lamb.
Fabaceae: Bean Family

Caragana arborescent

Very small trees or shrubs native to eastern Asia.
Hardy as far north as Zones C2, NA2. **Leaves**
deciduous, alternate, pinnately compound with
8–12 leaflets (no terminal leaflet) on a central stalk
4–8 cm long. Leaflets oval, 12–25 mm long, short-
pointed, toothless, stalkless. Stipules at the base
of each leaf becoming small spines. Terminal **bud**
present; buds with chaff-like scales. **Twigs** green;
dwarf shoots borne on previous years' branchlets.
Flowers pea-like, bright yellow, in small clusters on
the short shoots. **Fruit** a pod, 4–5 cm long,
swollen, splitting open with a twist when ripe. An
umbrella-shaped weeping cultivar is commonly
grown as a lawn tree. Easy to grow, very cold
hardy, and tolerant of drought, poor soil, salt, and
wind; used as a windbreak on the Canadian
prairies and in Russia.

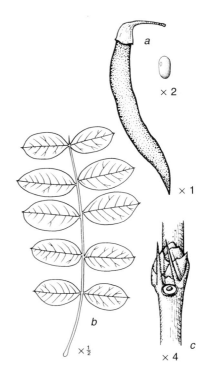

a. Fruit pod (left); seed (right). *b.* Leaf. *c.* Lateral
bud, leaf scar, and paired spines.

Laburnum*

Goldenchain tree

Laburnum anagyroides Medic.
and *Laburnum* ×*watereri* Dippel
Fabaceae: Bean Family

Cytise à grappes

Laburnums are very small trees native to southern
Europe. The common cultivar in Canada is
'Watereri' (*L. anagyroides* × *L. alpinum*). Hardy as
far north as Zones C6, NA5. **Leaves** deciduous,
alternate, compound with 3 leaflets on a central
stalk about 5 cm long. Leaflets elliptic, about 5 cm
long, pointed, toothless, stalkless. Terminal **bud**
present; buds silvery, with 2–4 scales. Dwarf
shoots occur on previous years' branchlets.
Flowers pea-like, showy, bright yellow, attractive,
in long drooping clusters, 50 cm long. **Fruit** a pod,
3–5 cm long. Seeds and other parts very
poisonous.

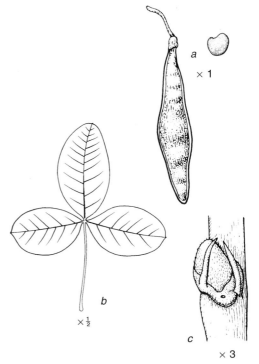

a. Fruit pod (left); seed (right). *b.* Leaf. *c.* Lateral
bud and leaf scar.

Siberian Pea-Tree: Flowers in small clusters on short shoots.

Laburnum: Flowers in long drooping clusters.

Yellow-Wood*
Virgilia

Cladrastis lutea (Michx.) K. Koch
[syn. *C. kentukea* (Dum. Cours.) Rudd]
Fabaceae: Bean Family

Virgilier à bois jaune

A small tree up to 18 m high; rare in its natural range, North Carolina, Kentucky, and Tennessee; planted in milder parts of Canada. Hardy as far north as Zones C4, NA3. **Leaves** deciduous, alternate, pinnately compound with 7–9 leaflets alternately arranged on a central stalk 15–20 cm long. Leaflets elliptic, large, 7–10 cm, blunt-pointed, rounded at base. No true terminal **bud;** buds naked, covered by the leaf stalk, then surrounded by the leaf scar. **Flowers** white, large, pea-like, 25–30 cm long, fragrant, in large, branched clusters at the tips of the new shoots. **Fruit** a flattened pod. **Bark** gray, beech-like. **Wood** yellow.

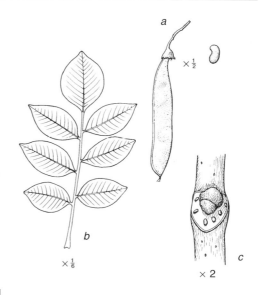

a. Fruit pod (left); seed (right). *b.* Leaf. *c.* Scaleless lateral bud and leaf scar.

Japanese Angelica-Tree*

Japanese aralia

Aralia elata (Miq.) Seem.
Araliaceae: Ginseng Family

Angélique du Japon

Small trees with a few coarse prickly upright
branches; introduced from Asia. Hardy as far north
as Zones C5, NA3. **Leaves** deciduous, alternate,
very large, doubly compound; with several pairs of
branches from a prickly central stalk 50–150 cm
long. Leaflets, numerous, lance-shaped, toothed,
short-stalked, often with prickles beneath. Terminal
bud conical; lateral buds smaller, pressed against
the stem, with few scales. Leaf scars narrow,
crescent-shaped, nearly encircling the twig, with
about 20 vein scars. **Twigs** very stout, grayish,
with many prickles; pith large. **Flowers** small,
white, in large clusters at the tip of the shoot.
Fruits plum-like, 6 mm across, black, with 5 ribs, in
large clusters. **Bark** straw-colored, ridged, with
prickles.

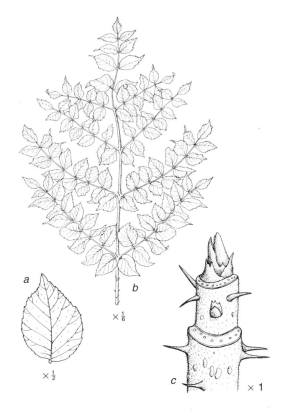

a. Leaflet. *b.* Doubly compound leaf. *c.* Winter
twig with lateral bud, leaf scars, and prickles.

Common Prickly-Ash

Toothache tree

Zanthoxylum americanum Mill.
Rutaceae: Rue or Citrus Family

Clavalier d'Amérique

Shrubs or very small trees; fairly common in
southern Ontario and Quebec. Occurs along
fencerows and forest edges; moderately shade-
tolerant. **Leaves** deciduous, alternate, pinnately
compound; composed of 5–11 leaflets on a prickly
central stalk 10–20 cm long; with 2 spines on the
shoot at the leaf base. Terminal and lateral **buds**
small, 4–6 mm long, rounded, more than 1 above
a leaf scar, covered with red woolly hairs; leaf
scars with 3 vein scars. **Twigs** stiff, dark brown,
with a spicy citrus odor when bruised. **Flowers**
small, greenish, in small clusters on previous
year's twig; appearing before the leaves; pollen
flowers and seed flowers on separate trees. **Fruit** a
capsule; rounded, small, 4–5 mm across, bright
red, with a spicy odor, containing 1 or 2 shiny black
seeds. Reproduces by root sprouts to form
colonies. **Bark** smooth, becoming furrowed, gray
or brown with lighter blotches. **Wood** hard, yellow.

9

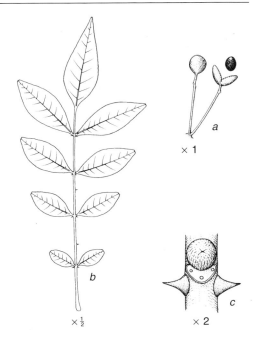

a. Fruit capsules; closed (left), open, with seed
(right). *b.* Leaf with prickly stalk. *c.* Lateral bud,
leaf scar and paired spines.

Mountain-Ash

Rowan trees, whitebeams, dogberries

Genus *Sorbus*

Rosaceae: Rose Family

Les sorbiers

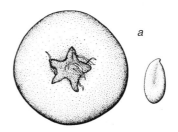

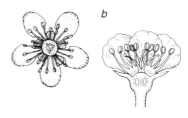

Worldwide, there are about 75 species of mountain-ash distributed throughout the North Temperate Zone in North America, Europe, and Asia; 4 tree-sized species are native to Canada. Frequently planted as ornamentals because of their handsome foliage, showy flowers, and bright red to orange fruit.

Leaves Deciduous, alternate, pinnately compound; composed of 9–17 similar leaflets on a central stalk 8–20 cm long; terminal leaflet present. Leaflets sharply toothed, pointed, short-stalked. Stipules small, leaf-like, usually shed before the leaves are fully grown. Central stalk often remains attached after the leaflets fall. [Whitebeams, which are sometimes planted in Canada, are also in the genus *Sorbus* but, have simple, lobed leaves.]

Buds Terminal bud pointed, often a flower bud, with about 5 scales. Lateral buds smaller. Leaf scars large, crescent-shaped, with 5 vein scars.

Twigs Stout. A pleasant bitter almond taste. Shoot growth often ends with the formation of a terminal flower cluster. Next year's shoot growth occurs from the first lateral bud below the flower cluster; it may develop into a new shoot during the same season or remain a bud over winter, becoming enlarged and resembling a terminal bud.

Flowers Small, with 5 white petals, in many-flowered, multibranched, flat-topped or rounded, showy clusters at the tip of a shoot. Appear in June or July after leaves are fully grown.

Fruits Small, 6–12 mm across, globular, apple-like, orange or red, 1- or 2-seeded, bitter; remaining on the tree after the leaves fall. In clusters at the shoot tip; may appear to be in a lateral position because a summer shoot has developed from a bud below the cluster. Dispersed by birds.

Bark Thin, smooth, light gray, fragrant, with conspicuous, horizontally elongated lenticels; becoming scaly with age.

a. Apple-like fruit showing calyx (left); seed (right).
b. Flower; radial view (left); longitudinal section, 2 petals removed (right).

Size and Form Small trees or shrubs with coarse ascending branches and full, rounded crowns.

Wood Moderately light, weak, pale brown. Diffuse-porous.

Habitat Occur on cool moist sites, often bordering swamps; also grow on poor soils; found mixed with other broadleaf trees.

Notes Fruit is an important source of food for wildlife, especially in winter when other food is scarce; browsing mammals eat the twigs.

Quick Recognition Leaves pinnately compound; leaflets sharply toothed. Terminal bud prominent. Flowers small, white; in large, erect, terminal clusters. Fruit small, red or orange, in drooping clusters; remaining after the leaves fall. Bark with elongated lenticels.

Showy Mountain-Ash

Dogberry

Sorbus decora (Sarg.) C.K Schneid.

Sorbier décoratif

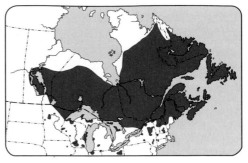

Leaves 13–17 leaflets on a central stalk about 20 cm long. Leaflets narrowly elliptic, 3–8 cm long, scarcely tapered, blunt-pointed; finely toothed from the tip to the middle or just below it; firm; upper surface blue-green, paler beneath; slightly hairy when young.

Buds Terminal bud narrowly cone-shaped, 10–14 mm long, sharp-tipped often with a curve, shiny, dark reddish-brown, sticky; outer scales hairless, inner ones hairy.

Twigs Reddish-brown to grayish, with a skin that weathers off, hairless.

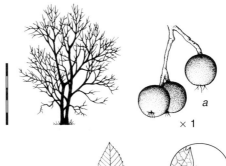

Flowers Petals round, 4–5 mm long; flowers on short, stout, hairy stalks, in dense, many-flowered, open clusters. Appear in May and June, a week or so later than American mountain-ash.

Fruits Shiny, red, 8–10 mm across, in many-fruited, rounded clusters. Mature in August and September; flesh thick.

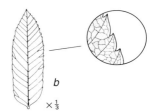

Bark Smooth, thin, light grayish-green to golden-brown; with age becoming slightly scaly.

Size and Form Small trees, up to 15 m high and 25 cm in diameter. Trunk straight, branch-free; crown short, rounded.

Habitat Typically on rocky shores of rivers and lakes.

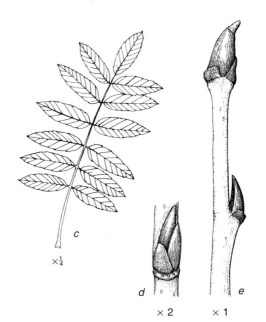

Notes Showy mountain-ash and American mountain-ash (both spread by birds) are absent from southwestern Ontario, even though present farther south, east, and west.

Quick Recognition [Contrasting features of American mountain-ash in brackets.] Leaves horizontally positioned ["on edge", arching]. Leaflets narrowly elliptic [lance-shaped].

a. Fruits. *b.* Leaflet with detail of finely toothed margin. *c.* Leaf. *d.* Lateral bud and leaf scar. *e.* Winter twig.

9

Flower clusters flat-topped and open.

Showy Mountain-Ash

Fruit clusters rounded; fruit shiny red.

Mature bark scaly and broken at tree base.

American Mountain-Ash

Dogberry

Sorbus americana Marsh.

Sorbier d'Amérique

Shrubs or very small trees, up to 10 m high. Not easy to distinguish from showy mountain-ash. **Leaflets** lance-shaped, taper-pointed, 5–8 cm long, thin, light green above, paler and hairless beneath; narrower than showy mountain-ash. **Bud** scales hairless. **Flowers** with petals broadest toward the tip, 3–4 mm long; on hairless stalks, in dense, flattish clusters; appearing in May and June, about 1 week earlier than showy mountain-ash. **Fruits** bright coral-red, 4–6 mm across, maturing in August; flesh thin. **Trunk** short, with spreading slender branches that form a narrow open round-topped **crown**. Occurs on moist sites bordering swamps and on rocky hillsides; also on dry soils.

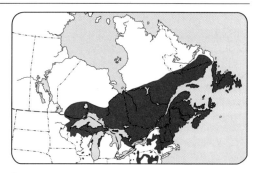

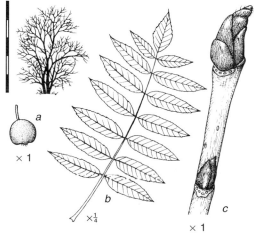

a. Fruit. *b*. Leaf. *c*. Winter twig, showing terminal bud, lateral bud, and leaf scar.

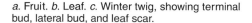

Sitka Mountain-Ash

Pacific mountain-ash

Sorbus sitchensis M.J. Roem.

Sorbier de Sitka

Shrubs or shrubby trees up to 6 m high, ranging
from British Columbia inland to western Alberta.
Reaches the alpine tree line; common on disturbed
sites. **Leaves** 7–11 leaflets on a red central stalk
8–15 cm long. Leaflets elliptic, 2–6 cm long,
toothed toward the tip, stalkless; dark green above,
paler beneath. Terminal **bud** cone-shaped, 6–15
mm long, pointed, sticky, reddish. **Twigs** reddish-
brown, with silvery dots; hairy when young.
Flowers small, 6–8 mm across, in erect, rounded
clusters. **Fruits** red, orange, or purple, 10–12 mm
across, in small clusters 5–10 cm across. **Bark**
smooth, light gray; separating into small curling
scales with age. **Crown** rounded.

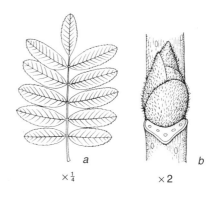

$\times \frac{1}{4}$ $\times 2$

a. Leaf. *b*. Lateral bud and leaf scar.

European Mountain-Ash**

Rowan tree, dogberry

Sorbus aucuparia L.

Sorbier des oiseleurs

Small trees native to Europe, up to 15 m high,
frequently planted in North America, naturalized
locally. Hardy as far north as Zones C3, NA3.
Leaves composed of 9–17 leaflets, 3–5 cm long,
very short-tipped, scarcely tapering, coarsely
toothed (except near the base), usually hairy on
both sides, whitish beneath. **Buds** not sticky,
densely covered with white woolly hairs. **Twigs**
hairy when young. **Flowers** on hairy stalks. **Fruits**
large, 10–12 mm across, orange to red; in round-
topped clusters. Another European species
planted in North America, whitebeam (*S. aria*), has
simple leaves with toothed or lobed margins and
7–10 pairs of veins.

9

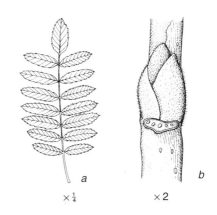

$\times \frac{1}{4}$ $\times 2$

a. Leaf. *b*. Lateral bud and leaf scar.

Sumacs

Genus *Rhus*

Anacardiaceae: Cashew Family

Les sumacs

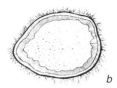

a. Fruit. *b.* Twig cross section showing large pith.

There are over 100 species of sumac, mostly found in southern Africa; 3 tree-size species are native to Canada.

Leaves Deciduous, alternate, pinnately compound; composed of many paired leaflets with a terminal leaflet.

Buds No terminal bud; lateral buds small, hairy; a withered twig tip or fruit cluster present beyond the last lateral bud; leaf scars crescent-shaped.

Twigs Stout, pith large.

Flowers Small, greenish-yellow, in clusters. Pollen flowers and seed flowers usually on separate trees. Appear after the leaves have developed.

Fruits Small, plum-like, with a single seed; in clusters. Dispersed by birds.

Vegetative Reproduction Frequently by root sprouts.

Size and Form Shrubs or very small trees growing in colonies.

Habitat Forest edges and old fields; often thriving on poor soil.

Notes Readily colonizes exposed sites. Important to wildlife for food and shelter; fruits are eaten by birds and small mammals.

> **Quick Recognition** Leaves compound, composed of many leaflets. Flowers and fruits in clusters.

Shining Sumac

Flameleaf sumac, winged sumac

Rhus copallina L.

Sumac brillant

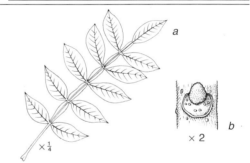

a. Leaf. *b.* Lateral bud and leaf scar.

Shrubs or shrubby trees, up to 8 m high; ranges north as far as southern Ontario. Occurs on sandy soils and rocky outcrops; tolerates heavier alkaline soils in cultivation. **Leaves** composed of 5–11 leaflets on a winged central stalk 10–20 cm long. Leaflets shiny above, hairy beneath, smooth-margined, stalkless; brilliant clear red in autumn. **Buds** with reddish-brown hairs. **Twigs** greenish-brown, hairy; branchlets ridged. **Fruits** red, hairy, in large terminal clusters.

Smooth Sumac

Rhus glabra L.

Sumac glabre

A tall shrub similar to staghorn sumac, but having hairless twigs and leaves and buds with whitish hairs. Ranges from British Columbia through the Prairie provinces to southern Ontario. Hybridizes with staghorn sumac where the 2 occur together.

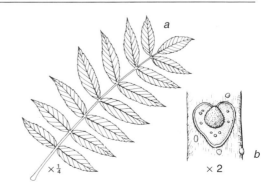

a. Leaf. *b.* Lateral bud and leaf scar.

Staghorn Sumac
Velvet sumac

Rhus typhina L.

Sumac vinaigrier

Ranges from the north shore of
Lake Superior east to Nova Scotia.

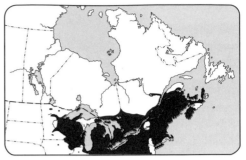

Leaves 11–31 leaflets on a reddish,
densely hairy central stalk 30–50 cm long.
Leaflets lance-shaped, 5–12 cm long, the
central ones largest, curved, long-pointed,
sharply toothed, almost stalkless; upper
surface dark green; paler beneath, with fine
hairs on the midvein and lateral veins; bright
scarlet or orange in autumn.

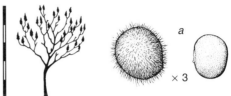

a
× 3

Buds No terminal bud. Lateral buds
rounded, 5–7 mm long, covered with long
pale brown hairs. Leaf scars crescent-
shaped, almost encircling the bud, with
3 groups of vein scars.

Twigs Very stout, densely covered with
dark velvety hairs, exuding a milky sap if
broken; ending in a cluster of fruits or a dead
stub. Growth resumes each year from
1–3 lateral buds some distance below the
dead tip, creating a distinctive branching
effect. Pith large, yellowish-brown.

Flowers Small, greenish-yellow, in large,
dense, erect clusters at the tip of a shoot.
Pollen flowers in clusters about 30 cm long;
seed flowers in smaller clusters, generally on
separate trees. Appear in July after the
leaves are fully developed.

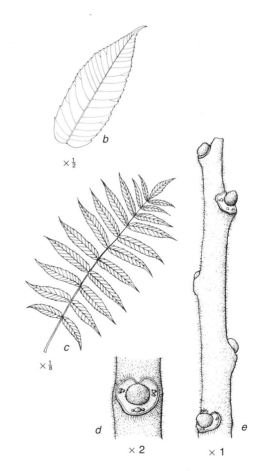

b
× ½

c
× ⅛

Fruits Small, red, juicy, with a single seed,
3–5 mm across, covered with glandular
reddish hairs; in large, dense, erect, cone-
shaped clusters at the tips of the twigs.
Remain on tree throughout most of the
winter.

Vegetative Reproduction Root sprouts
develop at the periphery of colonies.

Bark Thin, smooth, dark yellowish-brown,
with prominent lenticels; with age becoming
scaly.

Wood Light, soft, brittle; orange-green with
broad greenish rays.

Size and Form Shrubs or very small trees,
up to 6 m high, 10 cm in diameter, and 50
years old. Trunk forked; crown flat-topped,
spreading. Root system shallow, wide-
spreading.

d
× 2

e
× 1

a. Fruit (left); seed (right). *b.* Leaflet. *c.* Leaf.
d. Lateral bud and leaf scar. *e.* Winter twig.

Habitat Usually colonizes open areas,
characteristically on sandy or rocky soils;
typically found in large colonies.

9

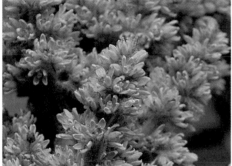

Pollen flowers (above) and seed flowers (below) usually on separate trees.

Young bark (inset) thin with prominent lenticels. Mature bark scaly.

Fruit in dense cone-shaped clusters, conspicuous.

Sumacs in autumn among showiest of trees and shrubs.

Notes In winter, the branches of staghorn sumac, with their distinctive branching pattern and stout woolly twigs, resemble the antlers of a deer in velvet; hence the name "staghorn". The fruit is a source of food for many species of birds; the twigs and leaves are browsed by rabbits, moose, and deer. The wood is occasionally used for decorative finishing and wood novelties.

Sometimes planted as an ornamental for its brilliant autumn foliage, but the numerous root sprouts cause problems.

> **Quick Recognition** Leaves compound, composed of 11–31 leaflets. Twigs very hairy, with a large yellowish-brown pith. Flowers and fruits in large, erect, cone-shaped clusters; fruits red. Occurs mostly in thickets.

Poison-Sumac
Swamp-sumac, poison-dogwood, poison-elderberry

Toxicodendron vernix (L.) Kuntze
[syn. *Rhus vernix* L.]
Anacardiaceae: Cashew Family

Sumac à vernis

The poison-sumacs are closely related to the sumacs; the 2 genera are often combined. *Toxicodendron vernix* is the only one of the 3 poison-sumacs native to Canada that attains tree size. Rare in Canada, occurring in southern Ontario and Quebec.
    The oils of poison-sumacs can cause a severe skin rash. See **Notes** for information on the 2 native shrubs, poison-oak and poison-ivy.

Leaves Deciduous, alternate, pinnately compound; composed of 7–13 leaflets on a central stalk 15–30 cm long. Leaflets 4–8 cm long, long-pointed, with a wedge-shaped base, smooth-margined, stalked; upper surface lustrous dark green, whitened and hairless beneath; stalk often reddish.

Buds Terminal bud conical, 10–19 mm long, with several purplish-brown scales. Lateral buds smaller. Leaf scars broad, shield-shaped with many dot-like vein scars more or less in 3 groups.

Twigs Slender, drooping; dark green, hairy when young; becoming mottled, brownish-yellow, hairless; lenticels prominent.

Flowers Small, yellowish, in open, drooping, branched clusters in the leaf axils.

Fruits Small, 10–13 mm across, rounded, berry-like, glossy white or ivory, hairless, 1-seeded, thin-fleshed; in loose, drooping clusters; usually remaining on the tree throughout the winter.

Bark Light gray, smooth.

Size and Form Shrubs or very small trees, up to 6 m high and 10 cm diameter. Trunk slender; crown small, rounded.

Habitat Occurs in open, swampy woodlands; mixed with other lowland species, such as willows, black ash, white elm, silver maple, and eastern white-cedar.

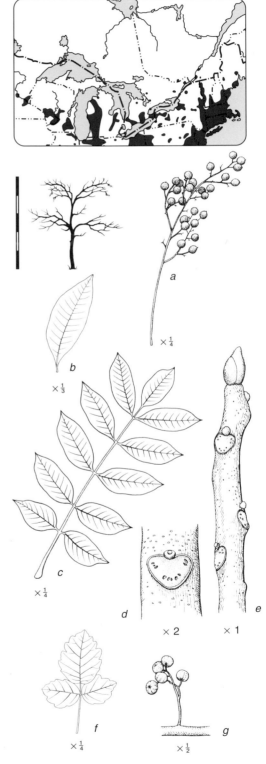

a. Fruit cluster. *b.* Leaflet. *c.* Leaf. *d.* Lateral bud and leaf scar. *e.* Winter twig. **Poison-oak:** *f.* Leaf. *g.* Fruit cluster.

9

Flowers small, in branched clusters.

Fruit glossy white; persist on tree in winter.

Poison-ivy: 2 lateral leaflets together, terminal leaflet stalked.

Poison-Sumac

Twigs (inset) show broad, shield-shaped leaf scars. Trunk slender; bark light gray, remains smooth.

Notes　The "black spot" test may be useful in the identification of all species of poison-sumac: crush a fresh leaf on white paper so that the juice wets the paper; the spot gradually turns brown and then black in about 24 hours.

Smoke from a burning plant is intensely irritating to skin, eyes, and breathing passages.

Poison-ivy, *T. radicans* (L.) Kuntze, ranges from British Columbia to Nova Scotia. Leaves composed of 3 leaflets on a central stalk 5–10 cm long. Leaflets ovate, 5–10 cm long, wavy-edged, irregularly toothed, or lobed; shiny green above; often red in the autumn; terminal leaflet stalked. Flowers and fruit similar to those of poison-sumac. The form of the plant varies greatly: single upright short stems, slender trailing vines, and stout climbing vines with aerial roots.

Poison-oak, *T. diversiloba* Torr. & A. Gray, occurs in British Columbia. Generally similar to poison-ivy; leaflets 5–10 cm long, oval, shallowly lobed; superficially resembling a white oak; often red in autumn.

Quick Recognition　(*T. vernix*) Leaves composed of 7–13 large stalked leaflets, with smooth margins. Terminal bud present. Leaf scars shield-shaped with many vein scars. Fruits glossy white, in drooping lateral clusters.

Common Hoptree

Wafer-ash, stinking-ash

Ptelea trifoliata L.
Rutaceae: Rue or Citrus Family

Ptéléa trifolié

Rare in Canada; occurs in southwestern Ontario on the north shore of Lake Erie; occasionally planted as an ornamental beyond its range (Zones C3, NA3). The genus *Ptelea* contains about 10 species of small trees and shrubs native to North America; of the 2 species reaching tree size, only common hoptree is found in Canada.

Leaves Deciduous, alternate, compound; composed of 3 leaflets on a central stalk 10–15 cm long. Leaflets 10–15 cm long, sharp-pointed, narrowing below the middle to a wedge-shaped base, nearly stalkless; margin smooth or with shallow irregular teeth; upper surface shiny dark green; much paler beneath, with many tiny translucent dots which can be seen by holding the leaf against a strong light. Pungent citrus odor when bruised.

Buds No terminal bud. Lateral buds very small, sunken, erupting through the leaf scar in spring. Leaf scars with 3 vein scars.

Twigs Slender, yellowish- to reddish-brown. Pith large, white. Pungent citrus odor when bruised.

Flowers Small, greenish-white, in clusters at the tips of shoots. Pollen flowers and seed flowers usually on separate trees. Perfect flowers occasionally present.

Fruits Flat, 1- or 2-seeded; central seed-case surrounded by a veined wing, about 25 mm across; in dense clusters; remaining on the tree through most of the winter.

Bark Reddish-brown, smooth becoming rough with age.

Wood Moderately heavy, hard, medium strength, yellowish-brown.

Size and Form Very small trees, up to 8 m high and 15 cm in diameter. Trunk often branched. Crown irregular, rounded, with many short ascending branches.

Habitat Occurs along shorelines, on dry, rocky soils bordering wooded areas, and in open woodlands. Tolerant of partial shade, but flowers only in full sunlight.

9

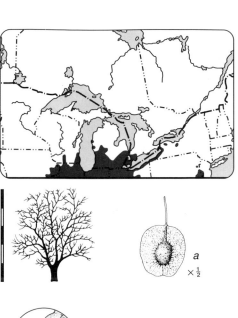

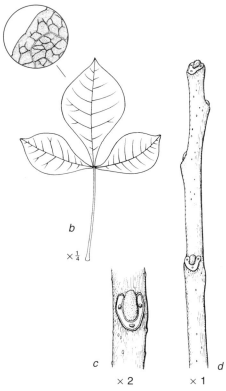

a. Fruit . *b.* Leaf with detail of shallow irregular teeth along margin. *c.* Lateral bud and leaf scar. *d.* Winter twig.

Seed flowers (above), pollen flowers, or perfect flowers may be together in a cluster.

Fruit wafer-like; persists on tree in winter.

Young bark (inset) reddish-brown with many conspicuous lenticels. Trunk short, branched; mature bark rough and scaly.

Notes The fruits have been used as a substitute for hops (*Humulus*) in flavoring beer; hence the name hoptree.

Quick Recognition Leaves compound, 3 leaflets with translucent dots. Buds small, partly buried. Twigs reddish-brown. Fruit flat, with an encircling wing. Strong citrus odor when bruised.

Ailanthus**

Tree-of-heaven, Chinese-sumac

Ailanthus altissima (Mill.) Swingle
Simaroubaceae: Quassia Family

Ailante glanduleux

Introduced from China; frequently planted for
ornamental purposes. Seed-producing trees
are usually planted because the pollen
flowers emit an unpleasant odor. Hardy as
far north as Zones C6, NA4.

Leaves Deciduous, alternate, pinnately
compound; composed of 11–41 leaflets on
a central stalk 25–75 cm long. Leaflets
5–15 cm long, pointed, widest toward the
base; a basal lobe or tooth on one side, with
a warty gland beneath. Unpleasant odor
when bruised.

Buds Rounded, small, brownish, hairy, with
2 or 4 scales; no terminal bud. Leaf scars
large, heart-shaped, with several groups of
vein scars.

Twigs Stout, blunt-tipped, yellowish- to
reddish-brown.

Flowers Small, yellowish-green, in large
erect clusters at the tip of the shoot; pollen
flowers and seed flowers on separate trees;
appearing after the leaves are fully grown.

Fruits A central seedcase with 1 dry seed,
surrounded by a long twisted wing, reddish-
or yellowish-brown.

Bark Thin, firm, greenish-gray; irregular
pale vertical lines with age.

Size and Form Medium-sized trees, up to
25 m high, 75 cm in diameter, and 90 years
old. Trunk straight or crooked, often forked;
crown open, broad, rounded.

Notes Reproduces readily by seeds on
exposed sites in urban areas. Intolerant of
shade; fast-growing when young (up to 2 m
per season).

9

a. Fruit. *b.* Leaf. *c.* Lateral bud and leaf scar.
d. Winter twig.

Quick Recognition Leaves large, compound;
many leaflets, often with basal lobe on one
side; unpleasant odor. Twigs stout, no terminal
bud. Fruits in large terminal clusters; seedcase
in middle of a long wing.

Young bark smooth, greenish-gray. Leaf scars prominent.

Pollen flowers (above) and seed flowers (below) in erect clusters on separate trees.

Fruit in dense clusters.

Mature bark shows pale, vertical lines.

Group 10

Leaves alternate, simple; edges lobed

The alternate arrangement of leaves is the most common. Alternate leaves are usually arranged spirally, but in a few cases they are in two ranks (rows) on opposite sides of the stem. All trees in Group 10 are deciduous with simple leaves (blade not divided into leaflets), alternately arranged.

Lobed leaves come in a variety of shapes. Some lobed leaves could be described as notched. Lobes may be arranged pinnately or palmately. Usually, there is a strong vein running through each lobe; ginkgo (Group 7) is an exception. There is no real difference between a large tooth and a small lobe. As a general rule, teeth are shorter than 1 cm; lobes are longer; there are seldom more than 9 lobes per leaf. The edge of a lobe may be toothed. Several species with lobed leaves are described in other groups because they belong in a genus where most species have unlobed leaves or the arrangement is opposite: European white poplar, European white birch, red alder, Scotch elm, hawthorns, apples (toothed, Group 11); maples, catalpas, and viburnums (opposite, Group 8). Ginkgo (Group 7) is with the conifers.

Some species may have cultivars with lobed leaves; they are not listed here, but the species may be identified by flowers, fruits, buds, bark, and other features.

Sycamore
American sycamore, American plane-
tree, buttonball-tree

Platanus occidentalis L.
Platanaceae: Sycamore Family

Platane occidental

One of the largest broadleaf trees in eastern
North America; occurs as scattered individuals in southern Ontario. Frequently planted
as an ornamental (Zones NA4, C5). The
sycamore genus contains 10 species; 3 in
the United States, 1 in Canada.

Leaves Deciduous, alternate, simple;
10–20 cm long, slightly wider; base wedge-
shaped to deeply indented, prominently
3-veined; coarsely toothed; usually 3 or
5 lobes; central lobe wider than it is long;
notches shallow. Stipules prominent in
spring.

Buds Bluntly cone-shaped, 6–10 mm long,
reddish, covered with a single scale; enclosed in the base of the leaf stalk until after
leaf fall. Leaf scars narrow, almost encircling
bud, with 5 or more vein scars. No terminal
bud; end bud originates as a lateral bud.

Twigs Zigzag, brownish, hairless, encircled
at each leaf scar by a line of stipule scars.

Flowers Pollen flowers and seed flowers on
the same tree, usually on separate shoots;
borne in leaf axils. Pollen flowers small,
yellowish-green, in clusters along the twigs.
Seed flowers larger, crimson, long-stalked, in
ball-like clusters near the shoot tips. Appear
with the leaves.

Fruits In a solitary ball-like aggregate,
20–35 mm across, hanging at the end of a
stalk, 8–16 cm long; each fruit (achene)
small, 1-seeded, elongated, with stiff brownish hairs at the base that spread apart when
the aggregate breaks up. Remain on the tree
throughout the winter, disintegrating gradually. Wind-dispersed. Seeds require light to
germinate.

Vegetative Reproduction Frequently by
stump sprouts.

Bark Smooth, brownish, flaking off in large,
irregular, thin pieces to expose the green,
cream-colored, or white inner bark, producing a striking mottled effect; dark brown
and scaly at the base of mature trees.

10

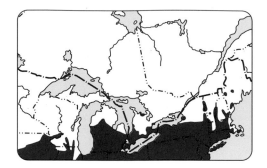

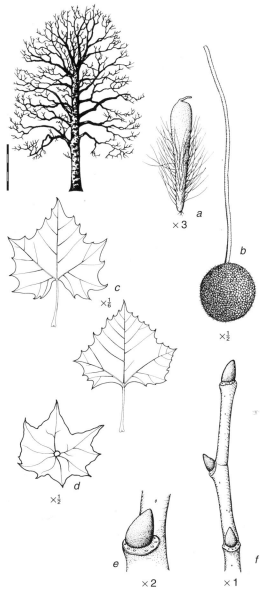

a. Fruit. *b.* Fruit aggregate. *c.* Leaf forms. *d.* Stipule.
e. Lateral bud and leaf scar. *f.* Winter twig.

Pollen flowers (left) and seed flowers (right) usually on different shoots of the same tree.

Fruit aggregate solitary, persists during winter.

Young bark (inset) brownish, flakes off to expose lighter inner bark. Mature bark brown and scaly.

Wood Medium heavy, hard, weak, brownish. Diffuse-porous, pores scarcely visible to the naked eye; rays wide, 2–4 mm high, prominent on radial and tangential surfaces.

Size and Form Large trees, up to 35 m high, 200 cm in diameter, and 250 years old. Trunk thick; crown open, spreading, with massive crooked branches. Root system shallow, spreading.

Habitat Grows on rich bottomlands; also on poorly drained soil; as scattered individuals mixed with other broadleaf trees. Moderately shade-tolerant; fast-growing.

> **Quick Recognition** Leaves 3- or 5-lobed; lobes broad. Buds conical, only 1 bud scale. Fruits in globular aggregates, solitary, drooping. Bark strikingly mottled.

London Plane-Tree*

Platanus ×acerifolia (Ait.) Willd.
(*P. occidentalis* × *P. orientalis*)

Platane à feuilles d'érable

Commonly planted in parks and along streets in Canada, the United States, and Europe; tolerant of urban conditions. Hardy as far north as Zones C6, NA4. Resembles American sycamore. **Leaves** maple-like; central lobe as long as or longer than it is wide; 2 or 3 **fruit** aggregates on each stalk.

a. Leaf. *b.* Fruit, typically with 2 aggregates on a stalk.

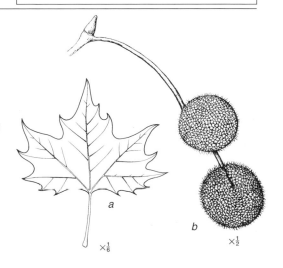

a

b

$×\frac{1}{6}$

$×\frac{1}{2}$

Red Mulberry
Morus rubra L.
Moraceae: Mulberry Family

Mûrier rouge

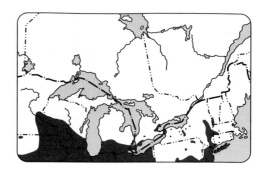

Rare in Canada; scattered throughout southern Ontario. The mulberry genus contains about 10 species; 2 are native to North America, 1 to Canada.

Leaves Deciduous, alternate, simple; ovate, various shapes on the same tree, 8–24 cm long; tip long-tapered; base broad, heart-shaped, asymmetrical, prominently 3-veined; coarsely toothed; unlobed or with 2 or 3 lobes; thin-textured; upper surface yellowish-green and rough like fine sandpaper, soft hairy beneath; yellow in autumn.

Buds Ovoid, asymmetrical, plump, brown, in 2 rows on the twig; 6–8 scales in 2 rows. No terminal bud; end bud originates as a lateral bud. Leaf scars raised, with 5 or more vein scars.

Twigs Slender, green becoming light brown; exuding a milky juice when cut.

Flowers Small, yellowish- to reddish-green. Pollen flowers and seed flowers occasionally in mixed catkins; usually in separate catkins on the same tree or on separate trees; borne in the leaf axils. Appear before and with the leaves.

Fruits Small, fleshy; in compact aggregates (resembling a blackberry), 22–30 mm long, red to dark purple, sweet, juicy, edible. Ripen in midsummer. Dispersed by birds and small mammals.

Vegetative Reproduction Occasionally by stump sprouts.

Bark Reddish-brown, separating into long flaky plates.

Wood Soft, weak, durable.

Size and Form Very small trees, usually up to 9 m high and 40 cm in diameter; forest-grown trees sometimes up to 20 m high and 75 cm in diameter. Trunk short, soon dividing into stout, spreading branches. Crown dense, rounded.

Habitat Grows rapidly on deep moist soils, forested floodplains and valleys; as scattered individuals mixed with other broadleaf trees. Shade-tolerant.

10

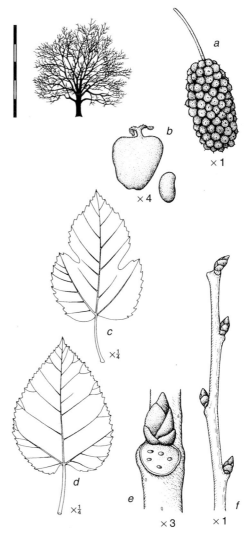

a. Fruit aggregate. *b.* Fruit (left); seed (right).
c. Lobed leaf. *d.* Unlobed leaf. *e.* Lateral bud and leaf scar. *f.* Winter twig.

Pollen flowers (above) and seed flowers (below) usually in separate catkins on same tree.

Mature bark breaks into flaky plates.

Young bark reddish-brown, smooth.

Fruit berry-like, sweet and edible.

Quick Recognition   Leaves large, long-tapered, very rough above; unlobed to variously lobed (2–3) on the same branch; base asymmetrical. Twigs grayish-brown, exuding a milky juice when cut. Bark grayish-brown.

White Mulberry**

Morus alba L.

Mûrier blanc

Small spreading trees native to eastern Asia, where the leaves are a major source of food for silkworms. Naturalized in Canada; occurs in waste areas, along fencerows and forest edges in southern Ontario. Thrives under urban conditions. Hardy as far north as Zones C3, NA4. Resembles red mulberry. **Leaves** variously lobed or unlobed; tip blunt, wedge-shaped; lustrous above, hairless beneath. **Fruits** in nearly globular aggregates, 10–20 mm long, white, reddish, or almost black. A cultivar with drooping branches is often planted as an ornamental.

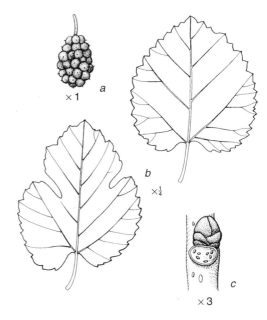

a. Fruit aggregate. *b*. Leaf forms. *c*. Lateral bud and leaf scar.

Sweetgum*

Redgum, starleaf-gum

Liquidambar styraciflua L.
Hamamelidaceae: Witch-hazel Family

Copalme d'Amérique

A large tree, native to the southeastern United States; planted in southwestern British Columbia because of its symmetrical shape and orange autumn leaves. Hardy as far north as Zones C7, NA5. **Leaves** deciduous, alternate, simple; star-shaped, maple-like, 10–18 cm long; 5 or 7 lobes, narrow, long-pointed, finely toothed; palmately arranged; leathery; resinous odor when crushed; leaf stalk almost as long as the blade. Terminal **bud** cone-shaped, 6–12 mm long, pointed, reddish-brown. **Twigs** with corky ridges. Pollen **flowers** and seed flowers on the same tree. **Fruits** in globular, drooping aggregates; each fruit with 2 woody horn-like projections; remaining on the tree during the winter; seeds winged. **Bark** gray, deeply furrowed, with scaly ridges. **Crown** conical.

10

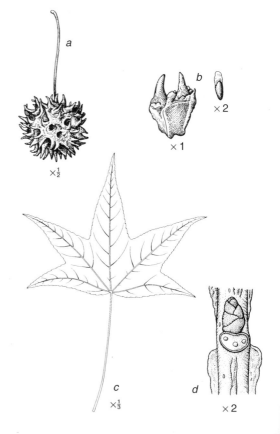

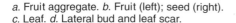

a. Fruit aggregate. *b*. Fruit (left); seed (right).
c. Leaf. *d*. Lateral bud and leaf scar.

White Mulberry

Pollen flowers (above) and seed flowers (below).

Fruit almost globular, white, reddish, or almost black.

Young bark (inset) brownish. Mature bark becomes deeply fissured.

Sweetgum

Star-shaped leaf and persistent spherical fruits.

Sassafras

Sassafras albidum (Nutt.) Nees
Lauraceae: Laurel Family

Sassafras officinal

Rare in Canada; occurs in southern Ontario, north of Lake Erie from southern Lake Huron to the west end of Lake Ontario. The sassafras genus consists of 3 species; 2 native to eastern Asia, 1 to North America.

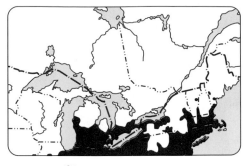

Leaves Deciduous, alternate, simple; 10–15 cm long; blunt-tipped; base wedge-shaped, prominently 3-veined above it; other veins pinnately arranged along the midvein, curving toward the tip of the leaf; smooth-margined; unlobed or with 2 or 3 lobes, all shapes occurring on same tree; under-surface hairless or slightly hairy; yellow to red in autumn; spicy odor when bruised.

Buds Terminal bud ovoid, 10–15 mm long, with several scales. Lateral buds smaller, plump, greenish. Leaf scars with 1 dash-like vein scar.

Twigs Stout, smooth, glossy, yellowish-green, brittle.

Flowers Small, greenish-yellow. Pollen flowers and seed flowers usually on separate trees; in loose clusters at the base of new shoots. Appear before the leaves unfold.

Fruits Berry-like, 10–15 mm long, dark blue, with a large, stone-like seed; in a red cup on a long red stalk; several to a cluster. Dispersed by birds.

Vegetative Reproduction Often by root sprouts, forming colonies.

Bark Dark brownish, deeply grooved with heavy soft corky ridges.

10

Wood Light, soft, weak, durable, coarse-grained; orange-brown; sapwood yellow; aromatic.

Size and Form Small trees, up to 20 m high and 50 cm in diameter; occasionally larger; often reduced to a shrub in dry sandy areas. Trunk bearing many abruptly spreading, crooked, brittle branches. Crown flat-topped, irregular, columnar. Branchlets have a stag-horn look because the first side shoot below the tip is usually much longer than the twig from which it originates. Root system sparse, spreading.

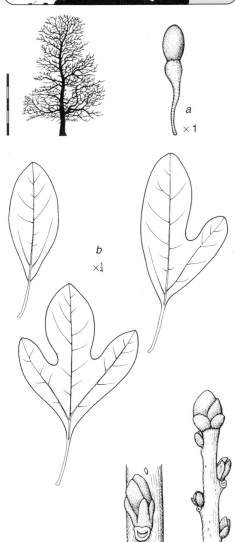

a. Fruit. *b.* Leaf forms. *c.* Lateral bud and leaf scar. *d.* Winter twig.

Pollen flowers (above) and seed flowers
(below) on separate trees.

Fruit in clusters; stalks turn red as fruit
ripens.

Young bark (inset) corky and rough. Mature bark
becomes deeply grooved.

Habitat Grows on a variety of sites from
nutrient-rich to poor; in forest openings and
edges; mixed with other broadleaf trees.
Shade-tolerant.

Notes The sassafras genus belongs to the
laurel family (Lauraceae), whose members
usually have a spicy fragrance (e.g., cinnamon and camphor) and are found in the
tropics and subtropics. All parts of the
sassafras have a pleasant spicy odor when
bruised. The shrub laurels (*Kalmia* spp.),
noted for their spectacular flowers, belong to
the heath family (Ericaceae).

> **Quick Recognition** Leaves lobed and unlobed
> on the same branch. Leaves, twigs, branches
> and bark with a spicy fragrance when crushed
> or broken. Twigs green, often with an abnor-
> mally long lateral shoot near the tip.

Tulip-Tree
Yellow-poplar

Liriodendron tulipifera L.
Magnoliaceae: Magnolia Family

Tulipier de Virginie

Occurs in southern Ontario, on the south shore of Lake Huron, the north shore of Lake Erie, and the Niagara Peninsula; planted beyond its range as an ornamental (Zones C5, NA4). The tulip-tree genus consists of 2 species; the other is a small tree in China.

Leaves Deciduous, alternate, simple; 7–12 cm long, as wide or wider; broad shallow notch at the tip so that leaf appears cut off at the top; smooth-margined; 4 (sometimes 6) lobes, basal lobes broader on juvenile leaves; light green, turning yellow in autumn. Stipules prominent in spring. Stalk slender, usually longer than the blade.

Buds Terminal bud flat, 12–14 mm long, with 2 outer scales meeting at their edges. Lateral buds much smaller, with a powdery coating. Leaf scars roundish, elevated, with several vein scars.

Twigs Stout, shiny, brownish, encircled at each leaf scar by a line of stipule scars. Pith solid, banded at intervals.

Flowers Large, showy, solitary, about 5 cm across, greenish-yellow; shaped much like tulip flowers; 6 petals, 4–6 cm long; at the tips of shoots. Appear after the leaves. Insect-pollinated; a source of nectar for bees.

Fruits Winged, 3–5 cm long, in cone-like aggregates 5–7 cm long, spindle-shaped; falling away from the central stalk when ripe, leaving it erect on the shoot tip. Good seed crops occur nearly every year. Seeds eaten by birds and small mammals; wind-dispersed.

Seedlings Cotyledons green, leaf-like, raised above the surface.

Bark Smooth, dark green when young; becoming brownish, with firm intersecting rounded ridges, separated by grayish fissures.

10

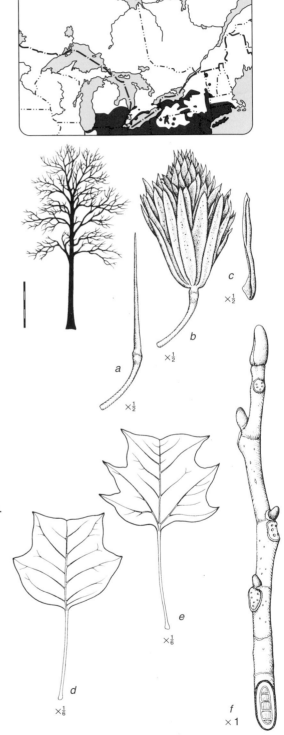

a. Persistent central fruit stalk. b. Cone-like fruit aggregate. c. Fruit. d. Typical leaf form. e. 6-lobed leaf form. f. Winter twig with section showing banded pith.

Flowers large, terminal on shoots.

Fruit aggregate will disintegrate when fruit ripens.

Young bark (inset) green with small whitish spots. Mature bark has deep intersecting ridges.

Wood Light, soft, weak, fine-grained, easily worked, yellowish- to greenish-brown, often with darker streaks. Diffuse-porous, pores too small to be visible; annual rings and rays visible.

Size and Form Large trees, up to 35 m high, 100 cm in diameter, and 150 years old; larger in the United States. Trunk tall, massive, straight, with little taper, branch-free for two-thirds or more of its height. Crown small, narrow, compact in the forest; long and irregular in the open. Root system deep, wide-spreading.

Habitat Occurs on deep, rich, moist soils along streams or around swampy areas; usually mixed with other broadleaf trees or eastern hemlock; rarely in pure stands. Intolerant of shade; fast-growing.

Notes Generally free of pests and diseases.

Quick Recognition Leaves notched at the tip, unlike any other. Terminal bud flat with 2 scales. Flowers large, tulip-like; fruits winged, in cone-like aggregates.

Oaks

Genus *Quercus*

Fagaceae: Beech Family

Les chênes

Oaks are prominent components of the deciduous forests of North America, Europe, and Asia. The oak genus comprises 500–600 species; about 60 occur in the United States; 11 in Canada.

Classification of the Canadian oaks into 2 groups aids in identification: the **red oaks** have leaves with bristle-tipped lobes (red, black, pin, northern pin, Shumard, and scarlet); and the **white oaks** have rounded lobes or large regular teeth (white, bur, swamp white, chinquapin, dwarf chinquapin, Garry, English, and chestnut).

Species in the red oak group resemble each other and may be difficult to separate into species. Leaves can vary in size and form with position on the twig and in the crown and are not as useful in identification as the buds, acorns (especially the cups), and bark. Species within the white oak group, however, can usually be identified by leaves and twigs. Most species hybridize with other species in the same group producing intermediate forms.

Features described below apply to both groups; where they differ, those of the white oaks are in brackets.

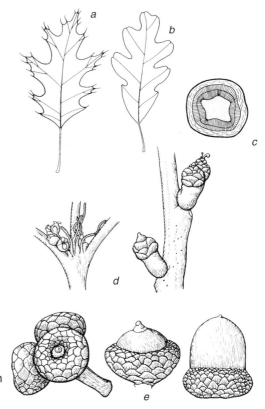

a. Leaf with bristle-tipped lobes (red oaks). b. Leaf with rounded lobes (white oaks). c. Cross section, 2-year-old twig. d. Seed flower (left); immature red oak acorns at end of 1st growing season (right). e. Maturation of acorns.

Leaves Deciduous, alternate, simple; firm-textured; often clustered at the ends of the twigs; pinnately lobed, with a prominent vein to each lobe; lobes bristle-tipped [lobes rounded, or margin toothed]. Dead leaves often remain on the tree over winter. Stipules deciduous soon after leaves appear.

Buds Terminal bud present; lateral buds smaller, in a group just below the terminal bud and along the twig; scales numerous, overlapping, in 5 rows. Leaf scars semi-oval to 3-sided, with 5 or more vein scars.

Twigs Moderately stout, stiff, straight, often with longitudinal ridges. Pith 5-pointed. First flush of growth preformed in the bud, ending with the formation of a new terminal bud; on vigorous shoots, a second and third flush of growth may take place, thus forming seasonal nodes.

Flowers Pollen flowers and seed flowers on the same tree. Pollen flowers small, in many-flowered drooping catkins; developing from buds on previous year's twigs or at the base of new shoots. Seed flowers small, solitary or in few-flowered clusters; in the axils of leaves on new shoots. Appear as the leaves are developing. Wind-pollinated.

Fruits Acorns, 1-seeded nuts with a tough leathery shell; tip rounded with an abrupt point; base enclosed by a scaly cup (involucre). Cup scales numerous, overlapping, flat [swollen or thickened]. Solitary, or in small clusters. Mature at the end of the second [first] season on previous year's twigs [on new twigs]. Small first-season and large second-season acorns may be on the same branch [all acorns the same size]. Shed in autumn soon after ripening. Nut usually separates from the cup [remains with the cup]. Shell hairy [hairless] on the inside. Good seed crops occur at intervals of 2–6 years. Dispersed by rodents, birds, wind.

Seeds Bitter, yellow [mostly edible, sweet, white]; viable over winter in cool moist conditions.

10

red oak
Seed flowers at time of pollination.

black oak
1-year-old acorns at the beginning of 2nd
growing season (red oak group).

Seedlings Newly germinated seedlings produce a scaly shoot with a cluster of 2–5 unlobed leaves at the tip. A terminal bud forms and may flush in a few weeks to produce a second shoot with another cluster of leaves. Later leaves increasingly resemble the adult type. Leaves on white oak seedlings may have bristle tips and superficially resemble red oaks. Cotyledons fleshy, pale (non-photosynthetic), retained within the shell. Germination of red oak takes place in spring [in autumn soon after seed fall, thus exposing seedling to winter damage]. Oak seedlings may produce a taproot up to 1 m long in the first year.

Vegetative Reproduction Sprouts readily develop on young stumps. Even 1-year-old seedlings have the ability to send out a new shoot if the original shoot dies.

Bark Mostly furrowed [scaly].

Wood For most species, hard, heavy, strong. Large rays and rings of large pores easily visible. Ring-porous; rays of 2 sizes. Water permeable because the water-conducting pores are open [pores plugged with intrusions (tyloses), hence impermeable]. Susceptible to decay [resistant to decay].

Size and Form Small to large trees. Crowns broad, full. Leading shoot upright. Side branches mainly in small clusters immediately below annual node. In some species, old horizontal branches may be noticeably long. Root system deep, wide-spreading, usually with a strong taproot.

Notes Oak wood is famous for its strength, durability, and beauty of grain. It is much in demand for high-quality furniture and flooring; oak lumber sawn in eastern Canada is mostly from red oak; white oak, which is impermeable, is used to make barrels for storing liquids. The bark of some species is a source of cork. The acorns are food for livestock and wildlife. The presence of acorns is a sure means of identifying oaks.

Leaves, especially those of red oaks, are high in lignin and resist decomposition; hence they are not recommended for use as compost.

Oak "apples" on the leaves are galls formed by leaf tissues in response to chemicals deposited by insects.

> **Quick Recognition** Leaves pinnately lobed or coarsely toothed. Terminal bud present, with a cluster of lateral buds around it; bud scales numerous, in 5 rows. Fruit an acorn. Wood ring-porous, rays very wide and high.

Red Oak
Northern red oak

Quercus rubra L.

Chêne rouge

The common oak of eastern Canada; ranging from east of Lake Superior to Nova Scotia. Planted as a landscape tree (Zones C3, NA4).

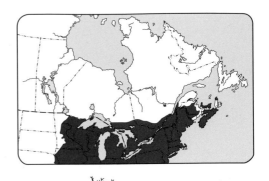

Leaves 10–20 cm long, base broadly wedge-shaped; 7–9 lobes, with several bristle-tipped teeth on the larger lobes; each lobe tapered from base to tip, separated by rounded notches; central lobe about as long as the width of the leaf between opposite notches; upper surface dull yellowish-green, paler beneath, with tufts of hairs in the vein axils. Stalk stout, 1–3 mm wide, 2.5–5 cm long. Leaves on young trees toothed rather than lobed.

Buds Terminal bud ovoid, 6–8 mm long, pointed, shiny reddish-brown, smooth except for a few brownish hairs at the tip.

Twigs Moderately stout, reddish-brown.

Fruits Acorns 12–25 mm long, almost as wide, short-stalked. Cup saucer-shaped, 15–25 mm across, enclosing about one-quarter of the nut; scales thin, hairless, reddish-brown.

Bark Smooth, dark gray when young; becoming grooved by wide shallow dark furrows into mostly unbroken, long, flattish pale gray ridges; inner bark pinkish.

Wood Pink to reddish-brown with an attractive grain; porous, not durable in moist situations.

Size and Form Medium-sized trees, up to 25 m high, 30–90 cm in diameter, and 150 years old; occasionally over 30 m high and 120 cm in diameter. Trunk straight. Crown symmetrically rounded. Root system deep, spreading, with a taproot on deep soils.

Habitat Grows mixed with other broadleaf trees and eastern white pine; stunted on dry rocky ridges toward the northern limit of its range. Intolerant of competition and shade, although moderately shade-tolerant when young.

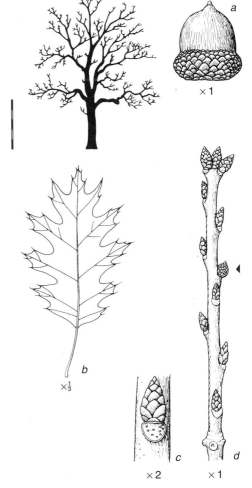

a. Acorn. *b.* Leaf. *c.* Lateral bud and leaf scar. *d.* Winter twig with 1-year-old acorn (◀).

Notes The provincial tree of Prince Edward Island.

10

Mature bark has unbroken vertical ridges.

Pollen flowers (above); seed flowers (below).

Young bark smooth, grooves beginning to develop.

Acorn almost as wide as long; cup encloses one-quarter of nut.

Quick Recognition Leaves with 7–9 lobes; each lobe wider toward the base, about equal in length to the constricted part of the leaf; notches V-shaped. Buds 6–8 mm, shiny reddish-brown, pointed, with a few hairs at the tip. Twigs reddish-brown, hairless. Acorns 12–25 mm long, almost as wide; cup shallow, enclosing about one-quarter of the nut; scales thin, hairless, tight-filling. Bark smooth, dark gray when young; dark wide grooves and long flat pale gray ridges with age; inner bark pinkish-red.

Black Oak
Quercus velutina Lam.
Chêne noir

Occurs in southern Ontario north of
Lakes Erie and Ontario.

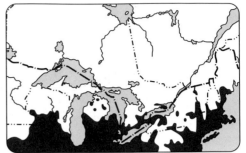

Leaves 10−20 cm long, base rounded;
5−7 lobes, with a few coarse bristle-tipped
teeth; each lobe oblique or at right angles to
the midvein, with parallel sides, separated by
deep U-shaped notches; central lobe may be
twice as long as the width of the leaf between
opposite notches; upper surface shiny dark
green, yellowish-brown and rather rough beneath; hairy early in the season, later a few
stellate hairs remain on the veins and in the
vein axils of the undersurface. Stalks stout,
7−15 cm long.

Buds Terminal bud ovoid, 6−8 mm long,
sharp-pointed, distinctly angled; grayish-
white woolly hairs obscure the surface of the
scales.

Twigs Moderately stout, dark reddish-
brown, hairless or slightly hairy.

Fruits Acorns small, 12−20 mm long,
almost as wide. Cup bowl-shaped,
12−24 mm across, enclosing about one-half
of the nut; scales loose-fitting, thin, slightly
hairy, dull brown, sometimes forming a fringe
to the cup.

Bark Smooth, dark gray when young; becoming almost black, and deeply furrowed
into irregular, rounded ridges divided into
squarish segments. Inner bark yellowish.

Wood   Light brown.

10

Size and Form Small trees, up to 20 m
high, 90 cm in diameter, and 200 years old;
larger farther south. Trunk distinct up to the
crown. Crown irregular, variable. Principal
branches horizontal in the lower part of the
crown, ascending in the upper part. Roots
deep, wide-spreading, with a deep taproot.

Habitat Occurs on dry sandy soils, also on
steep slopes with heavy soils; mixed with
other species. Intolerant of shade and competition.

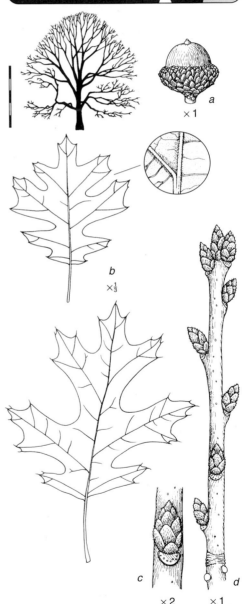

a. Acorn. *b*. Leaf forms and detail of hairs in vein
axils of leaf undersurface. *c*. Lateral bud and leaf
scar. *d*. Winter twig.

Pollen flowers (above); seed flowers (below).

Mature bark has vertical ridges horizontally broken into square segments.

Acorn cup with loose scales, encloses about one-half of nut.

Young bark dark gray.

Quick Recognition Leaves with 5–7 lobes; each lobe with parallel sides; notches U-shaped. Buds 6–8 mm long, dull grayish-brown, sharp-pointed, angled, densely hairy obscuring the surface of the scales. Twigs dark reddish-brown, hairless. Acorns 12–20 mm long, almost as wide; cup enclosing one-half of the nut; scales dull brown, hairy, loose-fitting, sometimes forming a fringe on the edge of the cup. Bark smooth, dark gray when young, becoming black, in squares; inner bark yellowish.

Pin Oak
Swamp oak

Quercus palustris Muenchh.

Chêne des marais

Occurs in southern Ontario at the east and west ends of Lake Erie; locally abundant. Frequently planted as an ornamental because of its symmetrical shape and attractive leaves (Zones C4, NA4); tolerant of urban conditions; easily transplanted. Small stiff dead branchlets often project like pins from the trunk and larger branches; hence the name "pin".

Leaves 7–15 cm long, base wedge-shaped; 5–7 lobes, with a few bristle-tipped teeth on the larger lobes, wide-spreading, separated by deep wide U-shaped notches; upper edge of lobes at right angles to the midvein; basal lobes often recurved; central lobe 3 times as long as the width of the leaf between opposite notches; upper surface shiny dark green, paler beneath, with a few tufts of hairs in the vein axils. Stalk slender, 2–5 cm long.

Buds Terminal bud ovoid, small, 2–4 mm long, sharp-pointed, light chestnut-brown, almost hairless.

Twigs Slender, hairless, reddish-brown.

Flowers One of the first oaks to bloom in spring.

Fruits Acorns small, 9–13 mm long, almost as wide, short-beaked at the tip. Cup shallow, saucer-shaped, 12–16 mm across, enclosing one-quarter of the nut; scales tight-fitting, thin, pointed, hairy, reddish-brown.

Bark Grayish-brown, thin, smooth; with age, dividing into narrow inconspicuous ridges. Inner bark reddish.

Wood Light brown.

Size and Form Small trees, up to 20 m high, 60 cm in diameter, and 100 years old. Trunk straight, with a gradual taper, distinct well into the crown. Principal branches slender (for an oak), ascending in the upper crown, horizontal in the center of the crown, and curving downward in the lower crown. Root system shallow.

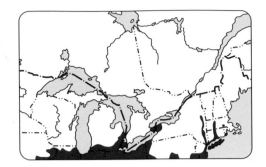

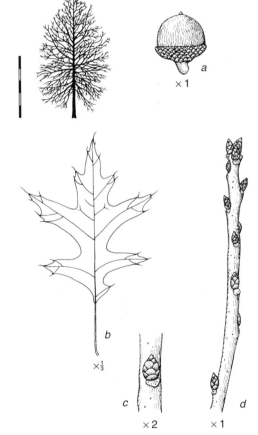

a. Acorn. *b.* Leaf. *c.* Lateral bud and leaf scar. *d.* Winter twig.

Habitat Occurs mainly on poorly drained soils, in swamps and along streams, yet flourishes on well-drained loamy soils; mixed with other oaks, elms, and willows. Early height growth rapid; intolerant of shade.

10

Pollen flowers (above); seed flowers (below).

Mature bark has low, scaly ridges and shallow fissures.

Acorn cup with tight scales, encloses one-quarter of the nut.

Young bark grayish-brown, smooth.

Quick Recognition   Leaves with 5–7 lobes; each lobe 3 times as long as the width of the constricted part of the leaf; notches wide, U-shaped; leaf stalks slender. Buds 2–4 mm long, light reddish-brown, almost hairless. Twigs slender, reddish-brown. Acorns 9–13 mm long; cup shallow, enclosing one-quarter of the nut; scales thin, tight-fitting, hairy. Bark smooth, grooves inconspicuous even on older trees.

Northern Pin Oak

Hills oak

Quercus ellipsoidalis E.J. Hill

Chêne ellipsoïdal

Uncommon in Canada; occurs in Ontario west of
Lake Superior and north of Lake Erie on sandy
soils in open disturbed habitats. Resembles pin
oak, but more likely to be growing on upland sites.
Leaves 7–12 cm long; lobes often constricted
toward the base; leaf stalk slender, 3.5–5 cm long.
Terminal **bud** 4–6 mm long. **Twigs** bright reddish-
brown, hairy when young. **Acorns** 12–18 mm
long; cup 10–15 mm across, enclosing one-third to
one-half of the nut, tapering to a stalk-like base.
Bark shallowly furrowed, with narrow vertical
plates; inner bark light yellow. Hybridizes with
black oak.

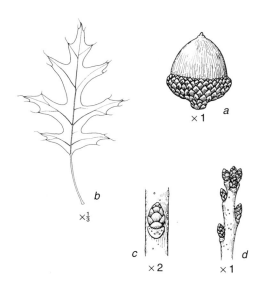

a. Acorn. *b*. Leaf. *c*. Lateral bud and leaf scar.
d. Winter twig.

Shumard Oak

Swamp red oak

Quercus shumardii Buckl.

Chêne de Shumard

Medium-sized to large trees, up to 40 m high,
occuring in a few locations in southern Ontario
north of Lake Erie; on lowlands, along streams, on
poorly drained, heavy soils. Resembles black oak.
Leaves larger, 12–20 cm long, base rounded;
7–11 toothed lobes, separated by deep narrow
notches; dark shiny green above, paler beneath;
remaining green after those of other oaks have
turned brown; rich red in autumn. Leaf stalk 5–7
cm long. **Buds** 6 mm long, very pale grayish-
brown. **Acorns** large, 15–30 mm long; cup
shallow, saucer-shaped, 15–30 mm across, gray,
slightly hairy, enclosing about one-third of the nut.
Trunk buttressed; **crown** open, wide-spreading,
with massive branches. **Bark** dark gray, deeply
furrowed.

10

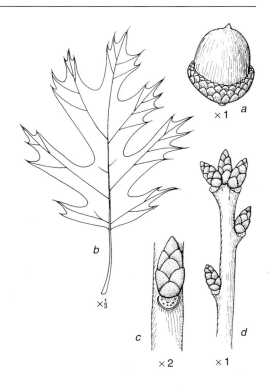

a. Acorn. *b*. Leaf. *c*. Lateral bud and leaf scar.
d. Winter twig.

Scarlet Oak*

Quercus coccinea Muenchh.

Chêne écarlate

Medium-sized trees, up to 25 m high and 60–90 cm in diameter. Native to the eastern United States; often planted in Canada because of the leaves, which are bright green in summer and scarlet in autumn. Hardy as far north as Zones C4, NA4. **Leaves** 10–15 cm long; 7–9 deeply indented lobes; stalks 4–6 cm long. **Buds** ovoid, 4–7 mm long, dark reddish-brown. **Acorns** 15–20 mm long; cup enclosing one-half to one-third of the nut; scales tight-fitting, thin, light reddish-brown. **Bark** dark brown to nearly black with age; shallowly furrowed, with irregular ridges. Grows on drier sandy soils.

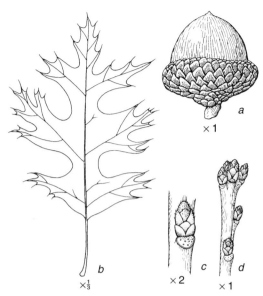

a. Acorn. *b.* Leaf. *c.* Lateral bud and leaf scar. *d.* Winter twig.

White Oak
Stave oak
Quercus alba L.

Chêne blanc

A characteristic tree of the broadleaf forests of southern Ontario and Quebec.

Leaves Usually widest above the middle, 10–20 cm long; 7–9 lobes, rounded, narrow, some with 1 or 2 large blunt teeth, separated by deeply cut notches; downy and pinkish when unfolding, later hairless; upper surface bright green, paler beneath; may turn reddish-purple in autumn.

Buds Terminal bud ovoid, 3–5 mm long, blunt, usually hairless, reddish-brown, not angled; lateral buds diverge from the twig.

Twigs Moderately stout, green to reddish-green when young, becoming red, then gray; mostly hairless.

Fruits Acorns 12–20 mm long, solitary or paired, stalkless or short-stalked. Cup broadly bowl-shaped, enclosing about one-quarter of the nut; scales thickened, warty, free at the tips but not forming a fringe to the cup.

Bark Pale gray, often with a reddish cast, scaly.

Wood Hard, strong, tough, light brown.

Size and Form Medium-sized to large trees, up to 35 m high, 120 cm in diameter, and several hundred years old. Trunk distinct well into the crown, often branch-free for two-thirds of the tree height. Crown of open-grown trees sometimes composed of large branches with many wide-spreading side branches that become gnarled and twisted. Roots deep, spreading, with a deep taproot.

Habitat Grows well on a variety of soils; usually mixed with other oaks, basswood, black cherry, hickories, sugar maple, white ash, eastern white pine, and eastern hemlock. Moderately shade-tolerant; persists as an understory tree, but grows well after the stand is opened.

10

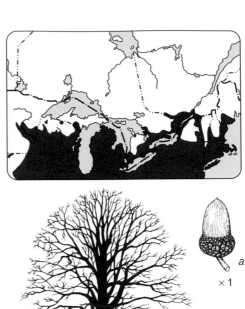

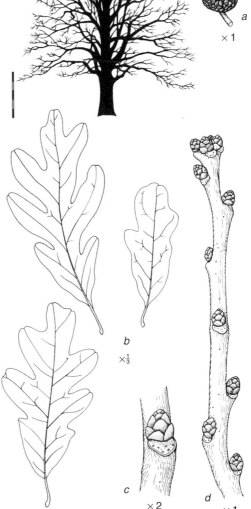

a. Acorn. *b.* Leaf forms. *c.* Lateral bud and leaf scar. *d.* Winter twig.

Pollen flowers (above); seed flowers (below).

Mature bark pale gray, scaly and narrowly ridged.

Young bark soon scaly.

Acorn cup with warty scales, encloses one-quarter of nut.

Quick Recognition [Contrasting features of bur oak in brackets.] Leaves hairless [hairy] underneath, never [often] with a large terminal lobe. Lateral buds diverge from [pressed against] the twig; all bud scales broad [a few narrow loose scales in the bud cluster]. Branchlets smooth [often with corky ridges]. Acorn cup enclosing one-quarter [one-half or more] of the nut; no fringe of scales [fringe of scales present]. Bark light [dark] gray, with thin [thick] scales.

Bur Oak

Blue oak, mossycup oak

Quercus macrocarpa Michx.

Chêne à gros fruits

The most common native white oak, ranging from southern Saskatchewan east to New Brunswick.

Leaves 15–30 cm long; shape variable; often with a broadly expanded toothed portion above 2 deep notches and a lower portion with a few short, rounded lobes; other leaves resemble white oak, with 7–9 deep lobes; upper surface shiny green, pale and hairy beneath. Stipules often present.

Buds Terminal bud 3–6 mm long, blunt-pointed, brown, hairy; often surrounded by a few elongated, pointed scales. Lateral buds pressed against the twig.

Twigs Stout, yellowish-brown, somewhat hairy; branchlets often with corky ridges.

Fruits Acorns 20–30 mm long, usually solitary, stalkless or short-stalked. Cup large, enclosing one-half or more of the acorn; scales knobby, long-pointed with narrow free tips, forming a conspicuous fringe to the cup.

Bark Rough, becoming deeply furrowed, with ridges broken into irregular thick dark gray scales.

Wood Light brown.

Size and Form Small trees, up to 15 m high, 60 cm in diameter, and 200 years old, occasionally much larger. Trunk straight, tall, distinct to the upper crown. Principal branches ascending in the upper crown, nearly horizontal in the lower crown. New shoots (epicormic branches) often occur along the trunk. On exposed shallow soils, trees may be stunted with a trunk that divides into radiating, crooked, and gnarled branches. Root system deep, wide-spreading, with a deep taproot.

Habitat Grows best on deep, rich bottomlands; also occurs on upland limestone soils, and at the northern limits of its range, on shallow soils over granitic bedrock; mixed with various other species. Drought-tolerant; moderately shade-tolerant.

Notes Fire-resistant because of its thick bark; this accounts for its presence in grasslands.

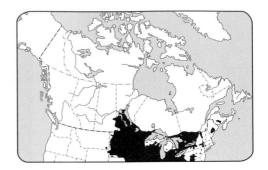

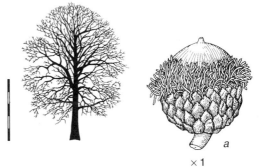

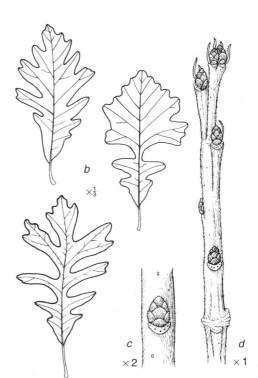

a. Acorn. *b.* Leaf forms. *c.* Lateral bud and leaf scar. *d.* Winter twig.

Transplants readily; tolerant of urban conditions.

Hybridizes with swamp white oak.

Pollen flowers (above); seed flowers (below).

Trunk straight, distinct to upper crown; mature bark deeply furrowed.

Acorn cup with fringe of knobby scales, encloses one-half or more of nut.

Young bark furrowed and rough.

Quick Recognition [Compare white oak]. Leaves hairy underneath, most with a large terminal lobe. Lateral buds pressed against the twig; a few narrow loose scales in the bud cluster. Branchlets with corky ridges. Half or more of the acorn covered with a fringed cup. Bark dark gray, with thick scales.

Swamp White Oak

Quercus bicolor Willd.

Chêne bicolore

Uncommon in Canada; occurs in southern Ontario and southern Quebec.

Leaves 12–17 cm long, widest above the middle, tapering to a wedge-shaped base; principal veins 4–6 per side, each ending in a rounded shallow lobe or tooth; upper surface shiny dark green; pale grayish-green beneath, with abundant white hairs; strong contrast in color between upper and lower surfaces.

Buds Terminal bud rounded, 2-4 mm long, blunt, reddish-brown, usually hairless; lateral buds diverging from the twig.

Twigs Stout, reddish-brown, usually hairless.

Fruits Acorns 20–30 mm long, solitary or paired, on stalks 2–10 cm long. Cup covered by swollen scales with recurved tips, enclosing one-third to one-half of the nut; fringed margin of cup usually evident.

Bark Light grayish-brown, scaly; fissured with flat ridges with age.

Size and Form Small to medium-sized trees, up to 22 m high, 90 cm in diameter, and 200 years old. Trunk short, forked. Crown broad, open, rounded, with relatively slender branches for an oak; upper branches ascending, lower ones usually drooping; larger branches and the trunk with many small, crooked, hanging branchlets, giving the lower part of the tree an untidy appearance.

Habitat Grows on moist bottomlands and at the edges of swamps. Moderately shade-tolerant.

10

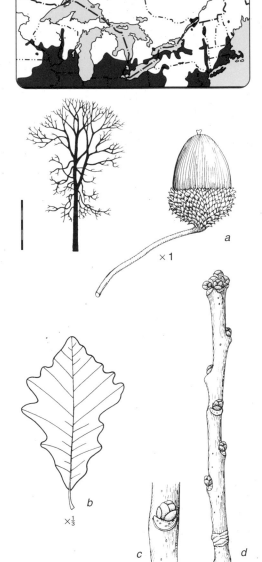

×1

×⅓

×2   ×1

a. Acorn on long stalk. *b.* Typical leaf form. *c.* Lateral bud and leaf scar. *d.* Winter twig.

Quick Recognition Leaves very pale and hairy underneath, contrasting with the green upper surface; shallowly lobed, widest above the middle. Acorns on long stalks. Buds blunt. Lower crown untidy in appearance. Occurs on moist sites.

Pollen flowers (above) not yet open; seed flowers (below).

Mature bark fissured with flat ridges.

Immature acorn showing swollen scales on cup.

Young bark scaly, on upper limbs peels in sheets.

Chinquapin Oak

Chinkapin oak, yellow chestnut oak

Quercus muehlenbergii Engelm.

Chêne jaune

Uncommon in Canada; occurs in south-western Ontario, and in southeastern Ontario near the Thousand Islands.

Leaves 10–18 cm long, narrow, tapered to both ends, tip narrow; coarsely toothed (rather than lobed); upper surface glossy green; pale grayish-green and downy beneath, with stellate hairs that have rays parallel to leaf surface; principal veins 10–15 per side, straight, parallel, each ending in a large pointed (or slightly rounded) tooth. Stalk slender, 1–3 cm long.

Buds Terminal bud ovoid, 4–6 mm long, rather sharp-pointed, light reddish-brown, mostly hairless.

Twigs Slender, hairless, green at first, with age becoming grayish-brown to orange-brown.

Fruits Acorns 12–25 mm long, short-stalked; cup deep, bowl-shaped, enclosing about one-third to one-half of the acorn, covered with small, hairy scales that are only slightly swollen; may form an inconspicuous marginal fringe.

Bark Pale gray to whitish, with thin narrow scales.

Size and Form Medium-sized trees, up to 30 m high (15 m when open-grown) and 60 cm in diameter; occasionally larger. Trunk usually straight, distinct well into the crown; strongly tapered with a swollen base. Crown narrow, rounded, made up of many short branches.

Habitat Occurs on dry rocky slopes and ridges, especially on limestone soils, and on sand dunes.

Notes Hybridizes with dwarf chinquapin oak.

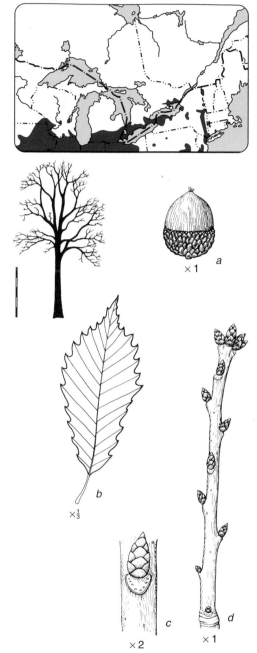

a. Acorn. b. Leaf. c. Lateral bud and leaf scar. d. Winter twig.

10

Quick Recognition Unlobed leaves with straight parallel veins each ending in a large tooth.

Pollen flowers (left); seed flowers (right).

Acorn cup with hairy scales, encloses one-third to one-half of nut.

Young bark (left) pale, scaly. Mature bark
rough and flaky with thin scales.

Dwarf Chinquapin Oak

Quercus prinoides Willd.

Chêne nain

Rare in Canada; occurs in southern Ontario near
Grand Bend on Lake Huron, Long Point on Lake
Erie, and Brantford on the Grand River. A very
small tree or spreading shrub, often occurring in
colonies on dry slopes; resembles and hybridizes
with chinquapin oak. **Leaves** 5–15 cm long;
principal veins 4–9 per side. **Acorns** 15–25 mm
long, with cup enclosing about one-half of the
acorn. **Bark** light brown, scaly.

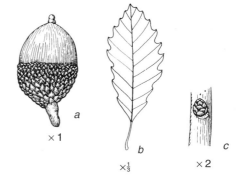

a. Acorn. *b*. Leaf. *c*. Lateral bud and leaf scar.

Garry Oak
Oregon white oak

Quercus garryana Dougl.

Chêne de Garry

The only oak native to British Columbia. A coastal species occurring on Vancouver Island and the Gulf Islands and in scattered locations along the lower Fraser River.

Leaves Thick, stiff, 7–10 cm long; 5–7 lobes, rounded, separated by deep narrow notches; larger lobes may be notched at the tip; upper surface shiny dark green, with a pebbled appearance; dull yellowish-green beneath, with brown hairs; may turn red in autumn. Stalk 1–3 cm long.

Buds Terminal bud elongated, 6–12 mm long, pointed; lateral buds diverging from the twig; scales narrow, rather loose, densely hairy, pale brown.

Twigs Stout, densely hairy the first season, becoming hairless and dark reddish-brown.

Fruits Acorns 25–30 mm long, solitary or paired, on short stalks up to 5 mm long. Cup shallow, saucer-shaped, enclosing one-third of the acorn; scales slightly thickened, hairy, free at the tips, but not forming a marginal fringe.

Bark Dark grayish-brown, scaly, with narrow shallow furrows.

Size and Form Small trees, up to 20 m high, 150 cm in diameter, and 200 years old; may be a shrub on poor rocky sites. Trunk short, stout, often forked within 5 m of the ground. Crown broad, rounded, composed of numerous twisted, gnarled, knobby branches.

Habitat A slow-growing tree even on deep, moist, loamy soils; often found at forest edges on south-facing slopes, on dry rocky knolls, and on seaside sites not subject to tidal flooding. Grows in small pure stands or mixed with Douglas-fir and arbutus. Intolerant of shade and competition.

Notes Similar in appearance to naturalized English oak.

10

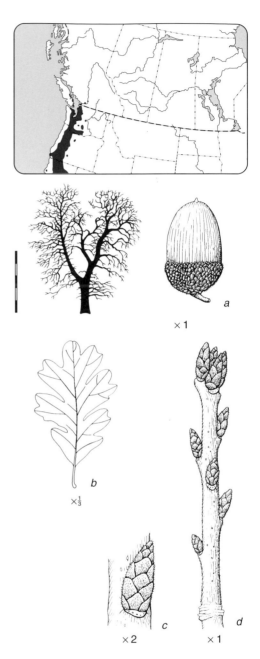

× 1

× ⅓

× 2   × 1

a. Acorn. b. Leaf. c. Lateral bud and leaf scar. d. Winter twig.

Pollen flowers (above); seed flowers (below).

Trunk short, stout; mature bark has narrow shallow furrows.

Acorn short-stalked; cup encloses one-third of nut.

Young bark dark grayish-brown, smooth, becomes scaly.

Quick Recognition [Contrasting features of English oak in brackets.] Leaves with 5–7 rounded lobes; upper surface shiny dark green [less shiny]; leaf stalks more than [less than] 1 cm long. Acorns solitary or paired [in small clusters of 2–5], on short stalks up to 5 mm [long stalks 30–80 mm] long. Trunk short, forked [tall, straight].

English Oak**

Pedunculate oak

Quercus robur L.

Chêne pédonculé

Large trees up to 35 m high and several hundred years old. Native to Europe. Naturalized locally in Canada and the United States. Hardy as far north as Zones C5, NA4. **Leaves** 5–12 cm long with 3-7 rounded lobes and 2 small ear-like lobes at the base; dark green above, bluish beneath; leaf stalk very short. Terminal **bud** 7–9 mm long, blunt. **Twigs** slender, greenish-brown, slightly powdery. **Acorns** 15–40 mm long, in clusters of 2–5, on stalks 30–80 mm long; cup encloses one-quarter to one-third of the nut. **Trunk** tall, straight, sturdy. **Crown** irregular. In Europe, this oak provided wood for shipbuilding and other uses. It has been planted in North America since colonial times. One of its common cultivars has strongly upright (fastigiate) branches.

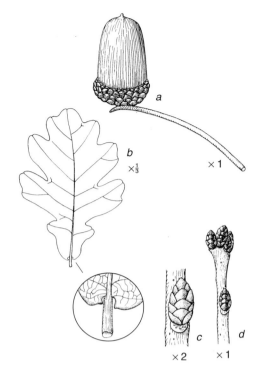

a. Acorn on stalk. b. Leaf with detail of basal lobes. c. Lateral bud and leaf scar. d. Winter twig.

Chestnut Oak*

Quercus montana Willd.

[syn. *Q. prinus* L.]

Chêne châtaignier

Native to the eastern United States; occurs on dry sandy soils and sometimes on bottomlands. Hardy as far north as Zones C5, NA4. A small slow-growing tree that can reach 400 years old. Resembles chinquapin oak, but **leaves** have rounded teeth and the rays on the stellate hairs on the undersurface spread away from the surface. Terminal **bud** large, about 10 mm long; lateral buds diverging widely from the twig. Scales of the **acorn** cup united at the base; rim of the cup without a fringe. **Bark** dark, firm and ridged, not scaly.

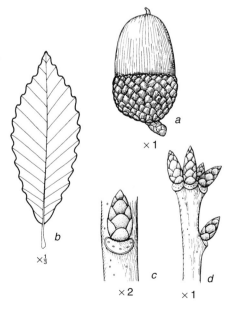

a. Acorn. b. Leaf. c. Lateral bud and leaf scar.
d. Winter twig.

10

Group 11.

Leaves alternate, simple; edges toothed

With 21 genera and about 110 species (over 75% native), Group 11 is the largest group in the book; the species are among the most difficult to identify. Most trees in Group 11 are deciduous with simple leaves (blade not divided into leaflets).

An alternate arrangement of leaves is by far the most common; alternate leaves are usually arranged spirally, but in a few cases they are in 2 ranks (rows) on opposite sides of the stem.

Teeth are projections on the edges of the leaf blade large enough to be seen with the naked eye; there is no well-defined distinction between a large tooth and a small lobe; refer to Group 10 if the teeth are 1 cm or more long and might be called lobes. Teeth may be rounded or sharp-tipped, sometimes hooked. A wavy margin is considered to be toothed; deeper waves grade into lobes (Group 10); shallow waves into a smooth edge (Group 12).

As with any classification system there are exceptions. The following genera in Group 11 have some species that sometimes have lobed leaves: hawthorn, poplar, apple, and birch. All willows are in Group 11, although some species, regularly or sometimes, have leaves without teeth. Some species of holly and Pacific bayberry are evergreen; leaf edges may be toothed or smooth. Mulberry and sycamore have leaves that are toothed and usually lobed (see Group 10). Chestnut oak, with rounded teeth, is described with the other oaks in Group 10.

American Chestnut

Castanea dentata (Marsh.) Borkh.
Fagaceae: Beech Family

Châtaignier d'Amérique

The chestnut genus comprises about 14 species of trees and shrubs; 5 are native to North America. American chestnut, the only species native to Canada, was once a prominent tree in the broadleaf forests of southern Ontario and the eastern United States; it is now rare due to the chestnut blight.

Leaves Deciduous, alternate, simple; 15–28 cm long, gradually tapering to both ends, short-stalked, yellowish-green; veins straight, parallel, 15–20 per side, each leading to a prominent tooth and extending beyond it to form a short curved bristle.

Buds Ovoid, 5–8 mm long, pointed, greenish-brown, with 2 or 3 hairless scales; end bud similar to the lateral buds. Leaf scars semi-oval, somewhat raised, with numerous vein scars.

Twigs Stout, shiny, reddish-brown, numerous light-colored lenticels. Pith 5-pointed.

Flowers Pollen flowers and seed flowers on the same tree. Pollen flowers on short stalks, in semi-erect catkins, 12–20 cm long, in leaf axils; seed flowers solitary, or in clusters of 2 or 3, at the base of some pollen catkins. Appear after leaves almost fully grown.

Fruits Edible nut; in small clusters of 1–5 within a spiny bur-like husk 5–8 cm across that splits into 4 parts; each nut ovoid, flat on 1 side, pointed, dull, brownish, smooth.

Seedlings Cotyledons white, fleshy, remaining in the seed coat under the surface.

Vegetative Reproduction By stump sprouts.

Bark Smooth, dark brown, separating into broad flat-topped ridges with age.

Wood Moderately hard and strong, oak-like grain on tangential face, straight-grained, reddish-brown, very decay resistant. Ring-porous, large pores easily visible with the naked eye; annual rings prominent; rays very small.

Size and Form Before the chestnut blight, large trees up to 35 m high and 100 cm in diameter; now seldom reaching a height of

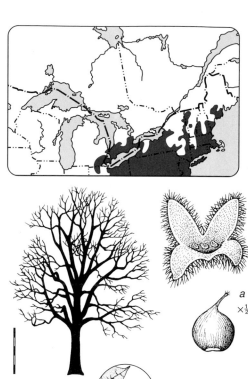

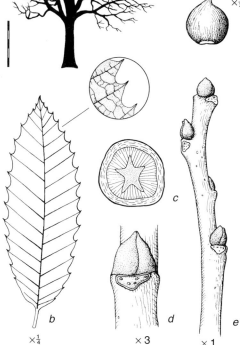

a. Fruit husk (above); nut (below). *b*. Leaf with detail of bristle-tipped teeth. *c*. Twig cross section. *d*. Lateral bud and leaf scar. *e*. Winter twig.

10 m, although a few specimens exceeding 50 cm in diameter exist in Ontario.

Habitat Occurs on a variety of sites; grows best on well-drained sands and gravels; usually found mixed with other broadleaf trees.

11

Pollen flowers (left) in catkins; seed flowers (right) solitary or in small clusters.

Fruit husks densely spiny; nuts edible.

Young bark (inset) smooth, dark brown. Large trees are rare; mature bark has prominent flat-topped ridges.

Notes Chestnut blight (*Endothia parasitica*), common in Europe and Asia, probably entered North America on stock imported from Asia in 1904. Although most North American trees have died, stump sprouts often grow to pole size. Attempts to select or breed for disease tolerance have been unsuccessful. The introduction of a less virulent disease strain to compete with the existing blight may reduce damage to the trees and permit more to survive. Wood formerly very valuable because of its durability.

> **Quick Recognition** Leaves alternate, long, narrow, with straight veins and large bristle-tipped teeth; fruit a large bur.

Chinese Chestnut*

Castanea mollissima Blume

Châtaignier de Chine

Native to China and Korea. Hardy as far north as Zones C6, NA4. Similar to American chestnut; somewhat resistant to the chestnut blight. **Leaves** with a rounded base and hairy undersurface. **Buds** hairy.

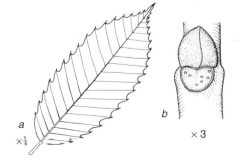

a ×¼

b × 3

a. Leaf. *b*. Lateral bud and leaf scar.

Witch-Hazel

Hamamelis virginiana L.
Hamamelidaceae: Witch-Hazel Family

Hamamélis de Virginie

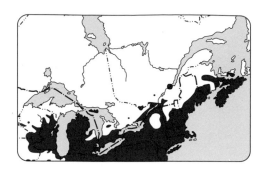

The witch-hazel genus comprises about 6 species; 2 are native to North America, but only *Hamamelis virginiana* attains tree size.

Leaves Deciduous, alternate, simple; irregularly oval, 6–15 cm long, base asymmetrical; margin irregularly wavy or coarsely toothed; texture thin; upper surface dark green; lighter, hairless beneath; veins straight, parallel, ascending, widely spaced, 5–7 per side. Yellow in autumn.

Buds Terminal bud flattened and curved, 10–14 mm long, stalked, without scales, densely covered with yellowish-brown hairs; lateral buds much smaller; commonly more than 1 bud above a leaf scar.

Twigs Slender, yellowish, hairy, becoming hairless and brown by autumn.

Flowers Bright yellow; 4 strap-like, twisted petals, each 15–20 mm long. In clusters of 3, in the leaf axils. Appear from September to October.

Fruits A capsule; 2-beaked, woody, orange-yellow, hairy. Maturing and splitting open in about a year, when the tree is in flower again; forcibly ejecting 2 small, shiny black seeds; empty capsules persisting on the tree for another year or more.

Bark Smooth or slightly scaly, light brown, often mottled; inner bark reddish-purple.

Size and Form Usually an upright shrub with many trunks, but may grow into a shrubby tree up to 8 m high and 15 cm diameter. Trunk leaning, crooked; crown spreading, irregular.

Habitat Occurs on moist, shaded sites, as an understory tree and at the forest edge.

Notes The bark, twigs, and leaves contain the astringent witch-hazel, used in making skin lotions and eyewashes.

11

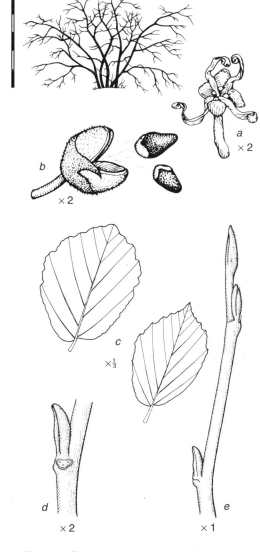

a. Flower. *b*. Fruit capsule ejecting seeds. *c*. Leaf forms. *d*. Lateral bud and leaf scar. *e*. Winter twig.

Flowers bright yellow, appear in autumn.

Trees usually in multistemmed clumps with trunks leaning.

Fruit capsules persist after seeds are shed.

Quick Recognition Leaves straight-veined with irregularly toothed margins and an asymmetrical base; buds without scales, yellow, stalked; flowers yellow, appear in autumn; capsules of several ages on the plant at any season.

Hollies

Winterberries, gallberries, dahoon,
yaupon, possum-haw

Genus *Ilex*
Aquifoliaceae: Holly Family

Les houx

The holly genus comprises nearly 400 species distributed throughout the tropical and
temperate regions of the world; of the 11
North American species that reach tree size,
one, common winterberry, is native to
Canada. Many introduced species and cultivars are planted for ornamental purposes.

Leaves deciduous or evergreen, alternate, simple; margin toothed or smooth.
Buds small, terminal bud present. Leaf scars
crescent-shaped, with 1 vein scar. Pollen
flowers and seed flowers on separate trees,
inconspicuous, appear early in spring. **Fruits**
berry-like, ripen in autumn, eaten by birds.

Common Winterberry
Black-alder

Ilex verticillata (L.) A. Gray

Houx verticillé

The largest native holly, but rarely reaching tree
size; occurring from Newfoundland to Lake
Superior, south to Tennessee and Georgia; found
in swamps and along streams. **Leaves** deciduous,
oval, 4–10 cm long, finely toothed, upper surface
dark green, may be hairy beneath. **Buds** small,
rounded, with several scales. **Twigs** light brown,
ridged. **Flowers** in small clusters in the leaf axils.
Fruits berry-like, red or orange, 5–8 mm across,
on short stalks, in small clusters in the axils of the
basal leaves; remaining on the tree after the leaves
fall. **Bark** mottled gray, with warty lenticels.

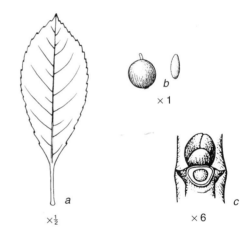

a. Leaf. *b.* Fruit (left); seed (right). *c.* Lateral bud
and leaf scar.

English Holly**

Ilex aquifolium L.

Houx commun

The Christmas holly, well-known for its shiny evergreen leaves and bright red fruits. Tall shrubs or
small trees up to 15 m high. Frequently planted in
southwestern British Columbia; occasionally naturalized there. Hardy as far north as Zones C7,
NA6. **Leaves** evergreen, shedding after 2 or more
years, oval, 3–7 cm long, thick, shiny green; teeth
coarse, tipped with spines in the lower part of the
tree, absent from leaves in the upper part of the
tree. **Buds** small, often more than 1 in a leaf axil.
Flowers small, greenish-white, in small clusters in
axils of leaves on 1-year-old twigs. **Fruits** bright
red, 7–10 mm across, remaining on the tree until
late winter.

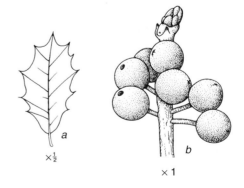

a. Leaf. *b.* Twig with terminal bud, leaf scar, and
fruit cluster.

11

American Holly*

Ilex opaca Ait.

Houx d'Amérique

Native to the eastern part of the United States; similar to the English holly, but the flowers occur on the new shoots, and leaves are larger. Hardy as far north as Zones C7, NA6.

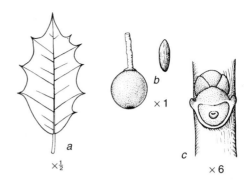

a. Leaf. *b.* Fruit (left); seed (right). *c.* Lateral bud and leaf scar.

Mountain-Holly

Nemopanthus mucronatus (L.) Trel.
Aquifoliaceae: Holly Family

Némopanthe mucroné

The genus *Nemopanthus* has only 1 species, mountain-holly. Tall shrubs, sometimes very small trees; occurring from Newfoundland to Lake Superior, south to Illinois and North Carolina; found in swamps and along streams. **Leaves** deciduous, alternate, simple; about 7 cm long, thin, abruptly pointed (mucronate); margin smooth or with a few scattered teeth; stalk purplish. **Buds** small, ovoid, with a thickened upper bud scale. **Twigs** slender, smooth, purplish; dwarf shoots a feature of twigs 1-year-old and older. Pollen **flowers** and seed flowers usually on separate trees, appear in spring, in the leaf axils; perfect flowers occur occasionally. **Fruits** berry-like, purplish-red, about 6 mm across, on a long slender stalk; shed in late summer. **Bark** smooth, gray.

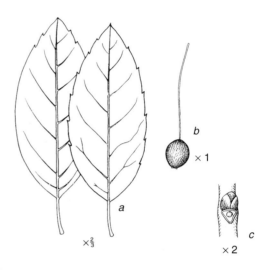

a. Leaf forms. *b.* Fruit. *c.* Lateral bud and leaf scar.

Pacific Bayberry

Myrica californica Cham.
Myricaceae: Wax-Myrtle Family

Myrique du Pacifique

Very small, usually shrubby, trees, up to 8 m high and 30 cm in diameter; uncommon in Canada; occurring near the coast of southern British Columbia. **Leaves** evergreen, alternate, simple; narrow, 5–10 cm long; upper surface bright green; glandular beneath; coarsely toothed to near base; fragrant when bruised. **Buds** 7–9 mm long, pointed, with several reddish-brown scales. Pollen **flowers** and seed flowers in separate clusters in the leaf axils. **Fruits** a globular aggregate 4–6 mm across, dark purple, a source of food for wildlife.

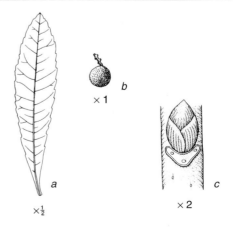

a. Leaf. *b.* Fruit aggregate. *c.* Lateral bud and leaf scar.

Buckthorns

Genus *Rhamnus*
Rhamnaceae: Buckthorn Family

Les nerpruns

Buckthorns are found chiefly in the temperate regions of the Northern Hemisphere. The buckthorn genus comprises about 100 species of trees and shrubs; 12 are native to North America; a shrub and cascara buckthorn are native to Canada. Two European species have been become naturalized.

Leaves deciduous, alternate, opposite, subopposite, simple; generally oval, tip pointed or blunt; veins characteristic, prominent, curving forward (arcuate) to form part of the leaf margin, several per side. **Flowers** mostly perfect, but some may bear only pollen or ovules; borne in the leaf axils of new shoots. **Fruits** berry-like, in small clusters in leaf axils toward the base of new shoots, or on dwarf shoots; eaten by birds, thus dispersing the seeds.

Cascara Buckthorn

Cascara

Rhamnus purshiana DC.

Nerprun cascara

A very small western tree. Young trees may retain green leaves into the winter.

Leaves Alternate (occasionally opposite), 4–16 cm long, usually widest just above the middle; tip short or blunt, base rounded; finely and evenly toothed; undersurface slightly hairy; veins very prominent, 10–15 pairs.

Buds Very small, without scales, hairy, usually in 4 rows along the twig; terminal bud present.

Twigs Slender, yellowish-green, hairy, becoming reddish-brown, smooth, bitter tasting. Leaf scars crescent-shaped, raised, with a few scattered vein scars.

Flowers Yellowish-green, inconspicuous, appear in June.

Fruits Berry-like, blackish, rounded, 8–14 mm across, containing 2–3 small nutlets.

Bark Thin, smooth, dark grayish-brown, becoming scaly with age.

Wood Light, close-grained, soft, brown tinged with red. Diffuse-porous, pores visible with a hand lens; annual rings and rays visible to the naked eye.

a. Fruit (left); seed (right). *b*. Leaf with detail of finely toothed margin. *c*. Lateral bud and leaf scar. *d*. Winter twig.

$\times \frac{1}{2}$ $\times 2$ $\times 1$

a × 1

b

c   *d*

11

Flowers in small inconspicuous clusters.

Fruits berry-like, when ripe almost black.

Trunk slender; mature bark becomes scaly.

Young bark thin, smooth.

Cascara Buckthorn

Size and Form A shrub or very small tree, up to 6 m high and 25 cm in diameter; occasionally much larger. Trunk slender; on young trees dividing into ascending branches; on older trees horizontal branches form a compact irregular crown.

Habitat Occurs mainly on moist organic soils; also on coarse, sandy, or gravelly soils.

Associated with alders and birches in colonizing burned-over areas and clearings. An understory tree in conifer forests.

> **Quick Recognition** Buds hairy, without scales. Leaves oval with fine teeth, veins curving forward, 10–15 pairs. Fruits berry-like, almost black.

European Buckthorn**

Common buckthorn, purging buckthorn

Rhamnus cathartica L.

Nerprun cathartique

Shrubs or small trees, up to 8 m high, and 10 cm in diameter; native to Europe. Hardy as far north as Zones C3, NA2. **Leaves** opposite, subopposite, or occasionally alternate; tip sharp-pointed, curved, folded; finely toothed, with 3–5 veins per side; remain green long after leaves of most other species have fallen. **Buds** scaly, almost black, appressed to the twig. Some dwarf shoots end in a thorn. **Fruits** black, in dense clusters; a laxative effect when eaten. **Bark** grayish-brown; like birch in that the lenticels are elongated horizontally and the loose ends of the smooth bark form tight curls.

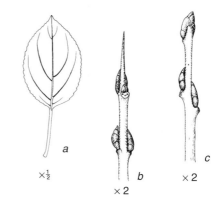

a. Leaf. *b*. Dwarf shoot, with terminal thorn, oppositely placed lateral buds, and leaf scar. *c*. Winter twig showing suboppositely placed lateral buds.

Glossy Buckthorn**

Alder buckthorn

Rhamnus frangula L.
[syn. *Frangula alnus* Mill.]

Nerprun bourdaine

Shrubs or small trees, native to Europe. Hardy as far north as Zones C3, NA2. **Leaves** alternate, smooth-margined, with 5–8 veins per side. **Buds** without scales. **Fruits** ripen throughout the summer; thus green, red, and black fruits may occur at one time.

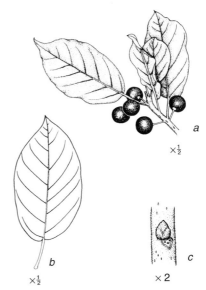

a. Twig with leaves and fruit. *b*. Leaf. *c*. Lateral bud and leaf scar.

11

Lindens/ Basswoods

Limes

Genus *Tilia*

Tiliaceae: Linden Family

Les tilleuls

Lindens are large trees found in the North Temperate Zone. Worldwide the linden genus comprises about 30 species; 4–7 are recognized in North America, 1 of which is native to Canada. Several non-native species are used for landscape purposes.

Leaves Deciduous, alternate, simple; heart-shaped, pointed, base asymmetrical, sharply toothed; lateral veins straight, intersecting margin; stalk almost half the length of the blade.

Buds Plump, pointed, asymmetrical, with 2 or 3 scales, in 2 rows along the twig; no true terminal bud. Leaf scars raised, with 5–10 vein scars.

Twigs Stout, zigzag, smooth; pith round in cross section. Spring flush of shoot growth is preformed in the bud; subsequent shoot growth is neoformed. Shoot growth ends when the tip aborts and falls off; occurs in late May on less vigorous shoots.

Flowers Yellowish, fragrant; 5 petals and 5 sepals. In small, loose, pendulous clusters (cymes), in the axils of leaves on new shoots. Stalks long, joined along the lower part to a distinctive long, narrow leaf-like bract. Appear in midsummer. Insect-pollinated.

Fruits Nut-like capsule; globular, woody, nonsplitting, usually containing 1-seed; in small long-stalked clusters with a prominent bract. Often remaining on the tree into the winter. Dispersed by wind and animals.

Seeds Deeply dormant; germination may not take place for several years after sowing. Remain viable for several years if stored air-dry at near-freezing temperatures.

Seedlings Cotyledons distinctively 5-lobed. Taproot well developed during the 1st year.

Vegetative Reproduction Often by stump sprouts.

Bark Inner bark tough, fibrous.

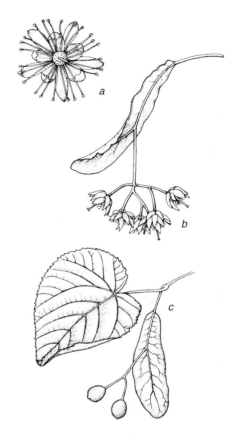

a. Flower. *b.* Flower cluster and leaf-like bract. *c.* Leaf, fruit cluster, and bract.

Wood Soft, texture uniform, weak, not durable. Diffuse-porous, pores visible with a hand lens; annual rings inconspicuous; rays visible with the naked eye.

Size and Form Medium-sized to large trees. Crown with a regular rounded outline; leading shoot oblique. Root system deep, wide-spreading.

Notes Wood suitable for carving. Flowers provide nectar for bees; fruits, bark, and twigs are food for wildlife.

Quick Recognition Flowers and fruits in clusters with a large distinctive bract joined to a long stalk. Leaves heart-shaped, single-toothed, asymmetrical at the base. Buds stout, asymmetrical with 2 or 3 bud scales. Flowers fragrant, parting in 5's; fruits nut-like, woody, roundish.

Basswood
American linden
Tilia americana L.
Tilleul d'Amérique

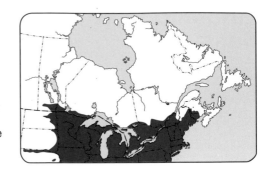

Leaves 12–15 cm long, thick; tip elongated; teeth sharp, gland-tipped; upper surface dull green; lighter beneath, with tufts of hair in the vein axils.

Buds 5–7 mm long, hairless, often reddish. Leaf scars semi-oval, with 5–10 vein scars.

Twigs Yellowish-brown, hairless.

Flowers Creamy yellow, fragrant, 11–13 mm across. Appear in July after the leaves are fully grown.

Fruits 8–12 mm across, coated with brownish hairs. Abundant seed crops occur almost every year.

Bark Thin, smooth, light greenish-brown when young, becoming dark grayish-brown, with many long, narrow, flat-topped ridges, transversely divided into short blocks.

Size and Form Large trees, up to 35 m high, 100 cm in diameter, and 200 years old. Trunk straight, distinct into the upper part of a symmetrical, smooth, rounded crown. Branches spreading out and ascending at tips.

Habitat Commonly occurs on moist slopes that face north and east. In mixed stands with other broadleaf trees such as sugar maple, American beech, white ash, and red oak, usually outgrowing them. Very shade-tolerant.

Notes Wood valued for hand-carving and turnery; used for interior trim, veneer, plywood, and furniture parts. Basswood considered an excellent source of nectar by bee-keepers.

Thin-barked and thus susceptible to fire damage; however, stump often remains alive and produces a clump of suckers.

The decaying leaves, rich in nitrogen and minerals, contribute to soil fertility.

11

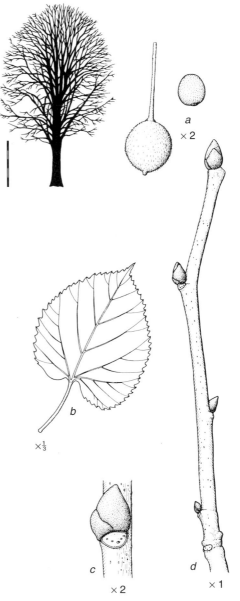

a. Fruit (left); seed (right). *b.* Leaf. *c.* Lateral bud and leaf scar. *d.* Winter twig.

Quick Recognition	See genus.

Flowers creamy yellow; appear in July.

Fruit coated with brownish hairs.

Basswood

Young bark (left) thin, smooth. Trunk straight and distinct; mature bark with long, narrow ridges.

Little-Leaf Linden*

Small-leaf European linden

Tilia cordata Mill.

Tilleul à petites feuilles

Native to Europe; commonly planted for landscape purposes; tolerates urban conditions. Hardy as far north as Zones C3, NA3. **Leaves** 4–8 cm long. **Fruits** gray, hairy.

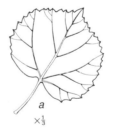

a
×⅓

b
× 2

a. Leaf. *b*. Lateral bud and leaf scar.

White Linden*

Silver linden

Tilia tomentosa Moench

Tilleul argenté

Native to southeast Europe–western Asia; tolerates urban conditions. Hardy as far north as Zones C5, NA4. **Leaves** 5–12 cm long; upper surface shiny dark green; white, densely hairy beneath. **Buds** and **twigs** hairy. **Fruits** warty, hairy, faintly 5-angled.

a
×⅓

b
× 2

a. Leaf. *b*. Lateral bud and leaf scar.

Birches

Genus *Betula*
Betulaceae: Birch Family

Les bouleaux

Birches are found in the North Temperate Zone, northward to the tree line. Worldwide, the birch genus comprises about 50 species of trees and shrubs. About 12 species are native to North America, 10 or 12 to Canada; 8 or more Canadian species reach tree size. The exact number depends on what populations are considered distinct species, subspecies, varieties, or hybrid swarms.

Birches are easily recognized by their bark; it forms in thin sheets marked by horizontally elongated lenticels. Birch species in Canada may be divided into white birches and yellow birches.

White birches are difficult to classify. White (paper) birch occurs in all forested regions of Canada. Gray birch, mountain paper birch, and blueleaf birch are found in eastern Canada; water birch, Alaska paper birch, and Kenai birch in western Canada. European white birch is frequently planted.

Yellow birches are represented by yellow birch and cherry birch. They differ from the white birches in that the leaves are oval with 8−12 veins per side [variously shaped with 3−5 veins per side]; the mature seed catkins are cone-like, erect [slender, pendulous]; and the inner bark and twigs have a wintergreen fragrance and taste.

Hybrids occur naturally, including some between tree-size species and shrubs. Polyploidy is very common; the somatic chromosome number varies from 28 to 84, and is not always constant within a species.

Leaves Deciduous, alternate, but often appearing to be in pairs or whorls on the dwarf shoots, simple; oval, ovate, or triangular; lateral veins prominent, pinnately arranged along the midvein, tending to be straight and parallel; each vein ends in a large sharp tooth, with 2−5 intervening teeth. Among the first trees to leaf out in spring; leaves usually turn yellow in autumn.

Buds Lateral buds with 3 scales. No true terminal buds on long shoots; solitary buds at the tip of dwarf shoots have 5−7 scales. Leaf scars with 3 vein scars.

Twigs Slender, with prominent lenticels, often glandular; pith oval or round in cross section. Two kinds of side shoots develop: normal long shoots (extension shoots)

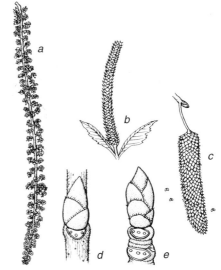

a. Pollen catkin at time of pollination. b. Seed catkin at time of pollination. c. Mature seed catkin shedding winged nutlets. d. 3-scaled lateral bud and leaf scar. e. 6-scaled bud on dwarf shoot.

toward the tip of the previous year's twig; dwarf shoots toward the base. Neoformed side shoots sometimes develop in the leaf axils of vigorous new shoots. Shoot growth ends when the tip aborts and falls off, leaving a lateral bud at the end of the twig.

Flowers Borne in many-flowered catkins; pollen and seed catkins on the same tree. Pollen catkins narrowly cylindrical, solitary or in small clusters at or near the tips of long shoots; formed in late summer, visible through the winter as stiff finger-like structures; in early spring, lengthening 2−4 times and becoming pendulous. Seed catkins stout, solitary (occasionally paired), on dwarf shoots; enclosed in the bud during the winter; appear in early spring before the leaves; wind-pollinated.

Fruits Borne in catkins. Each fruit a small flat 1-seeded 2-winged nutlet (often erroneously called a seed), located in the axil of a 3-lobed scale and bearing 2 withered thread-like stigmas at the tip. Mature in early autumn; scales and fruits fall away in the succeeding months, eventually leaving the bare central axis of the catkin attached to the tree. Can be blown long distances over glazed snow in winter. Seed production begins at about 15 years of age, with abundant crops from a stand, but not from every tree, almost yearly.

Seeds Exposure to light required for prompt germination. Viable for some years if stored air-dry at near-freezing temperatures.

11

white birch

Young bark shows conspicuous, horizontally elongated lenticels.

yellow birch

Immature pollen catkins visible on twigs in winter.

Seedlings Cotyledons small green. First true leaves small.

Vegetative Reproduction Commonly by stump sprouts, resulting in clumps of birches.

Bark Smooth when young, in thin papery layers, with conspicuous elongated transverse lenticels; in many species becoming white after the first few years; with age thickens and splits into irregular rough segments at the base of the trunk. Very durable (made of cork); forms part of the forest litter long after the wood has rotted away.

Wood Texture uniform, straight-grained, moderately hard and strong, light or medium reddish-brown, susceptible to decay. Diffuse-porous; pores and rays can be seen with a hand lens; annual rings indistinct.

Size and Form Crown open, relatively narrow, pyramidal. On young trees, branches slender, ascending; leading shoot upright. On older trees, branches bent down, becoming horizontal; leading shoot oblique.

Habitat Fast-growing; most birches are not shade-tolerant; often forming a large part of the new tree cover on areas disturbed by logging or fire; becoming overtopped by conifers and tolerant hardwoods after a few decades. Although forest fire favors regeneration of birch by removing competing vegetation and surface litter, fire usually kills the aboveground parts of existing trees.

Notes The birches, especially white birch, have experienced population decline occurring over extensive areas, referred to as birch dieback. It is characterized by a progressive dieback of twigs and branches, loss of vigor, and reduced tolerance to other stresses such as insects, diseases, drought, and frost. The precise causes have never been determined.

Commonly grow in association with conifers such as spruces and balsam fir; when these are harvested from the stand, the birch loses vigor and dies within a few years.

Used for wood pulp, lumber, and plywood. Buds, twigs, pollen catkins, and nutlets a source of food for birds and mammals. Often planted for landscape purposes.

In spring, birches can be tapped; sap flows from fresh stumps or wounds that penetrate the wood (xylem); the sap may be made into a fermented beverage, or boiled down into a sweet syrup.

Decaying birch leaves contribute to soil improvement.

Quick Recognition During any season by their papery bark and dwarf shoots, and when present, by their pollen catkins and seed catkins.

White Birch

Paper birch, canoe birch

Betula papyrifera Marsh.

Bouleau à papier

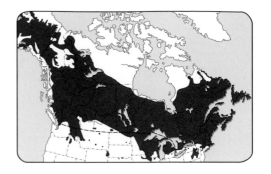

Occurs in all forested regions across Canada, north to the tree line. Frequently planted for landscape purposes (Zones C2, NA2).

Leaves Ovate or triangular, 5–10 cm long; tip pointed; base broadly wedge-shaped, rounded, straight, or cordate; double-toothed; upper surface dull green, lighter beneath; veins 9 per side or fewer, each ending in a large tooth, with 3–5 smaller intervening teeth; 35–55 teeth per side. Preformed leaves become essentially hairless; neoformed leaves remain hairy, especially at the vein axils.

Buds Slender, 5–7 mm long, tapering to a blunt point, resinous; scales greenish toward the base, brown toward the tip.

Twigs Slender, hairy during the first weeks, becoming dark reddish-brown, with sparse warty resin glands.

Flowers Pollen catkins in clusters of 1–3, 1–3 cm long and 2–4 mm wide in winter, about 9 cm long at pollination. Seed catkins 1–2 cm long, erect; stigmas pink or red.

Fruits Mature seed catkins 3–5 cm long, hanging from the dwarf shoots. Nutlets 1.5–2.5 mm long, half as wide; wings much wider than the nutlet. Scales variable, 2–3 mm long, usually hairy, with 2 rounded lateral lobes diverging from a short, pointed central lobe. Fruits and scales shed from September onwards, leaving the bare catkin axis on the tree.

Bark Thin, smooth, dark red to almost black on young stems, becoming reddish-brown and then bright creamy white; often shedding in large sheets. On removal of the outer bark, the reddish-orange inner bark soon dies, turns black, and divides into flakes. A new layer of bark develops inside the dead layer. Exposing large areas of the inner bark may kill the tree.

Wood Texture uniform, pale, no odor.

Size and Form In eastern Canada, small or medium-sized trees, up to 25 m high, 40 cm in diameter, and 120 years old. In western Canada, 2 forms are distinguished: western white birch, large trees up to 35 m high, with peeling orange-white bark; and northwestern

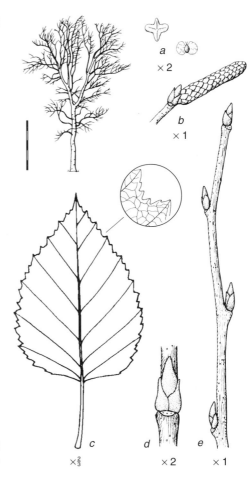

a. Fruit scale (above); winged nutlet (below). *b.* Pollen catkin in winter condition. *c.* Leaf with detail of double-toothed margin. *d.* Lateral bud and leaf scar. *e.* Winter twig.

white birch, small trees, up to 20 m high, with tight reddish-brown bark; however, differences relating to the leaves and fruits are not consistently associated with tree size and bark color. Trunk slender, often curved; usually distinct to midcrown or higher. Crown narrowly oval, open; branches ascending.

11

Seed catkins (left) and pollen catkins (right) at time of pollination

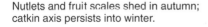

Nutlets and fruit scales shed in autumn; catkin axis persists into winter.

Young bark (top left) reddish brown; replaced by papery bark (top right); mature bark develops irregular, rough sections.

Habitat Found on forest edges, lakeshores, and roadsides. Grows on a wide variety of soils; in pure stands and mixed with various species such as other birches, pines, spruces, hemlocks, poplars, maples, balsam fir, northern red oak, and pin cherry. Not shade-tolerant. Among the first species to reforest areas that have been burned or cut.

Notes Isolated trees often die after the rest of a stand has been harvested. The tough pliable bark has long been used for making canoes and ornaments. Somatic chromosome number is 70(5x14) or 84(6x14), or occasionally 56(4x14).

Quick Recognition Leaves widest below the middle, double-toothed, base smooth-edged. Bud scales green and brown. Bark white, sheds in large sheets.

Mountain Paper Birch

Mountain white birch, eastern paper birch

Betula cordifolia Regel
[syn. *B. papyrifera* var. *cordifolia* (Regel) Fern.]

Bouleau à feuilles cordées

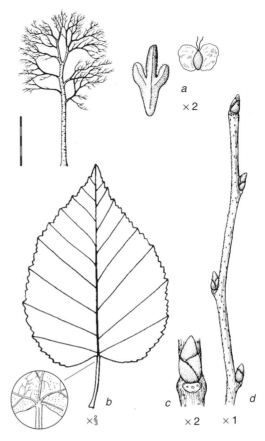

a. Fruit scale (left); winged nutlet (right). *b.* Leaf with detail of resin dots. *c.* Lateral bud and leaf scar. *d.* Winter twig.

An eastern species, sharing the cooler portions of the eastern range of white birch. Common in Newfoundland, New Brunswick, Nova Scotia, the Gaspé Peninsula and northern Quebec. Occurring in northwestern Ontario from Sault Ste. Marie westward to Algoma District and western Lake Superior, not reaching the Manitoba border; southward through northern New York state and New England. Not a true mountain species, as its name would suggest.

Leaves 6–12 cm long, tip taper-pointed, base usually heart-shaped; double-toothed; upper surface dull green, with numerous tiny resin dots; veins 7–9 per side.

Buds Ovoid, 4–6 mm long, blunt; scales resinous.

Twigs Yellowish-brown to dark brown, with light gray lenticels and warty resin glands.

Flowers Pollen catkins in clusters of 2–4, 2–4 cm long and 3–6 mm wide in winter, up to 10 cm at pollination. Seed catkins 1–2 cm long, often pendulous; stigmas light-colored or transparent.

Fruits Mature seed catkins 3.5–5 cm long. Nutlets 2–3 mm long, half as wide, wings much wider; stigmas may be as long as the nutlet. Scales 6–8 mm long, fringed with hairs, middle lobe with parallel sides and a rounded tip, much longer than the 2 upright, rounded lateral lobes.

Bark White or bronze white; separates into several thin layers, copper-colored on the inner surface; peels freely, giving trunk a ragged look. Lenticels up to 70 mm long, average 19 mm.

Size and Form Medium-sized trees, up to 25 m high and 70 cm in diameter. Crown broad; principal branches nearly horizontal.

Habitat Occurs generally in cooler habitats on upper elevations at the tree line in the southern part of its range and on cooler northern aspects and in depressions towards the northern part of its range; associates with a wide range of tree species in mixed broadleaf and conifer forests.

Notes The species epithet *cordifolia* is misleading, in that the leaf base is not always cordate (heart-shaped); it may be rounded or straight. Somatic chromosome number is 28.

Quick Recognition [Contrasting features of white birch in brackets.] An eastern [transcontinental] tree; leaves [not] dotted with resin glands; leaf-base usually [seldom] cordate; twigs [slightly] glandular; nutlets 2.4 [2.0] mm long, stigmas 1.5 [0.8] mm; middle lobes of fruit scale long [short] with a rounded [pointed] tip; bark white, with [without] a bronze tinge, separating into several thin sheets [1 thick sheet], resulting in a ragged [smooth] surface; lenticels often [seldom] over 30 mm; principal branches nearly horizontal [ascending]. More prevalent on moist [dry] sites.

11

Mature seed catkin.

Young bark.

Mountain Paper Birch

Mature bark peels freely, trunk appears ragged.

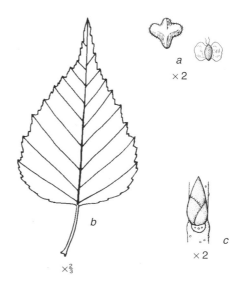

Blueleaf Birch

Betula ×caerulea Blanch.
[syn. *B. ×caerulea-grandis* Blanch.]
(*B. populifolia* × *B. cordifolia*)

Bouleau bleu

First identified in Vermont, this hybrid has now been found in eastern Canada; occurs in Labrador, the Gaspé Peninsula and Nova Scotia, south into northern New England and New York. Like both parents, the somatic chromosome number is 28.

Leaves 5–9 cm long, tip long-pointed, base variable; upper surface bluish-green, lighter beneath with some hairs in the vein axils; 8 veins and 38–50 teeth per side. **Buds** ovoid, 6 mm long, sharp-pointed, brownish; bud scales fringed with hairs. **Twigs** smooth, hairless, often glandular. **Pollen catkins** in small clusters of 1–3. **Mature seed catkins** 2.5–3.5 cm long; nutlets narrow 2.5–3.5 mm long, 1.2 mm wide, wings wider than the nutlet, stigmas 1.2–1.6 mm long; fruit scales 4–7 mm long, middle lobe short, lateral lobes diverging. **Bark** white, separable, peeling easily. Large trees, up to 35 m high. Often found with red spruce.

a Fruit scale (left); winged nutlet (right). *b*. Leaf. *c*. Lateral bud and leaf scar.

Gray Birch
Wire birch

Betula populifolia Marsh.

Bouleau gris

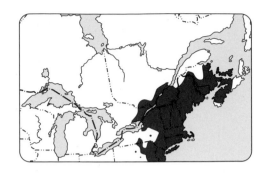

A small tree, commonly found in the Maritime provinces. May be extending its range northward and westward colonizing abandoned farmland.

Leaves Triangular, 4−7 cm long; tip long-pointed, tapering, base almost straight across; pendulous on slender stalk; upper surface rough, shiny dark green; shiny, lighter beneath; veins 6−9 per side, each ending in a large tooth; 40 (18−47) teeth per side, distinctly of 2 sizes, larger ones sometimes appear to be small lobes; toothless near the stalk. Preformed leaves hairless; neoformed leaves slightly hairy.

Buds Pointed, uniform light brownish-gray, gummy, often hairy.

Twigs Slender, dark brown to gray, hairless, with many resin glands.

Flowers Pollen catkins usually solitary, about 2 cm long in winter, 6−10 cm at pollination. Seed catkins 1−1.5 cm long.

Fruits Mature seed catkins 1.5−2 cm long, narrow, blunt-tipped, semi-erect. Nutlets almost hairless, wings much wider than the nutlet. Scales 2−3 mm long, densely hairy on the inner side, with 2 broad recurved lateral lobes and a very short central lobe. Shed in late autumn and early winter.

Bark Thin, smooth, dark reddish-brown when young; becoming dull chalky white, peeling with difficulty into small thin rectangular plates. Triangular black patches often form below the branches; caused by a fungus (*Pseudospiropes longipilus*) growing on excretions from the bark.

Size and Form Small trees, up to 12 m high, 15 cm in diameter, and 50 years old. Often in clumps. Trunk usually curved and leaning; distinct nearly to the top of the crown. Crown narrow, irregular, open, with many slender branches; older branches often S-shaped.

Habitat Occurs on sandy or gravelly soils of any moisture regime; in pure stands on abandoned pastures and areas recently cut or burned; eventually replaced by more tolerant and longer-lived species. Not shade-tolerant.

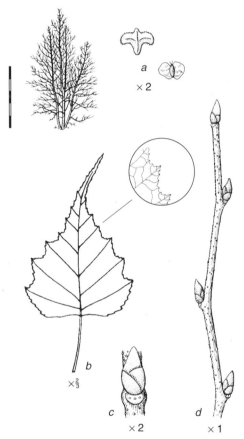

a. Fruit scale (above); winged nutlet (below).
b. Leaf with detail of venation and distinct double-toothed margin. *c.* Lateral bud and leaf scar. *d.* Winter twig.

Notes Leaf-out takes place about a week after that of white birch. The somatic chromosome number is 28.

Quick Recognition Leaf triangular, with a long tip; bark dull chalky white, nonpeeling, black patches below the branches; branches thin, often S-shaped. Seed catkin scales densely hairy on the inner side.

11

Seed catkins narrow, blunt-tipped.

Young bark dark, reddish-brown.

Gray Birch

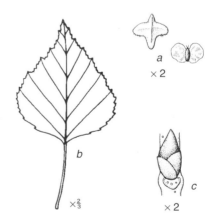

Mature bark dull, chalky-white, does not peel easily.

European White Birch**

Weeping birch, silver birch

Betula pendula Roth

[syn. *B. verrucosa* Ehrh.]

Bouleau verruqueux

Small trees, up to 15 m high; native to Europe and Asia. Often mistaken for gray birch. Frequently planted in Canada as a landscape tree. Many cultivars exist, some with deeply lobed leaves, some with strongly pendulous branches. Hardy as far north as Zones C2, NA2. [Contrasting features of gray birch in brackets.] **Leaves** 3–7 cm long, triangular or ovate, base wedge-shaped or straight, hairless; tip taper-pointed [long taper-pointed], strongly double- toothed; 20 (9–28) [40 (18–47)] teeth per side; 5–8 [6–9] veins per side; remaining on tree 3 or 4 weeks longer than native birches, a feature of many introduced species. **Twigs** hairless, with few [many] stalked glands, reddish-purple or reddish-brown [darker brown and grayish]. **Pollen catkins** in groups of 2–4 [1–2]. **Mature seed catkins** 2–4 cm long, slender-stalked, pendulous; fruit scales sparsely [densely] hairy on the inner surface, with a small central lobe, large lateral lobes, and an elongated base;

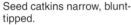

a. Fruit scale (above); winged nutlet (below).
b. Leaf . *c*. Lateral bud and leaf scar.

wings 3–4 times wider than the nutlet. **Bark** bright [dull] chalky white, peeling in long strands [peeling with difficulty into small thin rectangular plates]; with some [many] rough dark diamond-shaped marks just below the branches. Vertical black fissures [do not] develop at the base of mature trees. **Crown** broad [narrow].

Water Birch
Western birch, red birch, river birch,
black birch

Betula occidentalis Hook.
[syn. *B. fontinalis* Sarg.]

Bouleau fontinal

A small shrubby tree occurring from Hudson
Bay to interior British Columbia.

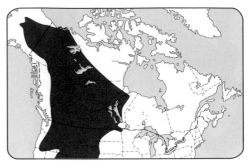

Leaves Small, broadly oval, or widest below
the middle, 2–5 cm long; tip wedge-shaped,
blunt or sharp; base rounded or wedge-
shaped with a smooth margin; stalk hairy;
upper surface shiny, deep yellowish-green,
lighter with tiny resin dots beneath; veins 4 or
5 per side, each ending in a large tooth, with
3 or 4 smaller intervening teeth; teeth sharp,
distinctly of 2 sizes.

Buds Pointed, greenish-brown, slightly
resinous.

Twigs Very slender, often hairy, reddish-
brown, with many reddish warty resin glands.

Flowers Pollen catkins about 2 cm long in
winter, 6 cm at pollination. Seed catkins
about 2 cm long.

Fruits Mature seed catkins 2.5–4 cm long,
slender, pendulous, on stalks 1–2 cm long.
Nutlet hairy, about as wide as the wings.
Scales hairy on the margin, with pointed
lobes; lateral lobes angled upwards, shorter
than the middle lobe.

Bark Thin, lustrous, dark to nearly black
when young, becoming purplish-brown with
age; not peeling readily. Lenticels long.

Size and Form Generally a tall shrub with
stems in clumps; may become a small tree
with a leaning crooked trunk, up to 12 m high
and 30 cm in diameter. Crown open, with upright branches drooping at the tips.

Habitat Grows on moist soils along
streams, mixed with poplars, willows, and
alders. Dense, pure thickets are common.

Notes A variable species; hybridizes with
paper birch.

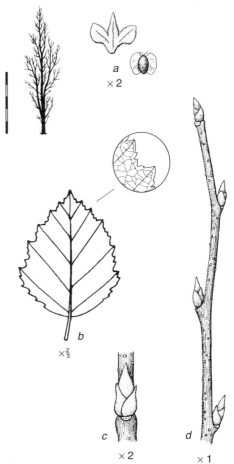

a. Fruit scale (above); winged nutlet (below).
b. Leaf with detail of distinct double-toothed
margin. *c.* Lateral bud and leaf scar. *d.* Winter
twig.

11

Seed catkins have hairy scales and nutlets.

Young bark nearly black.

Mature bark purplish-brown, does not peel freely.

Quick Recognition Small leaves with 4 or 5 veins per side; purplish-brown nonpeeling bark; reddish warty glands on branchlets.

Alaska Paper Birch

Alaska white birch, Alaska birch

Betula neoalaskana Sarg.
[syn. *B. resinifera* Britt.; *B. papyrifera* var. *neoalaskana* (Sarg.) Raup; *B. papyrifera* var. *humilis* (Regel) Fern. & Raup; *B. alaskana* Sarg.]

Bouleau d'Alaska

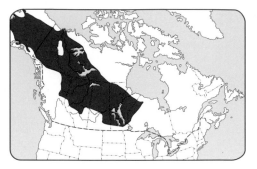

A northwestern species; similar to white birch, but smaller and occurring in bogs and on poorly drained sites.

Leaves Triangular to broadly oval, 4–7 cm long; tip long-tapered, sharp-pointed; base broadly wedge-shaped, with a smooth margin; teeth often gland-tipped; upper surface shiny dark green; light yellowish-green with tiny resin dots beneath, almost hairless; veins 4 or 5 per side, each ending in a tooth, with 3–5 smaller intervening teeth.

Buds Blunt, greenish-brown, resinous, slightly hairy.

Twigs Slender, bright reddish-brown, with an abundance of fine short hairs; covered with resin glands, sometimes large and crystalline and dense enough to conceal the twig surface.

Flowers Pollen catkins 2.5–4 cm long at pollination; seed catkins 1–2 cm.

Fruits Mature seed catkins about 2.5–3.5 cm long, blunt-tipped, drooping or slightly spreading from the dwarf shoots. Nutlets twice as long as broad; wings much wider than the nutlet. Scales with 2 rounded lateral lobes pointing away from the central lobe or curved slightly downward. Catkin disintegrates, shedding fruits and scales.

Bark Thin, smooth, dark reddish-brown when young, becoming creamy white or slightly pinkish with age; peels off in papery layers (but not as freely as white birch).

Size and Form Small trees, up to 15 m high and 20 cm in diameter. Crown narrow, oval. Branches slender, stiffly upright, some with drooping tips.

Habitat Characteristically occurs on bogs and poorly drained soils; in pure stands or mixed with other species, especially black spruce.

Notes Somatic chromosome number is 28.

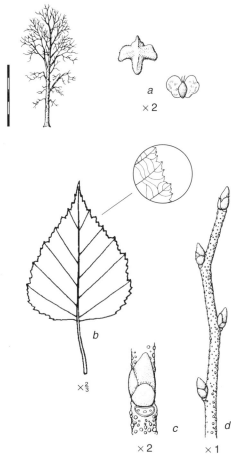

a. Fruit scale (above); winged nutlet (below).
b. Leaf with detail of resin dots on undersurface.
c. Lateral bud and leaf scar. *d.* Winter twig.

Quick Recognition A northwestern species; leaves long-pointed, twigs warty, bark white.

11

Seed catkins blunt-tipped.

Young bark dark reddish-brown.

Alaska Paper Birch

Mature bark does not peel as freely as that of white birch.

Kenai Birch
Kenai paper birch, red birch

Betula kenaica W.H. Evans
[syn. *B. papyrifera* var. *kenaica*
(W.H. Evans) L. Henry]

Bouleau Kenai

A small tree, rare in Canada, native only in Yukon, more common in Alaska. **Leaves** ovate to triangular, thick, short-pointed, coarsely toothed; upper surface dull dark green, hairy; lighter with resin dots beneath, hairy on the veins. **Twigs** reddish brown, hairy; becoming very dark and hairless. **Pollen catkins** 2.5 cm long, dark brown. **Seed catkins** 2.5 cm long on stiff stalks; nutlets with narrow wings; fruit scales with lobes of equal length. **Bark** papery, varies from light gray, to pink, to reddish-brown, to dark brown.

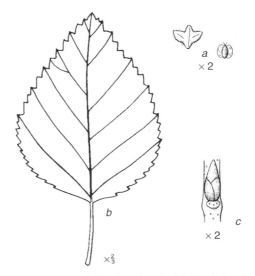

a. Fruit scale (left); winged nutlet (right). *b.* Leaf. *c.* Lateral bud and leaf scar.

Yellow Birch

Swamp birch

Betula alleghaniensis Britt.

[syn. *B. lutea* Michx. f.]

Bouleau jaune

The largest of the eastern birches.

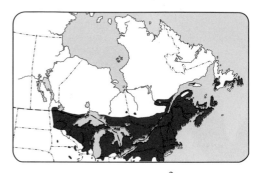

Leaves Oval, 8–11 cm long; tip slender, sharp-pointed; base rounded, indented; deep yellowish-green on the upper surface, lighter beneath; veins straight, parallel, 9 or more per side, each ending in a large tooth, with 2 or 3 smaller intervening teeth. Preformed leaves on dwarf shoots and the basal part of long shoots become hairless; neoformed leaves on the apical part of long shoots remain densely hairy.

Buds Sharp-pointed, often hairy, especially on young trees; 2 shades of brown on each scale.

Twigs Slender, usually slightly hairy, uniformly brown; wintergreen flavor when chewed.

Flowers Pollen catkins about 2 cm long and 2–3 mm wide by late autumn, about 8 cm long at pollination. Seed catkins 1.5–2 cm long, erect at pollination.

Fruits Mature seed catkins, about 3 cm long, cone-like, erect on spur shoots. Nutlets with narrow wings. Scales 5–7 mm long, hairy, with 3 narrow ascending lobes. Ripen in late September, shed during succeeding months; catkin axis with its scales often remains on the tree over winter.

Bark Thin, shiny reddish-brown when young, becoming dull yellow, with thin, papery shreds ending in tight curls, not peeling readily. Gradually darkens to a bronze color with age, separating into large ragged-edged plates on the lower part of the trunk.

Wood Heavy, often wavy-grained, hard, strong, golden brown to reddish-brown.

Size and Form Medium-sized trees, up to 25 m high, 60 cm in diameter, and 150 years old; occasionally older and larger. Trunk straight to sinuous, with little taper. Crown irregularly rounded; branches large wide-spreading, with drooping tips. Root system wide-spreading, often with some larger roots on or above the surface because the seed germinated on a rotting log or stump.

Habitat Occurs on rich moist soils; commonly mixed with beech, sugar maple, basswood, eastern hemlock, balsam fir, eastern

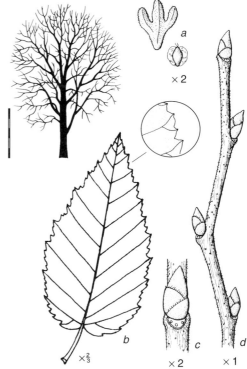

a. Fruit scale (above); winged nutlet (below).
b. Leaf with detail of double-toothed margin.
c. Lateral bud and leaf scar. *d.* Winter twig.

white pine, white spruce, and red spruce. Moderately shade-tolerant; the most shade-tolerant of the eastern birches.

Notes Wood can be stained; takes a high polish; used extensively for furniture, cabinetwork, flooring, doors, veneer, and plywood. An important source of hardwood lumber in eastern Canada.

    The provincial tree of Quebec.

> **Quick Recognition** Buds with 2 shades of brown. Twigs with a wintergreen flavor. Bark yellowish-gray with tight curls, broken into irregular plates on large trees.

11

Pollen catkins.
Seed catkin.
Mature seed catkin.
Young bark thin, not readily peeling.
Mature bark dark bronze.

Yellow Birch

Cherry Birch
Sweet birch, black birch

Betula lenta L.

Bouleau flexible

Closely resembles yellow birch; contrasting features of yellow birch are shown in brackets. A small tree seldom reaching 20 m high [frequently over 20 m]. **Buds** mostly hairless [mostly hairy], diverging from [pressed against] the twig. **Fruit scales** 6–12 mm long, hairless [5–7 mm, hairy]. **Bark** dark cherry red to almost black, becoming grayish with age [reddish-brown, becoming dull yellow]. **Wintergreen flavor** strong [moderate]. Canadian range limited to one confirmed site in southern Ontario, at Port Dalhousie near St. Catharines, on the south shore of Lake Ontario. Ranging through southern Maine and New England southward through Alabama, Georgia and Tennesee.

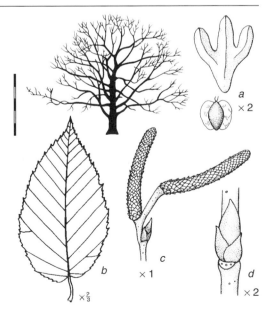

a. Fruit scale (above); winged nutlet (below). *b*. Leaf. *c*. Twig terminal with pollen catkins in winter condition. *d*. Lateral bud and leaf scar.

Alders

Genus *Alnus*
Betulaceae: Birch Family

Les aulnes

Worldwide, there are about 30 species of
alders, distributed through the north temperate regions and in Central and South
America. Of the 8 species recognized in
North America, 6 are native to Canada.

    The alders described can be divided into
2 subgenera: *Alnus* (red, speckled, mountain, hazel, and European black alder) and
Alnobetula (Sitka, green, and Siberian alder).
Where features of the subgenera differ in the
genus description below, those of *Alnobetula*
are shown in brackets.

    Subspecies are large populations within
a species complex that are separated geo-
graphically (sometimes with overlapping
edges) and have distinctive features. In this
treatment of alders, subspecies of *A. incana*
and *A. viridis* are recognized.

Leaves Oval, glandular beneath; teeth may
be of 2 sizes or on shallow undulations; veins
prominent, 5–12 per side, tending to be
straight and parallel, pinnately arranged
along the midvein, each ending in a large
tooth. Margin thickened. Green leaves retained late in the season. Stipules prominent,
glandular beneath.

Buds Large, purplish, irregularly shaped,
with 2–5 scales; stalked [stalkless]. Leaf
scars semicircular or triangular, with 3–5
vein scars; raised, on thickened portions of
the twig.

Twigs Slender to moderately stout, with a
3-angled pith. Neoformed side shoots may
develop in the leaf axils of vigorous shoots.

Flowers Borne in catkins; pollen catkins and
seed catkins on the same tree. Pollen catkins
narrowly cylindrical, in a few small clusters at
the shoot tip; formed in late summer, visible
from early autumn as stiff, finger-like structures; in spring, lengthening 2–4 times as
the pollen ripens and becoming pendulous.
Seed catkins in clusters of 3–5, on short
side shoots below the pollen catkins;
immature seed catkins visible during the
winter [enclosed in a bud]. Flowers open in
spring before the leaves appear [as the first
leaves reach full size]; wind-pollinated.

Fruits Borne in erect woody persistent
cone-like catkins. Each fruit a small, dry,

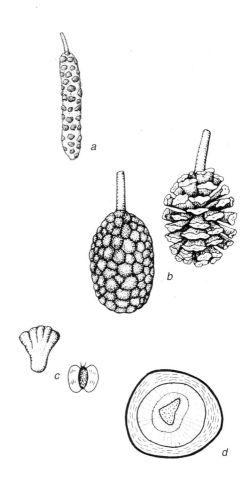

a. Pollen catkin in winter condition. *b.* Seed
catkin before (left) and after (right) nutlets are
released. *c.* Fruit scale (left); winged nutlet (right).
d. Twig cross section showing 3-angled pith.

flattened, 1-seeded, winged nutlet, located in
the axil of a 5-lobed woody scale on a stiff
central stalk and bearing 2 small thread-like
stigmas at the tip. Wings narrow [as wide as
or wider than the nutlet]; may encircle the
nutlet. Scales spread apart in autumn, releasing the fruit over the next few months;
catkin remains on the tree for a year or more.
Thus, in early autumn, alders may bear old
open catkins and new ones that have not
quite matured. Fruit production starts at
5–10 years old, with abundant crops about
every 4 years. Dispersed by wind and water.

Seeds Remain viable for several years if
stored air-dry in sealed containers. Germinate promptly on moist surfaces.

11

speckled alder

Alder fruit borne in woody, persistent conelike catkins.

speckled alder

Bark conspicuously patterned with horizontal lenticels.

Seedlings Cotyledons small, green, raised above the surface. First true leaves are small.

Vegetative Reproduction Young trees commonly reproduce by stump sprouts.

Bark Smooth, conspicuously patterned with enlarging lenticels; becoming rough with age. Inner bark tinged red.

Wood Light, soft, moderately strong, reddish-brown. Diffuse-porous; rays and annual rings indistinct.

Size and Form Mostly small trees or shrubs; red alder and European black alder are medium-sized trees. Leading shoot oblique.

Habitat Most species occur on wet sites and are intolerant of shade. Early growth is rapid.

Notes Alders are able to colonize exposed sites low in nutrients and improve soil fertility. Their rootlets bear clusters of nodules formed by bacteria that can convert atmospheric nitrogen into compounds that can be used by plants. Their fallen leaves also supply nitrogen to the soil.

Buds, twigs, bark, pollen catkins, and fruits of alders are eaten by birds and mammals.

Quick Recognition Fruits in small clusters of woody cone-like catkins; present year-round. Leaves toothed, oval, glandular below, stipulate, with straight pinnate veins. Buds large, purplish, irregularly shaped.

Red Alder

Oregon alder, western alder

Alnus rubra Bong.
[syn. *A. oregona* Nutt.]
Subgenus *Alnus*

Aulne rouge

Restricted in Canada to a narrow band within 150 km of the Pacific coast. The most common broadleaf tree in British Columbia.

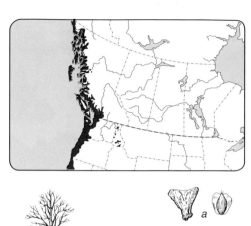

Leaves Oval to rhombic, 7–13 cm long, tapered from the middle to both ends, thick, not sticky; upper surface dull dark green; grayish and hairy on the veins beneath; double-toothed, the largest teeth almost lobe-size; margin clearly rolled under; veins 8–15 per side, impressed above, veinlets forming a ladder-like pattern. Considerable leaf-fall during the summer.

Buds Green, with 2 or 3 hairy scales that meet along their edges. Leaf scars with 3–5 vein scars.

Twigs Moderately stout, purplish-brown, triangular or roundish in cross section. Pith indistinctly triangular in cross section.

Flowers Pollen catkins in small clusters at the shoot tips; about 3 cm long in winter, 10–15 cm at pollination. Seed catkins about 5 mm long, visible in winter in small lateral clusters below the pollen catkins.

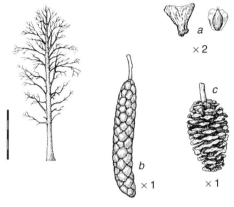

Fruits Mature seed catkins 15–25 mm long, short-stalked, abundant. Nutlet about 2 mm long, narrow-winged or encircled by wing. Ripen in early autumn; shed in late autumn and into the winter.

Bark Smooth, light gray, usually almost white with lichens; occasionally shallowly furrowed, separating into irregular flat plates with age; wounds turn red.

Wood Fine, even-textured, light brown, often turning reddish after cutting.

11

Size and Form Medium-sized trees, up to 25 m high, 60 cm in diameter, and 75 years old. In the forest, trunk straight, slightly tapered, distinct to the top of a narrow, rounded crown. In the open, crown extends to the ground, its spreading branches giving it a broadly conical shape. Root system shallow, wide-spreading.

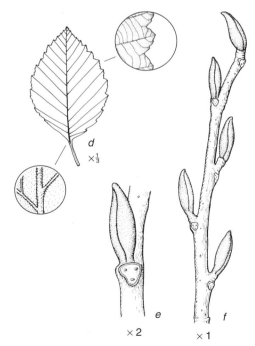

a. Fruit scale (left); nutlet (right). *b.* Pollen catkin in winter condition. *c.* Mature seed catkin after nutlets shed. *d.* Leaf with details of double-toothed margin (above) and hairy veins on lower surface (below). *e.* Lateral bud and leaf scar. *f.* Winter twig.

Young bark
smooth, light gray.

Pollen catkins (above) in clusters at shoot tips;
seed catkins (below) visible in winter.

Fruit ripens in early autumn.

Mature bark breaks into irregular flat plates.

Habitat Commonly grows in pure stands after colonizing cleared or burned land; also forms extensive stands on floodplains and along streams, mixed with black cottonwood, grand fir, Douglas-fir, western redcedar, western hemlock, western yew, bigleaf maple, vine maple. Like most pioneer species, it grows rapidly at first but is eventually over-topped by more tolerant longer-lived species. Intolerant of shade.

Notes Wood takes a stain well and is easily worked; used in wood carving, for making furniture, and for firewood. A form of this species with deeply lobed leaves has been found at Cowichan Lake, Vancouver Island, and locally in the states of Washington and Oregon.

Quick Recognition Leaf margin rolled under. The only native alder reaching 10 m high or more. White lichen blotches the bark.

Speckled Alder

Tag alder, gray alder, hoary alder

Alnus incana ssp. *rugosa* (Du Roi)
J. Clausen
[syn. *A. rugosa* (Du Roi) Spreng.]
Subgenus *Alnus*

Aulne rugueux

Small trees or tall shrubs; the common alder of stream banks and wet areas in central and eastern Canada.

Leaves Oval, 5–10 cm long, thick-textured, double-toothed, not sticky; upper surface wrinkled, dull green; hoary beneath; veins deeply impressed above, conspicuously projecting below, brownish-red; veinlets forming a ladder-like pattern. Considerable leaf-fall during the summer.

Buds Dark reddish-brown, blunt, stalked, with 2 or 3 scales that meet along their edges. Arranged in 3 ranks along the twig,

Twigs Moderately slender, reddish-brown.

Flowers Pollen catkins 1–2.5 cm long in winter, 5–8 cm at pollination. Immature seed catkins 2–5 mm long, visible in winter.

Fruits Mature seed catkins 13–16 mm long, ovoid to globular, short-stalked; in pendulous clusters. Nutlets very narrow-winged.

Bark Smooth, reddish-brown with conspicuous horizontal orange lenticels.

Size and Form Usually a coarse shrub with a clumped, crooked trunk, but occasionally a very small crooked tree up to 8 m high and 12 cm in diameter.

Habitat Occurs on moist sites such as stream banks and swamps; often mixed with black spruce or eastern white-cedar. Intolerant of shade.

Notes Hybridizes with mountain alder where their ranges overlap in Saskatchewan and westward.

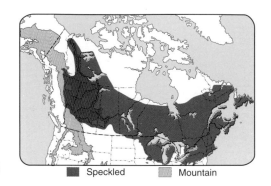

Speckled   Mountain

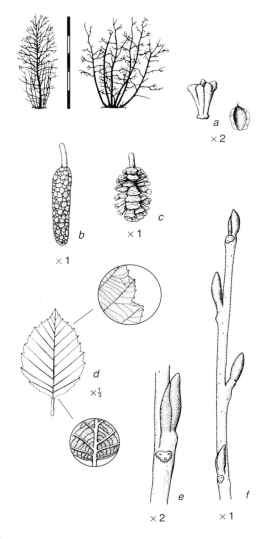

a. Fruit scale (left); nutlet (right). *b.* Pollen catkin in winter condition. *c.* Mature seed catkin after nutlets shed. *d.* Leaf with details of double-toothed margin (above) and prominent veins on lower surface (below). *e.* Lateral bud and leaf scar. *f.* Winter twig.

11

> **Quick Recognition** Leaves dull green above, hoary below, with undulating toothed margin; buds stalked, with scales that meet along the edges. Immature seed catkins visible in winter.

Seed catkins (left) and pollen catkins (right).

Seed catkins at time of pollination.

Mature seed catkins small, short-stalked.

Speckled Alder

Tree can be single-stemmed, more often multistemmed and shrubby.

Mountain Alder
Thinleaf alder

Alnus incana ssp. *tenuifolia*
(Nutt.) Breit.
[syn. *A. tenuifolia* Nutt.]
Subgenus *Alnus*

Aulne à feuilles minces

The common alder of stream banks in western North America from the Arctic southward, forming dense thickets, usually well inland. Resembles speckled alder and hybridizes with it where their ranges overlap. A larger tree up to 10 m high. **Leaves** thinner and smoother, with veins that are not impressed into the upper surface. **Buds** bright red, with fine hairs. **Twigs** slender, light green, hairy. Mature **seed catkins** smaller, 9–13 mm long; in groups of 2–5 on slender stalks. **Bark** gray or reddish-gray, not so conspicuously marked with lenticels; becoming scaly with age.

a
$\times\frac{1}{3}$

b
$\times 2$

a. Leaf. *b.* Lateral bud and leaf scar.

Sitka Alder

Alnus viridis ssp. *sinuata* (Regel)
Á. Löve & D. Löve
[syn. *Alnus sinuata* (Regel) Rydb.]
Subgenus *Alnobetula*

Aulne de Sitka

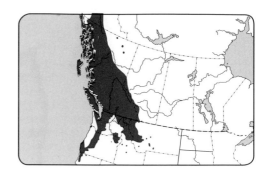

Occurs from the Pacific coast to the western
slopes of the Rocky Mountains. Resembles
birch.

Leaves Broadly oval, 7–15 cm long, thin;
shiny, deep yellowish-green; resinous and
fragrant when young; mostly hairless; base
broad, often asymmetrical; teeth sharp,
gland-tipped, of 2 sizes, none near the stalk;
6–8 veins per side , each ending in a large
tooth; veinlets irregularly branched.

Buds 12–14 mm long, curved, sharp-
pointed, purplish-brown, stalkless, with
3–5 overlapping scales. Usually arranged in
2 ranks along the twig. Leaf scars raised,
crowded along the twig.

Twigs Slender, light reddish-brown or
yellowish-brown with pale lenticels; sticky
when young. Dwarf shoots may develop.

Flowers Pollen catkins stalkless, in pairs;
about 1 cm long in winter, 10–14 cm at pollination. Seed catkins enclosed in a bud in
winter; opening with the leaves.

Fruits Mature seed catkins about
12–20 mm long with equally long stalks; several in an erect cluster. Nutlet about the same
width as wings.

Bark Reddish- to grayish-brown, smooth,
with long horizontal lenticels.

Size and Form Small trees or tall shrubs, up
to 15 m high and 20 cm in diameter. Trunk
slender; crown open, with short horizontal
branches. Root system shallow.

Habitat Found mainly on cool, moist sites in
the mountains; often occur where landslides
have created openings.

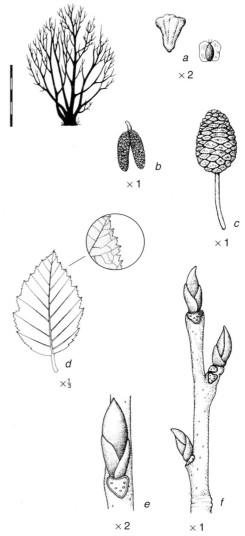

Quick Recognition Leaves sticky, fragrant
with sharp teeth of 2 sizes; buds stalkless, with
overlapping scales; seed catkins not visible
in winter; seed about the same width as the
wings.

a. Fruit scale (left); nutlet (right). *b.* Pollen catkin in
winter condition. *c.* Mature seed catkin. *d.* Leaf with
detail of double-toothed margin. *e.* Lateral bud and
leaf scar. *f.* Winter twig.

11

Pollen catkins at time of pollination.

Seed catkins just after emerging from buds.

Mature seed catkins.

Tree often multistemmed with low horizontal branches.

Young bark has very prominent lenticels.

Green Alder

Alnus viridis ssp. *crispa* (Ait.)
Turrill
[syn. *Alnus crispa* (Ait.) Pursh]
Subgenus *Alnobetula*

Aulne crispé

Large shrubs, ranging from Labrador to Alaska and southward. Intergrades with Sitka alder in Alberta. A northern species, often occurring on dry upland sites with jack pine; also found on wet sites; a coastal species in the Maritime provinces. More shade-tolerant than other alders. **Leaves** bright shiny green on the upper surface; yellowish-green beneath; margin with many small sharp teeth about equal in size, undulations may be present. **Buds** stalkless, with curved pointed tips; scales reddish or greenish overlapping. Immature **seed catkins** remain within the bud in the winter; mature seed catkins 10–15 mm long, long-stalked; nutlets broad-winged.

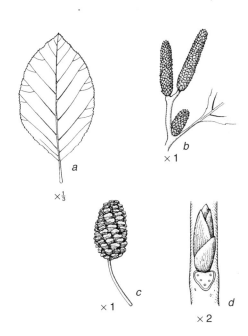

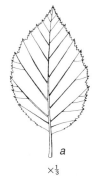

a. Leaf. *b.* Pollen catkins. *c.* Mature seed catkin.
d. Lateral bud and leaf scar.

Siberian Alder

Alnus viridis ssp. *fruticosa*
(Rupr.) Nym.
Subgenus *Alnobetula*

Aulne de Sibérie

A subartic Asian species common in coastal Alaska and northern British Columbia at elevations below 500 m; also southward as far as California. Distinguished from green alder by its larger, more coarsely toothed leaves, and from Sitka alder by its thicker, single-toothed leaves.

a. Leaf. *b.* Lateral bud and leaf scar.

11

Hazel Alder

Alnus serrulata (Ait.) Willd.
[syn. *Alnus incana* var. *serrulata* (Ait.)
Boivin]
Subgenus *Alnus*

Aulne blanc

Shrubs or very small trees of the coastal region of
eastern North America; native to but rare in New
Brunswick, Nova Scotia, and Quebec along the St.
Lawrence River. **Leaves** finely toothed and wavy.
Buds 7–10 mm long, reddish, rounded tip, with 2
or 3 scales. **Twigs** slender, smooth, brown. **Pollen
catkins** 2.5–3.5 cm long. Mature **seed catkins**
7–12 mm long, in erect clusters of 3–6.

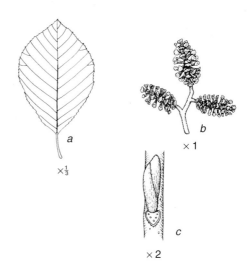

a. Leaf. *b.* Mature seed catkins. *c.* Lateral bud
and leaf scar.

European Black Alder**

Alnus glutinosa (L.) Gaertn.
Subgenus *Alnus*

Aulne glutineux

Medium-sized trees, up to 25 m high and 30 cm in
diameter. Frequently planted in eastern North
America; naturalized locally on floodplains. Hardy
as far north as Zones C4, NA3. **Leaves** almost
round, tip blunt, base broadly wedge-shaped, dark
green, double-toothed, resinous when young, only
5 or 6 veins per side. Mature **seed catkins** in
pendulous clusters.

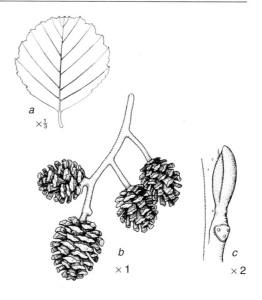

a. Leaf. *b.* Mature seed catkins. *c.* Lateral bud
and leaf scar.

Blue-Beech

American hornbeam, musclewood, ironwood

Carpinus caroliniana Walt.
Betulaceae: Birch Family

Charme de Caroline

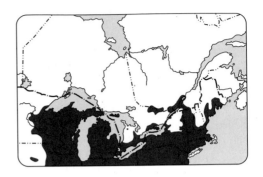

The genus *Carpinus* consists of about 25 species of trees and shrubs found in the North Temperate Zone; only blue-beech is native to North America.

Leaves Deciduous, alternate, simple; arranged in 2 rows on the twig, 5–10 cm long, becoming successively larger along the shoot; elliptic, tapering to a sharp tip, rounded at the base; texture firm; upper surface bluish-green, yellowish-green beneath, red in autumn; veins straight, parallel, each ending in a sharp tooth with 1–2 smaller intervening teeth.

Buds Ovoid, blunt, slightly hairy, pressed against the twig; scales several, in 4 rows, reddish-brown with a whitish margin. No true terminal bud. Leaf scars with 3 vein scars.

Twigs Slender, as narrow as the leaf buds, gray to reddish-brown.

Flowers Pollen flowers and seed flowers in separate clusters on the same tree. Pollen flowers in drooping catkins, 25–40 mm long, laterally on previous year's twigs. Seed flowers in elongated clusters (racemes) at the tips of new leafy shoots. Pollination takes place as the leaves begin to appear.

Fruits Small, ovoid, ribbed nut, 6–9 mm long, in the axil of a 3-lobed, leaf-like bract about 25 mm long, with 5–7 principal veins; in elongated pendulous clusters 10–15 cm long.

Bark Smooth, thin, unbroken, slate gray, with muscle-like longitudinal ridges giving a wavy outline in cross section.

Wood Very heavy, hard, strong. Diffuse-porous; pores visible with a hand lens, larger rays visible to the naked eye.

Size and Form Very small trees, up to 8 m high and 25 cm in diameter. Trunk usually short, crooked, fluted. Crown low, wide-spreading, with a few irregular, slightly zigzag branches bearing fine twigs in flat sprays.

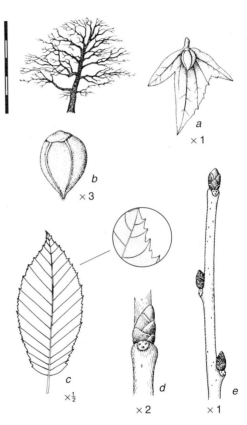

a. Fruit bract and nut. b. Nut. c. Leaf with detail of distinctly double-toothed margin. d. Lateral bud and leaf scar. e. Winter twig.

Habitat Generally occurs on deep, rich, moist soils on lower slopes in valleys and along the borders of streams and swamps. Very shade-tolerant; commonly found in the understory of broadleaf forests.

Notes "Ironwood" is a name applied locally to this species; "musclewood" refers to its fluted trunk. "Horn" means tough; "beam" means tree (old English). Its strong wood is used for tool handles.

11

Pollen flowers in drooping catkins.

Seed flowers in long clusters at tips of new shoots.

Fruit borne in axil of 3-lobed leafy bract, in long hanging clusters.

Mature bark remains smooth and unbroken.

Bark has wavy, muscle-like ridges.

Quick Recognition Bark smooth, gray; trunk fluted; buds pressed against the twig; each fruit in the axil of a leafy lobed bract.

Ironwood
Hop-hornbeam

Ostrya virginiana (Mill.) K. Koch
Betulaceae: Birch Family

Ostryer de Virginie

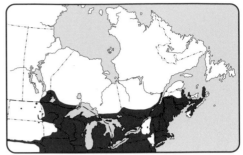

The genus *Ostyra* consists of 8 species; 3 are native to North America; only ironwood is native to Canada.

Leaves Deciduous, alternate, simple; arranged in 2 rows along the twigs, 7–12 cm long, successively larger toward the tip of the shoot; elliptic, usually widest near the middle, tapering from the middle to a long sharp tip, base narrowly rounded or indented; upper surface dark yellowish-green, similar but hairy beneath; dull yellow in autumn; texture soft; veins straight, parallel, each ending in a sharp tooth, with somewhat smaller intervening teeth; a branch vein may also end in a tooth.

Buds Ovoid, pointed, 3–4 mm long, spreading from the twig, greenish-brown, slightly hairy; scales several, striated, not in regular rows. No true terminal bud. Leaf scars with 3 vein scars.

Twigs Slender, not as wide as the buds, dark reddish-brown.

Flowers Pollen flowers and seed flowers in separate clusters on the same tree. Pollen flowers in drooping catkins, 15–50 mm long, in groups of 2 or 3 at the twig tips; visible, but short, stiff, erect in winter. Seed flowers in elongated clusters (racemes) at the tips of new leafy shoots. Pollination takes place as the leaves begin to appear.

Fruits Small flattened nut, 5–8 mm long, enclosed in an inflated sac about 15 mm long, covered with stiff hairs; in clusters at the tip of a short drooping axis. Sacs fall during winter, leaving the axis attached to the end of the twig.

11

Bark Grayish-brown; broken into short, narrow, longitudinal strips, loose at both ends; easily rubbed off.

Wood Heavy, very hard, tough; the densest Canadian wood; close-grained, light brown. Diffuse-porous; pores barely visible to the naked eye; rays and annual rings indistinct.

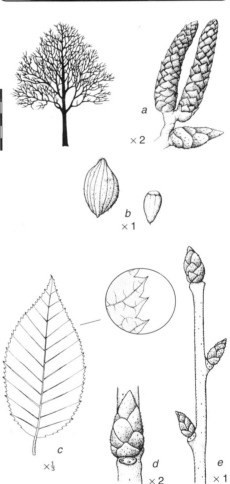

a. Pollen catkins and end bud in winter condition. *b.* Fruit sac (left); nut (right). *c.* Leaf with detail of indistinctly double-toothed margin. *d.* Lateral bud and leaf scar. *e.* Winter twig.

Size and Form Small trees, up to 12 m high and 25 cm in diameter; occasionally larger. Trunk upright, distinct almost to the top of the tree. Crown wide-spreading, with long slender branches.

Pollen catkins in groups of 2–3.

Seed flowers in elongated cluster at shoot tip.

Fruit cluster on short stem.

Young bark (inset) soon rough. Mature bark has narrow vertical strips that are loose at both ends.

Habitat Occurs on well-drained slopes and ridges. Very shade-tolerant; commonly found in the understory of broadleaf forests.

Notes The strong wood is used for tool handles. The name "ironwood" is also applied to blue-beech in Canada and to other species in various parts of the world. The cluster of fruit sacs is reminiscent of hops; hence the name, "hop-hornbeam".

> **Quick Recognition** Bark shaggy, in strips; buds spreading from the twig; fruit in inflated sacs.

Hazels

Hazelnuts, filberts

Genus *Corylus*

Betulaceae: Birch Family

Les noisetiers

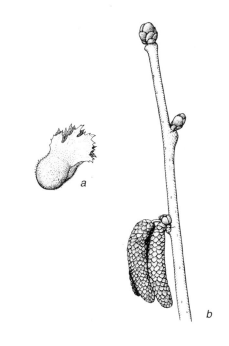

The genus *Corylus* comprises about
15 species of shrubs and trees native to the
North Temperate Zone. Several are valued
for their edible nuts. Both of the
North American species are shrubs, described here for comparison with the introduced tree species.

    Leaves deciduous, alternate, simple;
generally oval, pointed, somewhat asymmetrical, straight-veined, irregularly double-
toothed; folded lengthwise in the bud; stipules scale-like, soon deciduous. **Buds** ovoid,
stout, asymmetrical, with several scales.
Pollen flowers in catkins, often evident
during the winter, firm and stout, in clusters
of 1–3, on previous year's twigs; in spring,
elongated, pendulous. **Seed flowers** at the
tips of a leafy shoots; visible in early spring
as thread-like stigmas emerging from the
bud. **Fruit** an edible nut, surrounded or
entirely hidden by a green involucre, fringed
or bristly bracts.

a. Involucre. *b*. Winter twig and pollen catkins.

Beaked Hazel

Beaked hazelnut

Corylus cornuta Marsh.

[syn. *C. rostrata* Ait.]

Noisetier à long bec

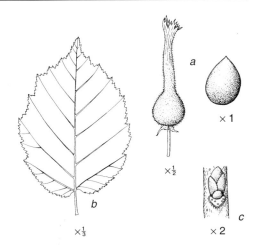

A native shrub ranging from British Columbia to
Newfoundland. **Leaf stalks** and **twigs** usually
hairless. **Involucre** somewhat fleshy, covered with
stiff hairs; completely hiding the nut and narrowing
into a long tube that extends beyond the nut. **Nuts**
about 12 mm long, solitary or in small clusters.

a. Involucre (left); nut (right). *b*. Leaf. *c*. Lateral
bud and leaf scar.

11

American Hazel
American hazelnut

Corylus americana Marsh.

Noisetier d'Amérique

A native shrub ranging from Saskatchewan to Quebec. **Leaf stalks** and **twigs** covered with gland-tipped hairs. **Involucre** composed of 2 leafy bracts, with large irregular teeth. **Nuts** about 12 mm long, in clusters, visible within the involucre.

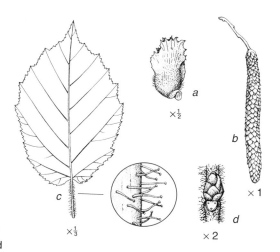

a. Involucre. *b.* Pollen catkin. *c.* Leaf with detail of gland-tipped hairs on leaf stalk. *d.* Lateral bud and leaf scar.

Turkish Hazel*
Turkish hazelnut

Corylus colurna L.

Noisetier de Byzance

Medium-sized trees native to southeastern Europe and western Asia. Hardy as far north as Zones C5, NA4. **Involucre** divided into narrow lobes that are bent back.

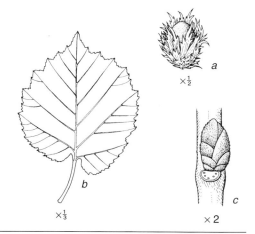

a. Involucre. *b.* Leaf. *c.* Lateral bud and leaf scar.

European Filbert*
European hazelnut

Corylus avellana L.

Noisetier commun

Very small trees or large shrubs, native to Europe and western Asia. Hardy as far north as Zones C5, NA4. **Involucre** irregularly toothed or lobed, about the same length as the nut.

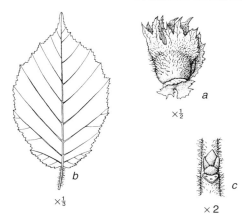

a. Involucre. *b.* Leaf. *c.* Lateral bud and leaf scar.

Willows

Genus *Salix*
Salicaceae: Willow Family

Les saules

Worldwide, there are about 400 species of willow; all are shrubs or trees, mostly confined to the Northern Hemisphere. The greatest diversity of species is in China. About 100 species occur in North America; 62 are native to Canada, with 15–20 of these attaining tree size.

Natural hybrids and polyploids occur. The somatic chromosome number is 38, but up to 6 times that many have been found in certain species.

Some species closely resemble each other and positive identification may require observations during both flowering and fruiting.

Leaves Deciduous, alternate, simple; 2–10 times as long as wide, tapering to the tip, margin finely toothed or smooth, short-stalked. Stipules are characteristic of many species. Leaves on vigorous shoots may be larger, more prominently toothed, and with larger more persistent stipules. Among the first trees to leaf out in spring and the last to shed their leaves in autumn.

Buds Pointed, pressed against the twig, closely spaced. Covered with a single scale; may form a closed cap or its edges may be free and overlapping. No true terminal bud; end bud originates as a lateral bud. Flower buds larger than leaf buds. Leaf scars V-shaped, with 3 vein scars; stipule scars on either side, usually inconspicuous.

Twigs Slender, tinted with green, brown, yellow, orange, purple, or red. New side shoots develop from buds on previous year's twigs; also in leaf axils on vigorous new shoots. Spring flush of shoot growth is pre-formed in the bud; subsequent shoot growth is neoformed. Shoot growth ends when the tip aborts and falls off; resulting scar inconspicuous. Twigs and branchlets frequently break off.

11

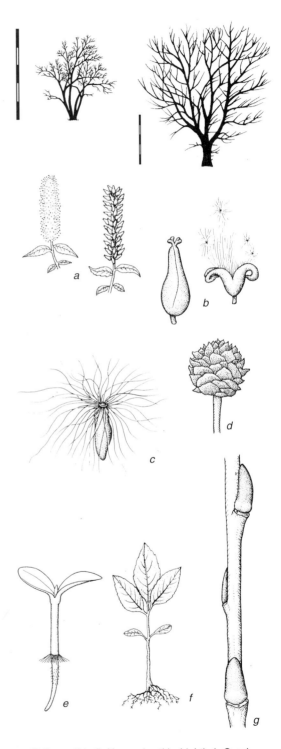

a. Pollen catkin (left); seed catkin (right). *b.* Seed capsule closed (left), open (right). *c.* Seed with silky hairs. *d.* Cone-like gall. *e.* Newly germinated seedling with fringe of hairs at base of hypocotyl. *f.* 1-year-old seedling. *g.* Lateral buds and leaf scars.

Hooker willow
Pollen catkin at time of pollination.

Scouler willow
Seed catkins at time of pollination.

pussy willow
Seed catkin before fruit capsules open.

pussy willow
Seed catkin at time of seed release.

Flowers Borne in many-flowered catkins; pollen catkins and seed catkins on separate trees. Pollen catkins erect or pendulous; seed catkins usually erect. Catkins pre-formed in buds; found at the end of short leafy side shoots, or on previous year's twigs; appear in spring before or with the developing leaves. Flowers small, greenish or yellowish, with 1 or more nectar-bearing glands at the base, no petals or sepals; set in the axil of a hairy, smooth-margined bract. Pollen flowers with 2 (rarely 1, 3, or 5) stamens. Insect- or wind-pollinated.

Fruits Many-fruited catkins bearing small, ovoid, pointed capsules that split into halves to release the seeds. Fruit crops occur nearly every year.

Seeds Small, 1–2 mm long, often greenish, covered by long white silky hairs. Light and moisture are required for germination; under natural conditions, germination or death occurs within a day or 2 after seedfall. Remain viable for years in ultra-cold storage. Dispersed by wind and water.

Seedlings During germination, a fringe of hairs grows out around the base of the hypocotyl, rendering the young seedling erect and positioning the new root to grow down into the soil. Cotyledons small, green, leaf-like, and raised above the surface. First true leaves small.

Vegetative Reproduction By stump sprouts; less frequently by root sprouts and rooting of detached branches. Many species are notable for the ease with which their stem cuttings take root; for some, even detached branches and pole-sized stems will root if they are put in contact with moist soil.

Bark Smooth or slightly rough on small stems, lenticels usually evident; scaly or furrowed on large stems. Inner bark has a bitter taste in most species.

Wood Light, texture uniform, straight-grained, soft, not durable, but tough and shock-resistant, odorless, pale. Diffuse-porous or early pores slightly larger; pores barely visible to the naked eye. Rays very narrow. Annual rings often not well defined.

Size and Form Vary from prostrate or upright shrubs to trees. Shrub willows usually have many stems. Tree willows may have a single trunk or several trunks in a group; crowns spreading; leading shoot oblique. Roots shallow, with many rootlets.

Habitat Occur on sites that vary widely with respect to available moisture and nutrients, most often where the soil is moist, such as wet meadows, swamps, and along rivers; colonize disturbed sites, such as burned-over and cut-over areas, and old fields. Shrubby willows frequently grow in large clumps with a single species covering a considerable area; tree-size species may occur mixed with other tree species. Most willows withstand seasonal flooding. Intolerant of shade.

Notes Willows have been cultivated for centuries, at least as far back as ancient Rome and Greece. Individual willow trees with special properties have been propagated by cuttings and so preserved for centuries up to the present time. The genus is ideal for tree breeding. There are over 400 species and great variation within species. Almost all species are intercrossable. Seedlings flower at an early age, and flowering continues almost every year. Resulting seedlings can be rapidly multiplied by stem cuttings.

Willows are useful for reclamation work, erosion control, and the stabilization of river banks because they thrive under moist conditions and quickly produce an extensive root system. Several clones and species are highly valued as landscape trees. Many of the willows that occur naturally on wet sites also thrive on well-drained upland sites under cultivation.

The wood of willows is used in the manufacture of paper pulp, boxes, crates, and small wooden products.

Many mammals and birds feed on the twigs and buds; moose depend heavily on willow twigs and branchlets for winter food. The flowers are a source of nectar and pollen for bees in early spring.

It has long been known that drinking an extract of willow bark would relieve pain; the active ingredient is salicin, a substance similar to acetylsalicylic acid, or ASA.

Galls shaped like pine cones are frequently found on the tips of willow twigs. These deformed shoots are caused by the willow pinecone gall midge (*Rhabdophaga stobiloides*).

Quick Recognition Buds with a single bud scale; crowded along the twig and pressed against it. In most species, leaves long, narrow, pointed, short-stalked; twigs long, slender. Fruit a capsule, on elongated catkins. Seeds small, green, covered by long white silky hairs.

Willow form with pendulous branches. golden weeping willow

Willow form with upright branches. crack willow

Peachleaf Willow

Salix amygdaloides Andersson

Saule à feuilles de pêcher

The tallest native willow in the Prairie provinces; ranges from British Columbia to Quebec, and southward to the United States and Mexico.

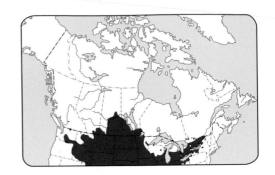

Leaves Lance-shaped, 5–14 cm long; thin; tip long-pointed, base rounded and uneven; finely toothed; upper surface green, whitish beneath; hairless; midvein prominent. Young leaves reddish, sparsely hairy. Stipules absent except on vigorous shoots.

Buds Ovoid, 2–4 mm, sharp-pointed, shiny yellowish-brown.

Twigs Smooth, slender, flexible, tough, yellowish-brown, becoming gray; lenticels pale.

Flowers Catkins loose, on leafy shoots; appear with the leaves. Pollen catkins 3–6 cm long, flowers in whorls; seed catkins 4–9 cm long, loosely flowered.

Fruits Capsules reddish or yellowish, 4–7 mm long, on stalks 1–2 mm long. Bracts shed before capsules ripen.

Bark Reddish-brown, becoming grayish-brown with age; irregularly furrowed, with broad, flat, shaggy ridges.

Size and Form Small trees up to 20 m high, often with several leaning trunks in a clump; single trunks sometimes 40 cm in diameter. Branches often curving upward, then arching over near the tip.

Habitat Grows on moist soil along rivers and lakes and in wooded swamps.

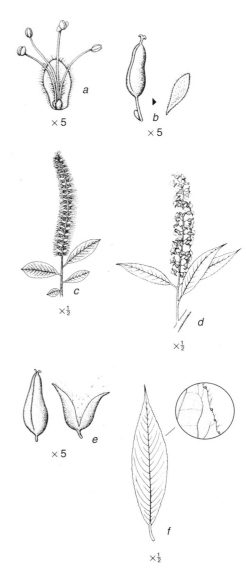

11

> **Quick Recognition** Leaf undersurface whitish; no stipules.

a. Pollen flower. *b.* Seed flower (left); deciduous bract (right). *c.* Pollen catkin. *d.* Seed catkin. *e.* Fruit capsule closed (left), open (right). *f.* Leaf with detail of finely toothed margin.

Seed catkin; fruit capsules on short stalks.

Young bark smooth, reddish-brown.

Mature bark grayish-brown, with broad shaggy ridges.

Peachleaf Willow

Sandbar Willow

Salix exigua Nutt.

[syn. *S. interior* Rowlee]

Saule de l'Intérieur

Shrubs or shrubby trees, up to 6 m high, ranging from New Brunswick to British Columbia. A pioneer species on river sandbars, where it spreads by means of root sprouts. **Leaves** very narrow, 5–15 cm long; teeth sparse, small, gland-tipped; both surfaces light yellowish-green; short-stalked. Stipules small, deciduous. **Twigs** upright, brown, smooth. **Catkins** on leafy shoots, often branched; appear after the leaves reach full size; bracts deciduous. **Fruit capsules** long, slender.

a. Pollen flower. *b.* Seed flower with deciduous styles (above) and deciduous hairy bract (left). *c.* Pollen catkin. *d.* Seed catkin. *e.* Leaf with detail of gland-tipped teeth.

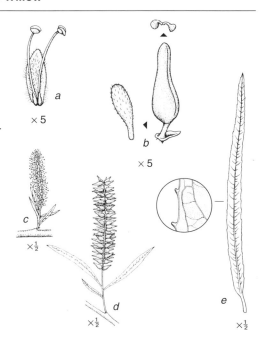

Bebb Willow

Diamond willow, beaked willow

Salix bebbiana Sarg.

Saule de Bebb

The most common and widespread of the tree-size willows native to Canada, ranging from the Atlantic coast to British Columbia.

Leaves Narrow to elliptic, 3–7 cm long, tapering to both ends; margin smooth or wavy, somewhat rolled under, may be toothed in the basal part especially on leaves toward the tip of the shoot; upper surface dull green, wrinkled; whitish, hairy beneath, with prominent meshing veins. Stalk 3–10 mm, hairy. Stipules small, 1–5 mm, soon falling away; larger and persistent on vigorous shoots. Young leaves hairy.

Buds Blunt, shiny brown.

Twigs Reddish-purple to orange-brown, becoming darker with age; hairy when young, becoming hairless.

Flowers Catkins on short leafy shoots; appear with the leaves. Pollen catkins 1–2 cm long; seed catkins about twice as long. Bracts narrow, pointed, 1–3 mm long, sparsely hairy, yellowish or straw-colored, tips may be reddish.

Fruits Capsules 6–8 mm long, stalks about as long, long-beaked, sparsely hairy.

Bark Reddish- or grayish-brown, furrowed on large stems. Diamond-shaped patches, caused by a fungal infection, often appear on a main stem.

Size and Form Shrubs or shrubby trees, up to 8 m high, 15 cm in diameter, and 20 years old. Usually multistemmed. Branches upright, then spreading.

Habitat Occurs on moist sites; an understory species on limestone flats, jack pine sand plains, and other forest types. Often one of the first species to appear after forest fires.

Notes Stems marked by "diamonds" have an attractive appearance when peeled; used for making rustic ornaments.

11

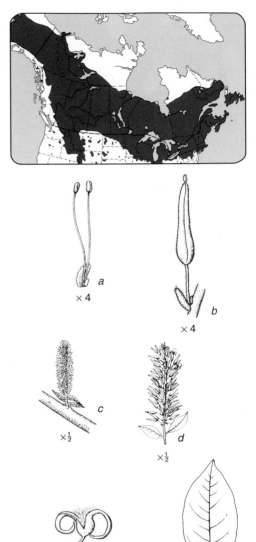

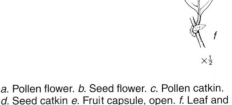

a. Pollen flower. *b.* Seed flower. *c.* Pollen catkin. *d.* Seed catkin *e.* Fruit capsule, open. *f.* Leaf and stipules.

Quick Recognition Upper leaf surface dull green, wrinkled, whitish hairs beneath. Capsules long-beaked, on long stalks; bracts yellowish.

Seed catkin; fruit capsules and stalks of equal length.

Young bark reddish- or grayish-brown.

Usually a mutlistemmed tree, mature bark furrowed.

Bebb Willow

Littletree Willow

Salix arbusculoides Andersson

Saule arbustif

Shrubs or shrubby trees, up to 9 m high; forming thickets, mostly along river banks and on recently burned-over areas, from northwestern Ontario to northern British Columbia, with a disjunct occurrence at Lake Mistassini, Quebec. **Leaves** elliptic, narrow, 6–7 cm long, pointed at both ends; teeth small, gland-tipped; upper surface shiny dark green, covered with silky white hairs beneath; veins straightish, parallel, closely spaced. Stipules small, deciduous. **Twigs** shiny, gray to reddish-brown. **Catkins** on very short shoots. **Fruit capsules** covered with white hairs. **Bark** with "diamonds" similar to Bebb and feltleaf willow.

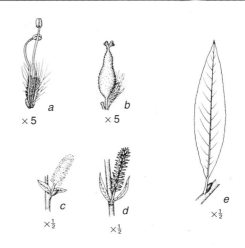

a. Pollen flower. b. Seed flower. c. Pollen catkin. d. Seed catkin. e. Leaf.

Pussy Willow

Salix discolor Muhl.

Saule discolore

Probably the best known native willow; twigs bearing immature catkins often used in floral decorations.

Leaves Oblong to narrowly elliptic, sometimes widest above the middle, 3–10 cm long; tip pointed or blunt, base tapered; toothed especially above the middle; upper surface bright green, with raised veins, whitish beneath; hairless; firm. Stipules small, much larger on vigorous shoots. Young leaves reddish, hairy.

Buds Flattened, pointed, reddish-purple, 7 mm long; flower buds up to 10 mm long.

Twigs Stout, shiny, dark reddish-brown; hairy when young, becoming hairless; lenticels pale.

Flowers Catkins on very short shoots with bract-like leaves; densely covered with silky hairs when immature; fully developed before the leaves appear. Pollen catkins 2–4 cm long; seed catkins 2–6 cm long. Bracts dark brown with white hairs.

Fruits Mature seed catkins up to 9 cm long. Capsules cylindrical, 7–12 mm long, long-beaked, hairy.

Bark Grayish-brown, tinged with red.

Size and Form Shrubs or shrubby trees, in clumps up to 6 m high, with upright branches.

Habitat Occurs in wet meadows, alder swamps, along shorelines.

Notes The catkins herald the coming of spring. Catkins can be induced to develop in late winter if twigs are brought indoors and set in water.

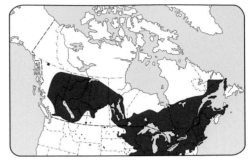

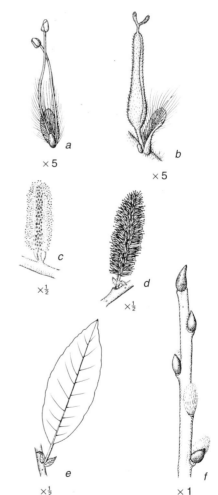

a. Pollen flower. *b.* Seed flower. *c.* Pollen catkin. *d.* Seed catkin. *e.* Leaf and stipules. *f.* Twig just after bud break.

11

Quick Recognition Strong contrast between green upper surface and whitish lower surface of leaves. Catkins densely hairy; appear very early in spring.

Seed catkin; fruit capsules long-beaked, hairy.

Young bark grayish-brown, tinged red.

Pussy Willow

Typical form upright, multistemmed.

Balsam Willow

Salix pyrifolia Andersson
[syn. *S. balsamifera* Barratt]

Saule baumier

Shrubs or very small trees up to 8 m high, found in spruce bogs and alder swamps, from Newfoundland and Labrador to British Columbia, and south into Michigan and Vermont. Odor of balsam from bruised leaves and twigs. **Leaves** oval, thick, 3–8 cm long; tip tapered to a point, base rounded; teeth short, gland-tipped; upper surface dark green, paler beneath, with a network of prominent veins; leaf stalks with glands near the leaf blade. Stipules small or missing. Young leaves thin, translucent, often tinged with red. **Twigs** yellowish or greenish in spring, becoming reddish-brown, shiny; larger stems gray or brown to purplish-red. **Catkins** on short, usually leafy shoots; appear with the leaves. **Fruit capsules** smooth.

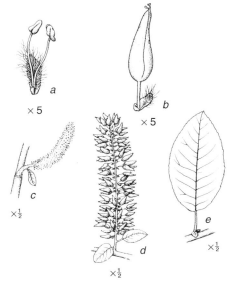

a. Pollen flower. *b.* Seed flower. *c.* Pollen catkin. *d.* Seed catkin. *e.* Leaf and stipules.

Shining Willow

Salix lucida Muhl. ssp. *lucida*

Saule brillant

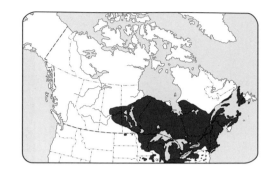

The tallest native willow in Newfoundland and Labrador; ranges west as far as Saskatchewan, and from Hudson Bay south to Indiana. Used for landscape purposes (Zones C1, NA2); resembles the introduced laurel willow.

Leaves Lance-shaped, 4–15 cm long, tip very long-pointed, base rounded; teeth gland-tipped; upper surface shiny dark green; paler, shiny beneath. Stalk glandular at junction with blade. Stipules 1–6 mm long with glandular margins. Young leaves red-tinged, with rusty hairs.

Buds Light brown.

Twigs Shiny, yellowish- to reddish-brown; with rusty hairs when young, becoming hairless.

Flowers Catkins on short leafy shoots; appear with the leaves. Pollen catkins 2–4 cm long; seed catkins slightly longer.

Fruits Capsules narrowly ovoid, 4–7 mm long, light brown, hairless. Bracts shed before capsules ripen.

Bark Brown.

Size and Form Shrubs or very small trees, up to 10 m high. Crown broad, rounded; branches upright.

Habitat Occurs on a variety of wet sites.

Quick Recognition Leaves shiny on both sides. Glands on leaf margins, leaf stalks, and stipule margins.

11

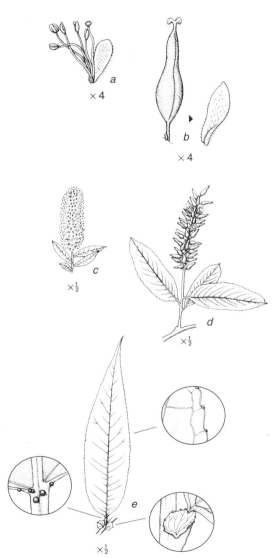

a. Pollen flower. *b.* Seed flower (left); deciduous bract (right). *c.* Pollen catkin. *d.* Seed catkin. *e.* Leaf with details of gland-tipped teeth on margin (above), stipule showng glandular margin (right), and glands at leaf stalk–blade junction (left).

Seed catkins on short leafy shoots.

Young bark shiny, becomes brown.

Mature bark irregularly furrowed.

Shining Willow

Laurel Willow*

Bayleaf willow

Salix pentandra L.

Saule laurier

Small trees, up to 20 m high, native to Europe and western Asia; thrives on acid soils in northern regions. Hardy as far north as Zones C2, NA3. Similar to the native shining willow whose leaves have a narrower tip. **Leaves** fragrant, ovate, 4–10 cm long, tip long-pointed, base rounded to wedge-shaped; margin with gland-tipped teeth; midvein yellow; upper surface very shiny green, paler beneath; leaf stalk glandular. Stipules small, glandular. **Buds** yellow; **twigs** shiny reddish-brown. **Catkins** golden yellow in spring, on short leafy shoots. Leaves can be used for flavoring foods.

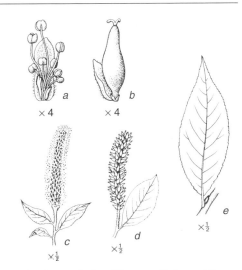

a. Pollen flower. *b.* Seed flower. *c.* Pollen catkin. *d.* Seed catkin. *e.* Leaf.

Black Willow

Salix nigra Marsh.

Saule noir

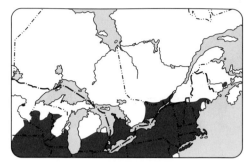

The largest native willow in North America; ranges from Ontario to the Atlantic coast, and southward to the Gulf of Mexico. Crack willow (*S. fragilis* and hybrids), naturalized in eastern North America, is often mistaken for black willow, and may be the more commonly encountered species.

Leaves Narrowly lance-shaped, 8–14 cm long; thin; tip long-pointed, curved back, base rounded; very finely toothed; upper surface green, paler beneath. Stalk short, 3–6 mm long, hairy. Stipules green, ear-like, toothed, persistent.

Buds Narrow, conical, 3–4 mm long, sharp-pointed, shiny, brown to yellowish-brown.

Twigs Light yellow to reddish- or purplish-brown, downy when young, becoming gray and hairless with age; longitudinal ridges run down from the leaf scars; brittle at the base.

Flowers Catkins erect, 2–7 cm long, on short leafy shoots; appear in early spring. Bracts 2–3 mm long, blunt-tipped, yellow, hairy on the inside.

Fruits Capsules ovoid, 4–5 mm long, light brown, hairless.

Bark Very dark, black or dark brown, deeply furrowed into scaly, flat-topped ridges.

Size and Form Small trees, up to 12 m high, 30 cm in diameter, and 70 years old; larger in the southern United States. Trunk crooked, often forked. Crown broad, irregular. Branches spreading, often breaking away from the tree.

Habitat Common on moist sites, along stream banks and in swamps; mixed with red and silver maple, cottonwood, green ash, white elm, other willows. Fast-growing; intolerant of shade.

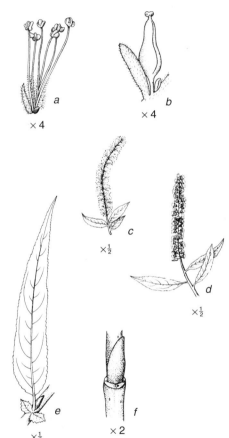

a. Pollen flower. b. Seed flower. c. Pollen catkin. d. Seed catkin. e. Leaf with persistent stipules. f. Lateral bud and leaf scar.

11

Quick Recognition Contrasting features of crack willow in brackets. Leaves green on both surfaces, finely toothed [whitish beneath, with irregular teeth], long-pointed, tip curved back. Stipules large, persistent [small or absent]. Buds shiny, narrow, sharp-pointed. Bark dark brown or black with scaly flat-topped ridges [gray with narrow ridges].

Seed catkin; fruit capsules ovoid, hairless.

Black Willow

Young bark becomes dark brown or black.

Mature bark furrowed into scaly flat-topped ridges.

Crack Willow**

Brittle willow

Salix fragilis L.

Saule fragile

Medium-sized trees, up to 30 m high and 100 cm in diameter, native to Europe. Hardy as far north as Zones C3, NA4. Named for its brittle branchlets, which break off and litter the ground following wind storms; branchlets may root. Hybrids with white willow occur as naturalized populations more frequently than the pure species. **Leaves** elliptic, 7–15 cm long, tip sharp, elongated; margin with large irregular gland-tipped teeth; upper surface green, whitish beneath; leaf stalk glandular at junction with blade. Stipules small or absent. **Buds** gummy. **Twigs** stiff, brittle, shiny yellowish-green to dark red. **Catkins** on short leafy shoots. **Bark** deeply furrowed, ridges narrow.

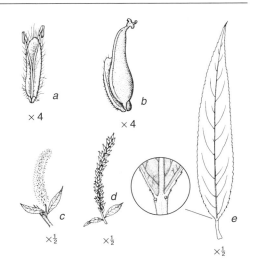

a. Pollen flower. *b.* Seed flower. *c.* Pollen catkin. *d.* Seed catkin. *e.* Leaf with detail of glands at leaf-stalk junction.

Golden Weeping Willow*

Salix alba L. var. *vitellina* Stokes cv. Pendula
[syn. *S. alba* var. *tristis* Gaudin; *S.* ×*sepulcralis* Simonk.; *S. chrysocoma* Dode]

Saule pleureur doré

The common willow planted as an ornamental in Canada; medium-sized trees with pendulous yellow branchlets. Hardy as far north as Zones C3, NA2.

Leaves Lance-shaped, 4–10 cm long; tip long-pointed, base tapered; finely toothed; stalk short, hairy, with glands; upper surface bright green; whitish, hairy beneath. Stipules lance-shaped.

Buds Flattened, about 5 mm long, tip rounded, smooth, hairy.

Twigs Very slender, flexible, pendulous, light yellow.

Flowers Catkins erect, on short leafy shoots; appear with the leaves.

Bark Light brown, corky, ridged, furrowed.

Size and Form Medium-sized trees, up to 25 m high. Crown very broad, with numerous yellow branchlets hanging vertically, often touching the ground. Root system deep, wide-spreading.

Habitat Thrives on moist to wet soils.

Notes An attractive ornamental, but subject to many diseases and insect pests; leaves, catkins, twigs, and branches litter the ground; aggressive root system may clog drains.
     The weeping willow native to China (*S. babylonica*) is much less hardy.

> **Quick Recognition** Easily recognized by the long slender pendulous yellow branches.

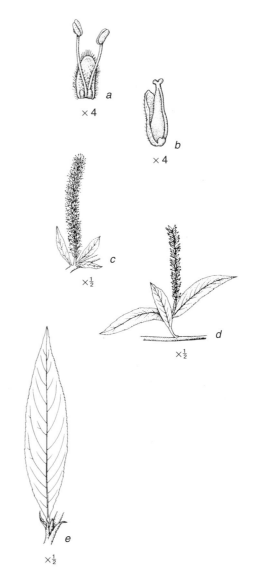

a. Pollen flower. *b.* Seed flower. *c.* Pollen catkin. *d.* Seed catkin. *e.* Leaf with lance-shaped stipules.

Golden Willow*

11

Salix alba L. var. *vitellina* Stokes

Saule jaune

Similar to the golden weeping willow but with upright branches; less commonly planted. Hardy as far north as Zones C3, NA2.

Seed catkins ripen in early summer.

Young bark dull, becomes light brown.

Mature bark ridged and deeply furrowed.

Golden Weeping Willow

Hybrid White Willow*

Salix alba × *S. fragilis* L.
[syn. *S.* ×*rubens*]

Saule blanc

Most of the planted trees called white willow are hybrids between golden willow and crack willow. Hardy as far north as Zones C3, NA2. Medium-sized trees, up to 25 m high, with open, spreading crowns. **Leaves** slender, 4–10 cm long, tip long-pointed, base tapered; finely toothed; short-stalked; grayish-green; hairy when young, becoming smooth. Stipules small or absent. **Buds** blunt, with silky hairs. **Twigs** yellowish-brown. **Catkins** erect, on short leafy stalks; appear with the leaves. **Fruit capsules** ripen about 1 month later.

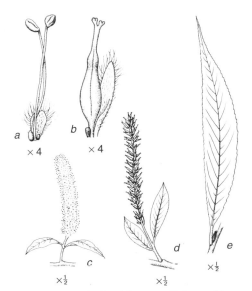

a. Pollen flower. *b.* Seed flower. *c.* Pollen catkin.
d. Seed catkin. *e.* Leaf.

Scouler Willow

Mountain willow, fire willow

Salix scouleriana Barratt

Saule de Scouler

Shrubs or very small trees, up to 9 m high; common on recently burned-over areas and in young upland forests, from Manitoba westward. The first willow to flush in spring. **Leaves** 5–12 cm long, usually widest above the middle, tapering to tip and base, blunt-pointed; margin smooth or sparsely toothed toward the base; upper surface dark green; whitish, covered with rusty hairs beneath. Stipules 1–3 mm long, longer on vigorous shoots. **Buds** red, pointed, sharp-edged. **Twigs** tough, yellowish- to greenish-brown; covered with dense hairs when young. **Catkins** on short shoots with bract-like leaves. **Fruit capsules** hairy. **Bark** dark brown, with ridges on larger stems; diamond-shaped patches similar to those on Bebb willow appear on main stems.

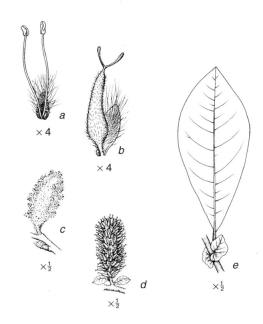

a. Pollen flower. *b*. Seed flower. *c*. Pollen catkin. *d*. Seed catkin. *e*. Leaf with stipules.

Sitka Willow

Salix sitchensis Sanson

Saule de Sitka

Shrubby trees, up to 7 m high, common along the Pacific coast. **Leaves** 4–9 cm long, widest above the middle, tip blunt-pointed, base narrow, tapering; margin wavy-toothed or smooth; upper surface dark green, with a few short hairs; paler beneath, with a sheen from silky hairs. Stipules small, about 1 mm long. **Twigs** brittle, dark reddish-brown, with appressed hairs. **Catkins** on short leafy shoots. **Fruit capsules** with silky hairs. **Bark** slightly furrowed, scaly. Roots reddish.

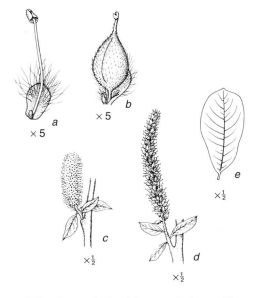

a. Pollen flower. *b*. Seed flower. *c*. Pollen catkin. *d*. Seed catkin. *e*. Leaf.

11

Feltleaf Willow

Salix alaxensis (Andersson) Coville

Saule feutré

Shrubs or shrubby trees up to 5 m high, common
in northwestern Canada; ranges west into Alaska
and Siberia; a pioneer species on river flats.
Leaves elliptic, 5–10 cm long, often widest above
the middle, tip pointed, base tapered; margin
smooth, rolled under; upper surface green, felt-like
covering of white woolly hairs beneath. Stipules
4–15 mm long, hairy, pointed. **Buds** hairy. **Twigs**
covered with woolly hairs during the 1st year, then
smooth, reddish-brown. **Catkins** appear before the
leaves, on previous year's twigs; pollen catkins
3–4 cm long, seed catkins about twice as long.
Fruit capsules 4–5 mm long, hairy. **Bark** furrowed, scaly, with "diamonds" similar to Bebb and
littletree willows.

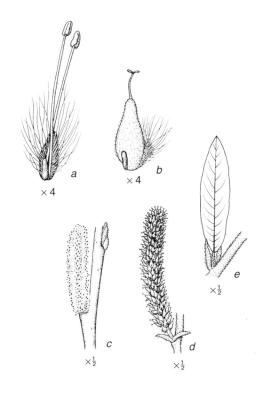

a. Pollen flower. *b.* Seed flower. *c.* Pollen catkin.
d. Seed catkin. *e.* Leaf with pointed hairy
stipules.

Mackenzie Willow

Salix prolixa Andersson
[syn: *S. mackenzieana* (Hook.) Andersson;
S. rigida var. *mackenzieana*]

Saule du Mackenzie

Shrubs or shrubby trees up to 5 m high; common
along streams and lakeshores, aspen woodlands,
and lower mountain slopes, from the Northwest
Territories to southeast Yukon, south through
Alberta and British Columbia into the United
States. **Leaves** 5–15 cm long, narrow, widest
above the middle, tip long, base rounded, margin
finely toothed and/or glandular; lower surface
whitened; stipules round-tipped; leaf stalks and
young leaves reddish. **Twigs** brownish, sometimes
with long shaggy hairs; larger stems light brown to
reddish-brown. **Flowers** appear with the leaves;
bracts brown, with curling hairs.

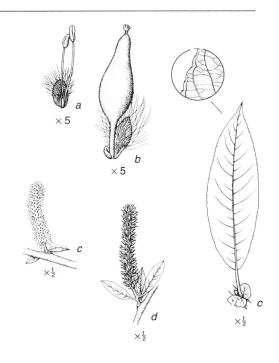

a. Pollen flower. *b.* Seed flower. *c.* Pollen catkin.
d. Seed catkin. *e.* Leaf with round-tipped
stipules and detail of glandular leaf margin.

Meadow Willow
Slender willow
Salix petiolaris Sm.
[syn. *S. gracilis* Andersson]
Saule à long pétiole

Shrubs or shrubby trees, up to 5 m high, ranging
from New Brunswick to northern British Columbia.
One of the most common willows in southern
Canada. Occurs on wet sites, but also on jack pine
sand plains. **Leaves** numerous, pointing up,
overlapping, 2–7 cm long, thin; pointed at both
ends; edged with small gland-tipped teeth toward
the tip; upper surface shiny green, hairless; paler
beneath; leaf stalk yellowish; young leaves
reddish, hairy. Stipules small or absent. **Twigs**
yellowish-green to olive-brown, becoming very
dark brown. **Catkins** appear with the leaves on
leafy dwarf shoots. Seed catkins up to 4 cm long.
Fruit capsules 3–7 mm long, elongated with
slender beaks; on slender hairy stalks. **Bark**
grayish-green or reddish-brown, becoming dark
brown. Often heavily browsed by deer.

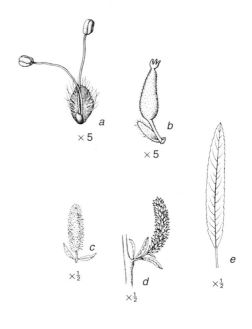

a. Pollen flower. b. Seed flower. c. Pollen catkin.
d. Seed catkin. e. Leaf.

Heartleaf Willow
Salix eriocephala Michx.
[syn: *S. rigida* Muhl.]
Saule à tête laineuse

Shrubs or shrubby trees, up to 6 m high, ranging
from Nova Scotia to Yukon and south into the
United States. **Leaves** lance-shaped, 10–15 cm
long, base rounded to heart-shaped; finely
toothed; upper surface dark green tinged with
reddish-purple, very pale beneath; hairy in spring,
becoming smooth in summer; red in autumn.
Stipules 10–20 mm long, somewhat toothed.
Buds reddish-brown. **Twigs** reddish-brown to
yellowish-green; hairy in spring, becoming smooth.
Catkins erect, on short shoots. **Fruit capsules**
smooth, long-pointed, reddish-brown. A highly variable species.

11

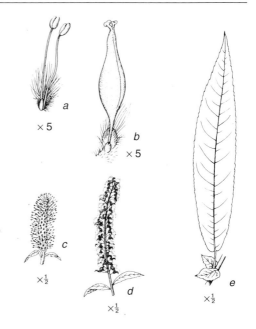

a. Pollen flower. b. Seed flower. c. Pollen catkin.
d. Seed catkin. e. Leaf with toothed stipules.

Satiny Willow

Salix pellita Andersson

Saule satiné

Shrubs or shrubby trees, up to 5 m high, ranging from Newfoundland to Saskatchewan. **Leaves** narrow, 4–12 cm long, tapering to both ends; thick; margin somewhat wavy, rolled under; upper surface dark green, pale with many satiny hairs beneath; midvein a prominent ridge on the lower surface; lateral veins depressed, somewhat straight and parallel. Stipules small or missing. **Twigs** brittle, yellowish- to dark reddish-brown; often coated with a waxy powder. **Catkins** elongated, on short shoots with bract-like leaves; appear before or with the developing leaves. **Fruit capsules** slender, densely hairy.

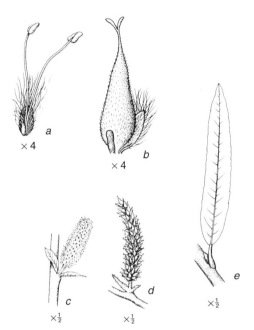

a. Pollen flower. *b.* Seed flower. *c.* Pollen catkin.
d. Seed catkin. *e.* Leaf.

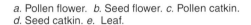

Basket Willow**

Common osier

Salix viminalis L.

Saule des vanniers

Small trees, up to 12 m high, with long erect yellowish branchlets; native to Europe and Asia; grown since ancient times to provide withes for basketry. Hardy as far north as Zones C5, NA4. **Leaves** very narrow, 5–20 cm long, long-pointed; margin smooth and rolled under; upper surface dull green, with silky hairs beneath; leaf stalk short, swollen at the base. **Twigs** flexible, greenish-yellow to reddish-brown, hairy. **Catkins** on short stalks; appear before the leaves. **Fruit capsules** densely hairy.

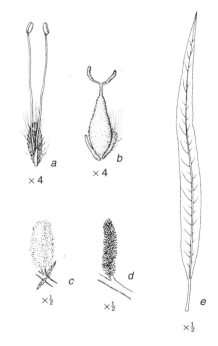

a. Pollen flower. *b.* Seed flower. *c.* Pollen catkin.
d. Seed catkin. *e.* Leaf.

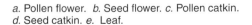

Purple-Osier Willow*

Salix purpurea L.

Saule pourpre

Shrubby trees, up to 5 m high; native to Europe and Asia; grown since ancient times to provide withes for basketry; attractive because of its purplish twigs. Hardy as far north as Zones C4, NA4. **Leaves** subopposite, widest above the middle, finely toothed toward the tip; upper surface dull green, paler beneath. Stipules small or absent. **Buds** subopposite, small, purplish. **Catkins** stalkless; appear before the leaves.

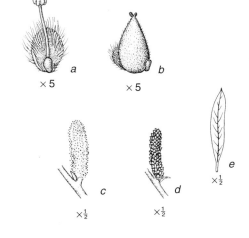

a. Pollen flower. *b.* Seed flower. *c.* Pollen catkin. *d.* Seed catkin. *e.* Leaf.

Violet Willow*

Salix daphnoides Villars

Saule daphné

Shrubby trees, up to 10 m high, native to the mountainous regions of central Europe; found on dry sites. Drought-tolerant; useful for stabilizing exposed soils. An attractive ornamental because of the bluish-white waxy powder on the branchlets. Hardy as far north as Zones C2, NA3. **Leaves** narrow, 6–10 cm long, pointed; margin finely toothed or smooth; upper surface shiny dark green, paler beneath. Stipules large.

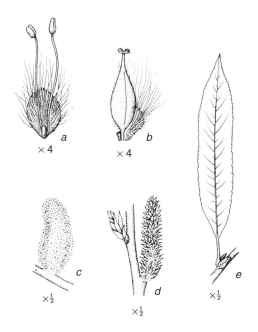

a. Pollen flower. *b.* Seed flower. *c.* Pollen catkin. *d.* Seed catkin. *e.* Leaf with stipules.

11

Poplars, Aspens, and Cottonwoods

Genus *Populus*
Salicaceae: Willow Family

Les peupliers

Poplars are distributed throughout the Northern Hemisphere, chiefly in the temperate zone. About 40 species of poplar are recognized; 6 are native to Canada. Natural hybridization between species commonly occurs. In addition, a number of cultivars derived from various species and their hybrids have been frequently planted throughout Canada.

The natural populations of poplar are difficult to classify into species. Identification may require observations at three different times of the year: during flowering, during fruit-ripening, and just before leaf fall (to examine preformed and neoformed leaves and winter buds).

The poplar genus is usually divided into sections, represented in Canada by the species in parentheses: Tacamahaca, or balsam poplars (black cottonwood, balsam poplar, narrowleaf cottonwood, and the introduced Simon poplar); Aigeiros, or cottonwoods (eastern cottonwood, plains cottonwood, and the introduced Lombardy poplar); and Leuce, or aspens (trembling aspen, largetooth aspen, and the introduced European white poplar). It is confusing that some species of the Tacamahaca are called cottonwood, a name usually applied to Aigeiros.

Several of the hybrids found in Canada are also briefly described: Jack's hybrid poplar, lanceleaf cottonwood, and Carolina poplar. Most characteristics of hybrids are intermediate between the parents.

Leaves Twice as long as wide in most species; tip pointed or taper-pointed; single-toothed; lateral veins pinnately arranged along the midvein, branching irregularly near the margin. Leaves preformed in the bud usually differ from those neoformed during the growing season. Also, neoformed leaves on seedlings, sprouts, and suckers often differ from those on adult trees. Stipules soon become deciduous. Warty glands occur on some species where the base of the leaf blade meets the leaf stalk.

Buds Terminal bud pointed. Lateral leaf buds similar but usually shorter, with 5-7 bud scales, the lowest directly above the leaf scar. Flower buds also lateral, but larger. Leaf

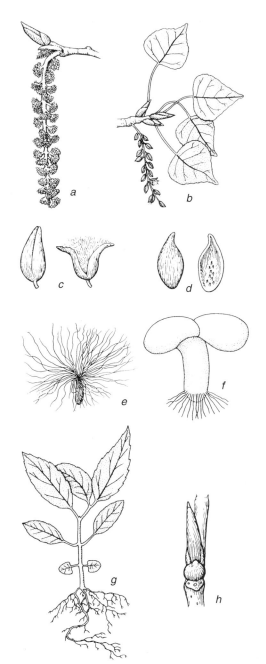

a. Pollen catkin. *b.* Seed catkin on leafy shoot. *c.* Seed capsule closed (left), open (right). *d.* Capsule scale outer surface (left), inner surface (right). *e.* Seed with silky hairs. *f.* Newly germinated seedling. *g.* 1-year-old seedling. *h.* Lateral bud; lower scale directly above leaf scar.

scars raised, 3-sided or 3-pointed; 3 vein scars. Stipule scars evident.

Twigs Medium to moderately stout; dwarf shoots may be present as well as the usual long shoots. Pith 5-pointed. Lenticels prominent on young stems.

Flowers Borne in catkins; pollen catkins and seed catkins on separate trees, rarely on the same tree. Catkins pendulous; preformed in buds; found on previous year's twig; appear early in spring, before the leaves unfold. Flowers small; set in the axil of a deciduous bract, with irregular long teeth and/or hair-like projections at tip. Wind-pollinated; spent pollen catkins fall off more or less intact.

Fruits Mature seed catkins pendulous, 10–15 cm long, bearing small pointed capsules that split in 2, 3, or 4 parts, releasing several seeds; maturing in 4–6 weeks, before the leaves are fully grown. Seed production begins at about 8–10 years of age; some seeds are produced most years.

Seeds Small, 1–3 mm long, half as wide, very pale, bearing a tuft of long, white, silky hairs (the "cotton" for which some species are named) by which they may be carried long distances by the wind. Under natural conditions, germination or death occurs within a few days after seedfall; light, continual moisture, and a favorable medium such as fine mineral soil are required for germination. Viable for some years when dried and kept very cold in air-tight containers.

Seedlings During germination, a fringe of hairs grows out around the base of the hypocotyl, rendering the young seedling erect and positioning the new root to grow down into the soil. Cotyledons small, becoming green, raised above the surface. Because the conditions required for seed germination are rare in nature, poplar seedlings are scarce; they are found on shorelines, sandbars in rivers, old gravel pits, and other freshly exposed areas.

Vegetative Reproduction Aspens chiefly by root sprouts. Cottonwoods by stump sprouts; detached branches can take root. Balsam poplars by root and stump sprouts; detached branches can take root. All species can be grafted readily; cottonwoods and balsam poplars can be propagated by stem cuttings; aspens and balsam poplars by root cuttings.

Species that can reproduce from long roots running parallel to the surface of the ground are important in the natural regeneration of poplar on cut-over and burned-over areas.

Bark Smooth, whitish-green, yellowish-green, or grayish-green, when young; with age much darker, furrowed.

Wood Light, straight-grained, soft, often with a disagreeable odor when moist, pale. Diffuse-porous or semi-ring–porous; pores visible with a hand lens; rays very narrow; annual rings indistinct.

Size and Form Medium-sized to large trees; fast-growing and short-lived. Root systems wide-spreading.

Habitat Intolerant of shade. Pioneer species, among the first to invade and recolonize open areas following fire, land clearing, or forest harvesting.

Notes Used for wood pulp, composition board, core stock for veneer and furniture, lumber, boxes, match sticks, and small woodenware.

Frequently planted for windbreaks and as ornamentals. Species that reproduce by root suckers may be undesirable in urban settings because of the rapid spread of suckers; also, the aggressive root systems of poplar trees may clog water drains.

Buds, twigs, and bark provide food for a great many species of birds and animals.

Susceptible to fire damage, but trees regenerate easily after fire by root or stump sprouts and seedlings.

Quick Recognition Terminal bud present. Lateral buds with the lowest bud-scale placed directly above the leaf scar. Flowers in drooping catkins, appearing in spring before the leaves. Fruits in catkins, ripening in early summer; seeds with a tuft of long white hairs. Pith 5-pointed.

11

largetooth aspen

eastern cottonwood

Pollen catkin (left) and seed catkins (right) at time of pollination

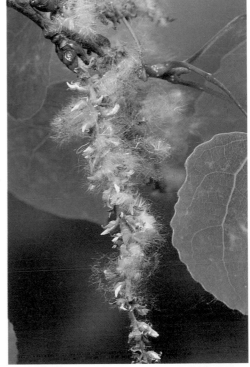

trembling aspen

trembling aspen

Seed catkins before (left) and after (right) opening

Black Cottonwood
Western balsam poplar

Populus trichocarpa Torr. & A. Gray
[syn. *P. balsamifera* ssp. *trichocarpa* (Torr. & A. Gray) Brayshaw]

SECTION TACAMAHACA

Peuplier de l'Ouest

The largest of the 3 balsam poplars native to Canada, and the largest broadleaf tree native to British Columbia. Similar to balsam poplar, hybridizing with it where their ranges overlap.

Leaves Broadly ovate, 7–12 cm long; tip long drawn-out, base usually rounded, sometimes wedge- or slightly heart-shaped; finely toothed with many short, rounded teeth that turn inward at their tips (margin sometimes partially smooth); upper surface dark green; silvery-green beneath, usually stained with brownish resin blotches, mostly hairless. Stalk 3–4 cm long, green, hairy; round in cross section. No warty glands at leaf base.

Buds Terminal bud round or slightly angled, 17–20 mm long, slender, long-pointed, orange-brown, very resinous, fragrant; 6–7 scales, sometimes fringed with hairs. Lateral leaf buds parallel to the twig. Pollen flower buds stout with thickened scales. Leaf scars small, triangular.

Twigs Moderately stout, hairy, orange-brown, round to slightly angular in cross section; lenticels large, sparse.

Flowers Pollen catkins 4–5 cm long; seed catkins 6–8 cm long.

Fruits Mature seed catkins 12–15 cm long, hairy, bearing many closely spaced capsules. Capsules nearly globular, 3–4 mm long, tip short; covered with short hairs; splitting into 3 parts when mature. Seeds 2 mm long.

Bark Smooth, grayish-green or yellowish-gray when young; becoming dark grayish-brown, separated into flat-topped ridges by irregular V-shaped furrows.

11

Size and Form Large trees, up to 35 m high, 120 cm in diameter, and 165 years old. Trunk long, straight. Crown narrow, somewhat columnar, open, with a few stout, ascending branches. Root system widespreading, with a few large roots, penetrating obliquely to the water table or a layer of hardened soil.

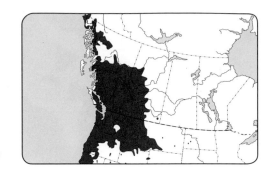

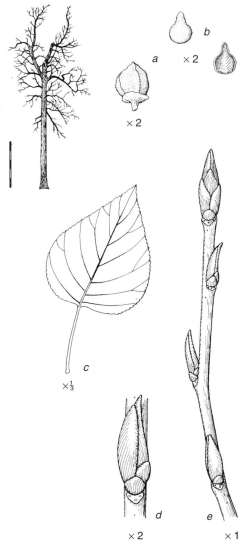

a. Fruit capsule. *b.* Fruit capsule scales; outer surface (left), inner surface (right). *c.* Leaf. *d.* Lateral bud and leaf scar. *e.* Winter twig.

Seed catkins; fruit capsules will split into 3 parts.

Mature bark has flat-topped ridges divided by V-shaped furrows.

Young bark grayish-green, lenticels prominent.

Habitat Grows at lower and middle elevations; typically found on bottomlands where it reaches its greatest size; also occurs on loose, porous, sandy, or gravelly soils; mixed with other species.

Quick Recognition Leaves ovate, with whitish resin-blotched undersurface and short round leaf stalks. Buds gummy, round, parallel to the rounded twigs. Fruit capsule splits into 3 parts.

Balsam Poplar
Eastern balsam poplar

Populus balsamifera L.

SECTION TACAMAHACA

Peuplier baumier

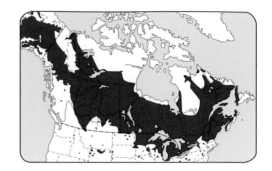

Very similar to black cottonwood; differences are noted below.

Leaves Leaf stalk longer, 7–10 cm; warty glands may be present at the leaf base.

Buds Terminal bud larger, up to 25 mm; 5 scales, margins hairless. Buds and twigs both round.

Twigs Bright reddish-brown becoming dark orange then gray with age.

Flowers Pollen catkins longer, 7–10 cm.

Fruits Mature seed catkins shorter, 10–13 cm; capsules ovoid, larger, hairless, 6–7 mm long, splitting into 2 parts when mature.

Bark Greenish-brown when young, becoming gray.

Size and Form Medium-sized trees, up to 25 m high, 50 cm in diameter, and 70 years old, occasionally much larger and older. Often the largest tree in the northern and western parts of its range.

Habitat Found most often in river valleys; also on any moist, rich, low-lying ground; grows in pure stands or mixed with alders, willows, balsam fir, black spruce, white spruce, white birch, and other boreal species.

Notes Frequently planted in rural areas for shelterbelts and windbreaks. Hybrids with eastern cottonwood are called Jack's hybrid poplar (*P.* ×*jackii*).

11

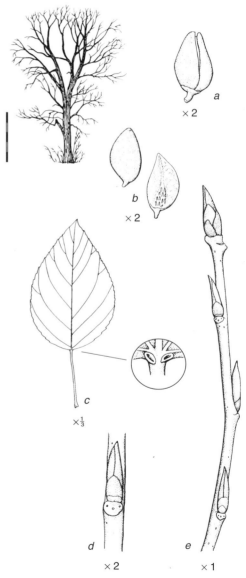

a. Fruit capsule. *b.* Fruit capsule scales; outer surface (left,) inner surface (right). *c.* Leaf with detail of glands at leaf base. *d.* Lateral bud and leaf scar. *e.* Winter twig.

Seed catkins; fruit capsules will split into 2 parts.

Balsam Poplar

Young bark (inset) greenish-brown. Mature bark gray with scaly ridges.

Jack's Hybrid Poplar

Populus ×jackii Sarg.

(*P. balsamifera* × *P. deltoides*)

Peuplier hybride de Jack

Medium-sized trees, up to 30 m high and 100 cm in diameter; occurring on floodplains, along lakeshores, and in other wet places; common where cottonwood and balsam poplar grow together. **Leaves** triangular-ovate, 7–10 cm long, tip short taper-pointed; base heart-shaped, with 2 or 3 glands; upper surface bluish-green, paler beneath; stalk 2.5–5 cm long, slightly flattened; preformed leaves with 15–30 teeth per side, neoformed leaves with 25–50. **Buds** ovoid, 9–13 cm long, reddish, resinous. **Twigs** reddish-brown. Flowering **catkins** 5–15 cm long; fruiting catkins up to 20 cm long, stalk hairy. Commonly reproduces vegetatively by root sprouts. **Bark** grayish-brown with an orange cast, furrowed on older trees.

    The cultivar Balm-of-Gilead is a (sterile) female clone that originated many years ago;

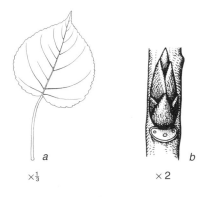

a. Leaf. *b.* Lateral bud and leaf scar.

possibly a hybrid progeny of eastern cottonwood fertilized with pollen of balsam poplar; frequently planted for landscape purposes; differs from most hybrids of the same 2 species in that the leaves are larger (up to 17 cm long), heart-shaped, with hairs on the stalks and lower veins.

Narrowleaf Cottonwood

Narrowleaf balsam poplar

Populus angustifolia James

SECTION TACAMAHACA

Peuplier à feuilles étroites

The smallest of the 3 balsam poplars native to Canada. Uncommon in Canada; found in a few river valleys in southwestern Saskatchewan and adjacent Alberta. Easily mistaken for a willow because of the leaf shape.

Leaves Narrow and willow-like, 5–9 cm long, widest near or below the middle, finely toothed to the tip, narrowed to a wedge-shaped base; upper surface light yellowish-green; slightly paler and often resin-stained beneath. Stalk very short, stiff, hairless, glandless; almost circular in cross section, but flattened near the base of the blade.

Buds Terminal bud 6–12 mm long, long-pointed, shiny, brownish, resinous. Lateral buds slightly smaller, diverging from the twig, with 5 scales.

Twigs Slender, yellowish-brown becoming a bright whitish or ivory by the 2nd year.

Fruits Mature seed catkins 6–8 cm long. Capsules broadly ovate, 6–8 mm long, tip short; hairless; splitting into 2 parts when mature.

Bark Whitish or yellowish-green; smooth, becoming furrowed at the base of mature trunks.

Size and Form Small trees, up to 15 m high and 30 cm in diameter. Typically a small slender tree with a narrowly conical crown of ascending slender branches and the appearance of a willow.

Habitat Often one of the first plants to become established on new sandbars.

Notes Frequently outnumbered by its hybrids. Hybrids with plains cottonwood called lanceleaf cottonwood (*P.* ×*acuminata*). Hybrids with balsam poplar less common; called **Brayshaw's poplar** (*P.* ×*brayshawii* Boivin); leaves dark green on the upper surface, whitish beneath.

11

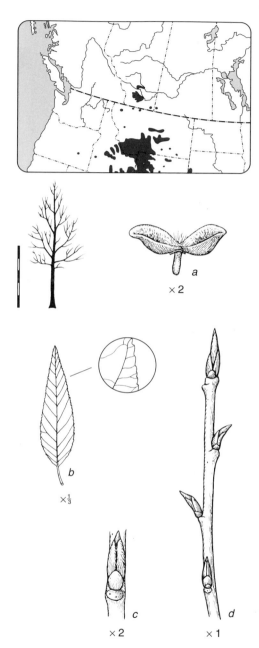

a. Fruit capsule, open. *b.* Leaf with detail of finely toothed margin. *c.* Lateral bud and leaf scar. *d.* Winter twig.

Quick Recognition In summer, the narrow leaves distinguish it from other poplars; in winter, the white branches and twigs distinguish it from all other species.

Leaves willow-like.

Narrowleaf Cottonwood

Young bark whitish-green.

Mature bark furrowed into broad, flat ridges.

Lanceleaf Cottonwood

Populus ×acuminata Rydb.

(*P. angustifolia* × *P. deltoides*)

Peuplier à feuilles acuminées

Common along streams in southern Alberta and south into the United States. Small trees, up to 20 m high and 50 cm in diameter, with narrow, flat-topped crowns. **Leaves** narrow, ovate to somewhat lance-shaped, 5–10 cm long, tip long-pointed, light green; preformed leaves with 15–25 teeth per side, neoformed leaves with 30–45. **Buds** slender, 9–12 mm long, reddish, resinous, slightly hairy. **Twigs** orange-brown. Flowering **catkins** 6–9 cm long; fruiting catkins up to 16 cm. **Fruit capsules** with 2 or 3 parts, 5–7 mm long. **Bark** light yellowish-brown, furrowed.

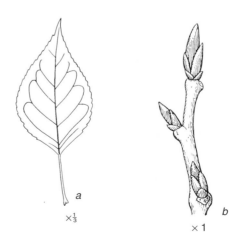

a. Leaf. *b.* Terminal bud, lateral buds, and leaf scar.

Eastern Cottonwood

Populus deltoides Bartr. ex Marsh. ssp. *deltoides*

SECTION AIGEIROS

Peuplier deltoïde

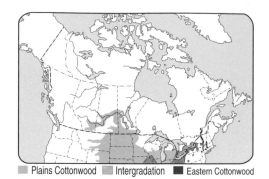

Plains Cottonwood   Intergradation   Eastern Cottonwood

Found in Quebec in the Montréal area and in southern Ontario near Lakes Erie and Ontario; not a common tree. In Ontario, eastern and plains cottonwoods intergrade; together, these 2 subspecies of *P. deltoides* range from Quebec to Alberta.

Leaves Triangular, 5–10 cm long, tip taper-pointed; 20–25 rounded teeth per side, no teeth on the tip or near the stalk; margin narrow, translucent; upper surface bright shiny green, slightly paler beneath; often pendulous on a long, smooth, hairless, yellowish-green, flattened stalk; usually 3–5 warty glands on the base of the leaf blade. Neo-formed leaves broadly ovate, hairy when unfolding; teeth smaller and more numerous.

Buds Terminal bud 3-sided, about 20 mm long, slender, long-pointed, yellowish-brown, very resinous, hairless. Lateral leaf buds slightly smaller, diverging from the twig; flower buds stout. Leaf scars large, 3-lobed, with eyelash-like hairs at juncture of bud and scar.

Twigs Stout, smooth, hairless, yellowish-brown, angular in cross section; narrow ridges extend down from either side of the bud. Lenticels sparse, linear.

Flowers Catkins 5–7 cm long.

Fruits Mature seed catkins 15–25 cm long; stalk hairless. Capsules oval, 8–12 mm long, tapering to both ends; splitting into 3 or 4 parts when mature.

Bark Smooth, yellowish-gray; with age becoming dark gray, deeply furrowed.

Size and Form Medium-sized to large trees, up to 30 m high, 100 cm in diameter, and 50 years old. In the open, trunk short, massive; often divided near the ground into a few large, wide-spreading limbs to form a broad, irregular-shaped, open crown. In the forest, trunk long, straight, with a small, rounded crown. Root system usually shallow, wide-spreading; may be deep in deep soils. The fastest growing native tree; often planted where fast growth is the main requirement.

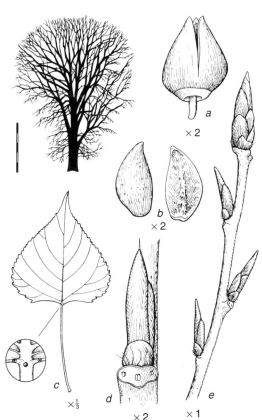

a. Fruit capsule. *b.* Fruit capsule scales; outer surface (left), inner surface (right). *c.* Leaf, with detail of glands at junction of leaf stalk. *d.* Lateral bud and leaf scar. *e.* Winter twig.

Habitat Occurs on rich moist sites, in open stands mixed with other species; may form pure stands on the banks of streams.

Notes Unrooted stem cuttings (30–50 cm long) can be planted. Hybrids with European black poplar are well-known in North America and Europe; see Carolina poplar.

> **Quick Recognition** Triangular leaf with coarse teeth. Large resinous angled buds, diverging from angular, ridged twig.

11

Seed catkin; fruit capsules will split into
3 or 4 parts.

▲
Plains Cottonwood
Eastern Cottonwood ▶

Young bark (inset) smooth, yellowish-gray. Trunk short,
massive; mature bark deeply furrowed.

Plains Cottonwood

Populus deltoides Bartr. ex Marsh.
ssp. *monilifera* (Ait.) Eckenw.
[syn. *P. deltoides* var. *occidentalis* Rydb.;
P. sargentii Dode]

Peuplier deltoïde de l'Ouest

Occurs along water courses in the southern
parts of the Prairie provinces; south to Texas.
Frequently planted on the prairies for
shelterbelts (Zones C2, NA2); naturalized
along many streams and rivers outside its
natural range. Differs from eastern
cottonwood in that the **leaves** have fewer
teeth (5–15 per side), a more rounded base,
a longer tip, and only 1 or 2 basal glands;
buds covered with minute hairs; **twigs** light
yellow.

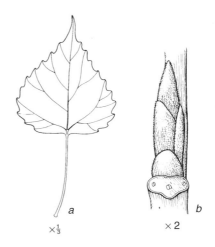

a. Leaf. b. Lateral bud and leaf scar.

$\times \frac{1}{3}$ $\times 2$

Trembling Aspen
Quaking aspen

Populus tremuloides Michx.

SECTION LEUCE

Peuplier faux-tremble

One of the most widely distributed species in North America; occurs throughout the forested areas of Canada.

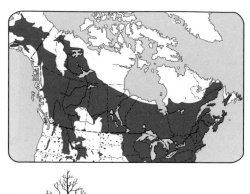

Leaves Broadly oval to kidney-shaped, 3–7 cm long; tip abrupt, short, sharp; base rounded; teeth fine, irregular (20–30 per side); upper surface deep green, paler beneath, usually hairless; stalk flattened, slender, usually longer than the blade, causing leaves to tremble in the breeze. Appear 7–10 days earlier than largetooth aspen. Preformed leaves usually broadly oval; neoformed leaves heart-shaped, often larger or smaller, with more numerous teeth, more readily visible basal glands, and round stalks.

Buds Terminal bud 6–7 mm long; flower buds larger; leaf buds smaller. Conical, pointed, appressed, with the tips curving inward; 6–7 scales, dark reddish-brown, shiny, slightly resinous, not fragrant. Leaf scars small, triangular.

Twigs Slender, shiny, dark green or brownish-gray, round in cross-section. Lenticels oval.

Fruits Mature seed catkins 10 cm long; stalk hairy. Capsules narrowly conical, 5–7 mm long, hairless; splitting into 2 parts when mature; with about 10 seeds. Seeds ripen 4–6 weeks after flowering. Good seed crops occur every 4–5 years.

Bark Smooth with a waxy appearance, pale green to almost white when young; becoming darker and furrowed into long flat-topped ridges; diamond-shaped marks about 1 cm across sometimes occur.

Size and Form Medium-sized trees, up to 25 m high, 40 cm in diameter, and 80 years old; larger and older on favorable sites. Trunk long, cylindrical, smooth, with little taper; branch-free in the lower part through self-pruning. Crown short, rounded. Root system shallow, very wide-spreading.

Habitat Occurs on a great variety of soils; prefers sheltered sites. Often grows in pure

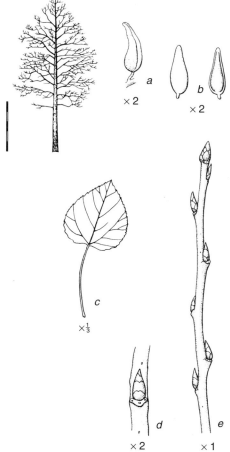

a. Fruit capsule. *b.* Fruit capsule scales; outer surface (left), inner surface (right). *c.* Leaf. *d.* Lateral bud and leaf scar. *e.* Winter twig.

stands, especially as a young tree; also mixed with white spruce, black spruce, balsam fir, white birch, balsam poplar, and jack pine. Considered a "nurse crop" because other broadleaf species and conifers enter a pure stand and eventually replace the aspen.

11

Seed catkins; fruit capsules will split into 2 parts.

Trunk cylindrical, smooth; mature bark becomes dark and furrowed.

Young bark smooth, pale green to whitish.

Notes Aspen wood has been increasingly used in the manufacture of "oriented strand board" and other forms of chipboard.

Groups of clones derived from a single aspen seedling by repeated generations of root sprouts have been found that occupy up to 80 hectares and consist of thousands of trees. These clones may have originated on land exposed soon after the Pleistocene ice sheet melted, making them among the largest and oldest organisms in the world.

Quick Recognition Small round leaves on long stalks, trembling in the breeze; reddish-brown buds; whitish bark.

Largetooth Aspen
Bigtooth aspen

Populus grandidentata Michx.
SECTION LEUCE

Peuplier à grandes dents

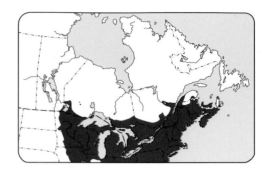

Similar in form and aspect to trembling aspen; range does not extend as far west, east, or north; usually found on drier sites.

Leaves Shape varies from ovate with a short sharp tip to broadly oval with a blunt tip; 5–10 cm long; 7–15 large uneven teeth per side; upper surface dark green, paler beneath, with white hairs when unfolding in spring (7–10 days later than trembling aspen); stalk flattened, shorter than the blade. Neoformed leaves on vigorous shoots heart-shaped, large, teeth smaller, more numerous; stalk less flattened; dense white woolly hairs on stalk and underside of leaf.

Buds Terminal bud ovoid, 7–8 mm long, blunt, dull brown, finely hairy, not gummy or fragrant. Lateral leaf buds similar but more slender. Flower buds stout.

Twigs Moderately coarse, dull brownish-gray, downy when the leaves are unfolding, some hairiness often remaining; round in cross section; lenticels orange.

Fruits Mature seed catkins 10–12 cm long. Capsules narrowly conical, 6–7 mm long, downy, closely spaced on flexible stalks; splitting into 2 parts when mature. Seeds ripen 4–6 weeks after flowering. Good seed crops occur every 4–5 years.

Bark Smooth, pale green to yellowish-gray, usually with an orange cast, darker than trembling aspen; becoming dark gray and furrowed with age; diamond-shaped marks about 1 cm across frequently occur.

Size and Form Small trees, up to 20 m high, 30 cm in diameter, and 60 years old; larger on favorable sites. Trunk long, cylindrical, smooth, with little taper; branch-free in the lower part through self-pruning. Crown short, rounded. Root system shallow, very wide-spreading.

Habitat Tree well-formed on moist fertile soils; scrubby and small on dry poor soils. Small pure stands are found, usually a single clone formed by root sprouts; more often mixed with trembling aspen, white birch, eastern white pine, balsam fir, white spruce, willows, and alders.

11

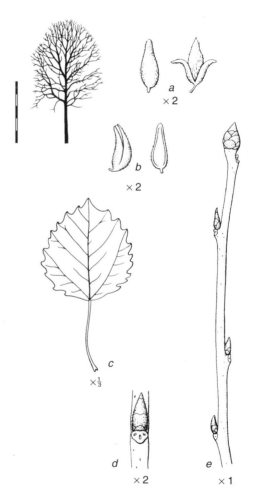

a. Fruit capsule, closed (left), open (right). b. Fruit capsule scales; outer surface (left), inner surface (right). c. Leaf. d. Lateral bud and leaf scar. e. Winter twig.

Quick Recognition Unfolding leaves grayish-white, hairy, 7–10 days later than trembling aspen; oval leaves with few coarse teeth; buds with fine hairs.

Seed catkins; fruit capsules will split into 2 parts.

Young bark darker than that of trembling aspen.

Mature bark on old trees dark gray and furrowed.

Simon Poplar*

Populus simonii Carrière
SECTION TACAMAHACA

Peuplier de Simon

Small attractive hardy trees with a narrow crown; native to eastern Asia. Hardy as far north as Zones C2, NA2. **Leaves** somewhat 4-sided, widest above the middle, 6–12 cm long, tip pointed, upper surface bright green, whitish beneath; stalk 1–2 cm long, red. **Buds** pointed, resinous; **twigs** angular, hairless, reddish-brown. Mature **seed catkins** 10–15 cm long. **Bark** grayish-green. A columnar-shaped cultivar is common.

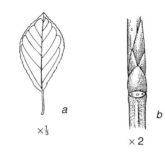

a. Leaf. b. Lateral bud and leaf scar.

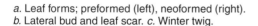

Lombardy Poplar*

Populus nigra L. cv. Italica
SECTION AIGEIROS

Peuplier noir d'Italie

A cultivar from Europe; a very common ornamental, planted for its slender, tapering crown. Hardy as far north as Zones C4, NA3. Subject to a number of diseases that may result in early death. Preformed leaves 4-sided (rhomboidal), smaller than those of cottonwood; neoformed leaves broader than long with an elongated tip.

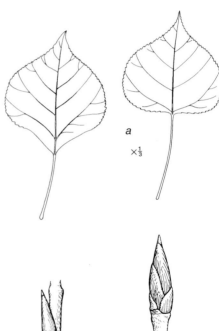

a. Leaf forms; preformed (left), neoformed (right).
b. Lateral bud and leaf scar. c. Winter twig.

11

Carolina Poplar*

Canada poplar

Populus ×*canadensis* Moench
cv. Eugenei

(*P. nigra* × *P. deltoides*)
[syn. *P.* ×*euramericana* (Dode) Guinier]

Peuplier de Caroline

First noted in France around 1750; a spontaneous hybrid between European black poplar (*P. nigra* Linnaeus) and eastern cottonwood, which had been introduced into Europe from North America in the 18th century; long cultivated in Canada, the United States, and Europe. Hardy as far north as Zones C3, NA3. In recent decades the cross has been repeated. The clones selected generally resemble cottonwood; however, the leaf base tends to be wedge-shaped rather than straight, and basal glands are usually absent; the trunk is more likely to be straight and distinct to the top of the tree; annual nodes are about 1 m apart, marked by a whorl of large ascending branches. Often propagated by stem cuttings for intensive culture of biomass (wood and bark).

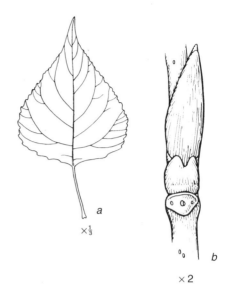

a. Leaf. b. Lateral bud and leaf scar.

European White Poplar*

Silver poplar

Populus alba L.

SECTION LEUCE

Peuplier blanc

A large handsome tree, among the first species introduced to North America from Europe. Hardy as far north as Zones C2, NA3. **Leaves** with white woolly hairs beneath; preformed leaves ovate to elliptic with a few rounded teeth, 4–6 cm long; neoformed leaves larger, 6–12 cm long, often palmately 3–5 lobed (like maple leaves) with large coarse teeth. **Buds** and **twigs** covered with short white hairs. Bark smooth, whitish gray, with dark lenticels; rough and dark at the base of old trees. Often pollinated by the 2 native aspens producing hybrid swarms. Unwanted root sprouts may be a problem when this tree is planted for landscape purposes.

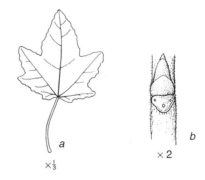

a. Leaf. b. Lateral bud and leaf scar.

Elms

Genus *Ulmus*
Ulmaceae: Elm Family

Les ormes

Worldwide the elm genus comprises about 18 species; 6 are native to eastern North America and 3 to Canada: white elm, slippery elm, and rock elm. Three introduced species have become naturalized and are planted as ornamentals.

In eastern North America, elms are among the best-known broadleaf trees. Unfortunately, Dutch elm disease has decimated the elm population. Only the introduced species, Siberian elm, is disease-resistant.

Leaves Deciduous, alternate, simple, in 2 rows along the shoot; oval, tip taper-pointed, base asymmetrical, thick, usually rough or hairy, short-stalked; veins prominent, straight, each ending in a large tooth, usually with smaller intervening teeth. Whether or not veins fork is helpful in species identification. Stipules small, thin, soon shed. Preformed leaves are smaller, more likely to have forked veins.

Buds Ovoid, asymmetrical, off-set above a leaf scar, in 2 rows along the twig; scales numerous, overlapping, in 2 rows. Flower buds plump, toward the base of the twig. No true terminal bud; end bud originates as a lateral bud. Leaf scars prominent, semi-circular, with 3 (3–6 in rock elm) sunken vein scars. Stipule scars obscure. Position of aborted shoot tip marked by a twig scar beside the end bud opposite the leaf scar.

Twigs Slender, often zigzag; pith solid, round. Side branches develop in 2 rows along previous year's twig, each shorter than the one above it. Occasionally, neoformed side shoots develop in leaf axils of vigorous shoots during the growing season. Shoot growth ends when the tip aborts and falls off, during late summer on vigorous shoots, much earlier on weaker ones. In some species, weak branchlets may break apart at the annual nodes and fall from the tree.

Flowers Small, inconspicuous, without petals; in small clusters, along previous year's twig. Appear early in spring before the leaves. Wind-pollinated.

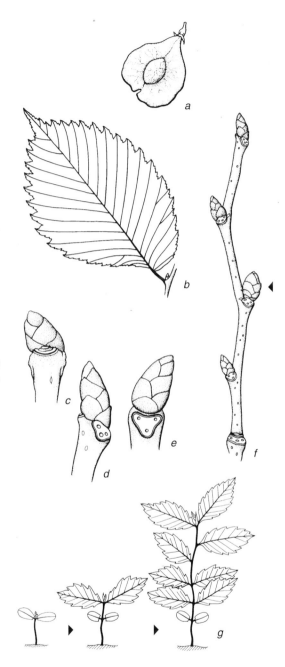

a. Fruit. *b.* Leaf. *c.* End bud showing twig scar. *d.* End bud showing twig scar (left), leaf scar (right). *e.* End bud showing leaf scar. *f.* Winter twig with leaf buds and flower bud (◀). *g.* Seedling development; cotelydon stage (left), primary leaf stage (center), seedling at end of 1st year (right).

11

white elm

Classic form of mature elm an increasingly rare sight.

Fruits Winged (a samara); small, flat, dry; a broad membranous wing surrounding a seedcase with 1 seed; in small slender-stalked clusters. Mature and fall away in spring before the leaves reach full size. Dispersed by wind and flowing water. Fruit production begins at 15–20 years of age; some fruits produced nearly every year, with abundant crops at intervals of 2–4 years.

Seeds Flat, dry; remain in the fruit; germinate soon after being shed. Viable for a year or more if stored air-dry at freezing temperatures.

Seedlings Cotyledons leaf-like, green, less than 1 cm long, with 2 basal lobes, raised above the surface of the ground. First true leaves may be in opposite pairs.

Vegetative Reproduction All species by stump sprouts; some by root sprouts.

Bark Gray to brownish, furrowed, with elongated, oblique, flat-topped, rough, irregular ridges.

Size and Form Medium-sized to large trees. Trunk often forked below the crown;

root flares sometimes prominent at the base of the trunk. Branches spreading, often arched; side branches 2-ranked. Leading shoot oblique.

Wood Heavy, attractive grain pattern on the tangential face, tough, hard, flexible, light brown; difficult to split because of interlocking grain. Ring-porous, rays inconspicuous.

Habitat Occur on a variety of soils from well-drained to wet; mixed with broadleaf species such as maple, basswood, and butternut; also as single trees along fences and in pastures. Fast-growing. Moderately-shade tolerant.

Notes Dutch elm disease is the fungus *Ophiostoma ulmi,* also known as *Ceratocystis ulmi* and *Ceratostomella ulmi.* It was first studied in Holland; hence the name "Dutch". The disease was accidentally introduced to North America in the late 1920s; it is spread by bark beetles (the native *Hylurgopinus rufipes* and the introduced *Scolytus multistriatus*) and also through root grafts. The disease disrupts the water-conducting system of the tree, causing the leaves to wilt and die; within a few years the whole tree succumbs. No satisfactory method has been found to control the fungus in woodlands, although removing and burning infected trees delays the spread. Individual trees can be protected by injecting the trunk with a systemic fungicide, or spraying the whole tree with an insecticide (to kill the bark beetles). Both methods are expensive. Efforts to select and breed resistant individuals have been unsuccessful.

Elms are among the easiest species to grow from seed. Seeds are produced almost every year; they ripen early and germinate readily. Seedlings are large enough to transplant by the end of the first season.

Because of its toughness and flexibility, elm wood was used to make such items as wheel hubs, hockey sticks, fruit baskets, and boat frames.

Quick Recognition Leaves thick, oval, straight-veined, short-stalked; tip pointed, base asymmetrical (most species). Side branches in 2 rows. Buds in 2 rows along the twig; bud scales numerous, in 2 rows. Flowers small, in few-flowered clusters, appearing before the leaves. Fruits slender-stalked, with a membranous wing surrounding a single seed; ripen in spring. Bark with coarse oblique ridges.

White Elm

American elm

Ulmus americana L.

Orme d'Amérique

The largest and most graceful of the elms. Once a characteristic feature of city streets, parks, and rural landscapes in eastern Canada, but Dutch elm disease has eliminated many of the large trees.

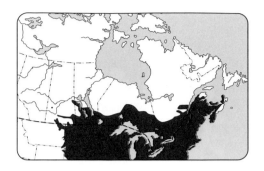

Leaves 10–15 cm long; upper surface dark green, moderately rough (occasionally smooth); paler, slightly hairy beneath; yellow in autumn; 15–20 veins per side, not more than 2 or 3 forked.

Buds Somewhat flattened, about 5 mm long, blunt; end bud bent, others pressed against the twig. Scales 6–9, reddish-brown, sometimes slightly hairy; margins darker, hairy.

Twigs Often decidedly zigzag, grayish-brown, hairless or slightly hairy; lenticels inconspicuous.

Flowers Each on a separate stalk, in loose, few-flowered, tassel-like clusters (fascicles).

Fruits Oval, 8–10 mm long, veined, fringed with hairs, otherwise hairless; wing deeply notched at the tip; seedcase distinct.

Bark Dark grayish-brown, deeply furrowed with broad, oblique intersecting ridges; becoming mottled ash-gray and scaly with age. Outer bark shows alternating layers of light orange-brown and dark brown.

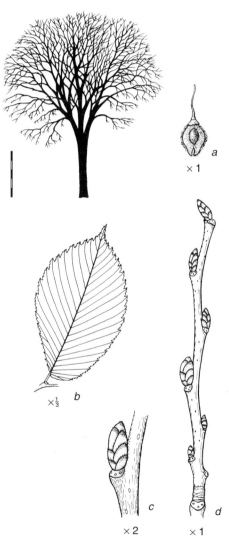

Size and Form One of the largest trees of eastern Canada, up to 35 m high, 175 cm in diameter, and 200 years old. Trunk buttressed at the base with prominent root flares; dividing into a few large, upright limbs and many outwardly fanning branches to form a graceful, spreading, vase- or umbrella-shaped crown easily recognized from a distance. Branch tips often droop, even on relatively young trees. Root system shallow, wide-spreading.

Habitat Occurs mainly on wet sites and on alluvial flats where water often remains in the spring; grows largest on rich, moist, sandy or gravelly loams with good drainage. Moderately shade-tolerant, but grows better in full sunlight.

11

a. Fruit. *b.* Leaf. *c.* Lateral bud and leaf scar. *d.* Winter twig.

Each flower on a separate, undivided stalk.

Mature bark deeply furrowed, mottled and scaly.

Fruit hairless except along wing margin, deeply notched at the tip.

Young bark dark brown with shallow intersecting ridges.

Notes Few large trees have survived Dutch elm disease, but saplings large enough to produce seeds are abundant; seedlings are often numerous on recently disturbed areas. Some individual trees are less susceptible to Dutch elm disease than others. Elm clones with good form and some tolerance to the disease are available for landscape purposes.

Quick Recognition No more than 2 or 3 forked veins per leaf. Buds flattened, pressed against the twig. Fruits oval; wing deeply notched; margin hairy. Bark shows alternating light and dark layers. Tree silhouette vase-shaped with drooping branch tips.

Rock Elm

Cork elm

Ulmus thomasii Sarg.
[syn. *U. racemosa* Thomas)

Orme liège

A rare tree. Its form is the least elm-like of the native elms; easily recognized by the irregular corky ridges on the branches.

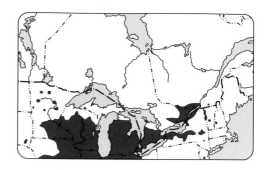

Leaves 5–10 cm long, leathery; upper surface very dark green, shiny, smooth; paler, somewhat hairy beneath; bright yellow in autumn; teeth incurved; about 20 veins per side, rarely forked.

Buds Conical, about 5 mm long, sharp-pointed, plump at the base; diverging from the twig. Scales dark reddish-brown, margins hairy. Vein scars 3–6 per leaf scar.

Twigs Slender, light yellowish-brown, often covered with fine hairs during the 1st year, becoming smooth and dark reddish-brown or ash-gray. Branchlets usually developing irregular corky ridges in the 2nd season.

Flowers In clusters with a central stalk (racemes).

Fruits Oval, 10–15 mm long, hairy, especially on the margin; wing pointed at both ends, narrow, not clearly distinct from seed-case, tip shallowly notched.

Bark Dark gray, tinged with red, shaggy; separated into broad, flat-topped ridges by wide, irregular, interrupted furrows; surface broken into large irregular scales, often with a mottled appearance similar to white elm.

Size and Form Medium-sized trees, up to 25 m high, 75 cm in diameter, and 175 years old. Trunk distinct almost to the top of the tree. Crown cylindrical or somewhat oval. Principal branches comparatively short; lower ones drooping, often crooked and gnarled, with thick ridged bark. General appearance rough and shaggy. Root system deep, wide-spreading.

11

Habitat Most often occurs on the heavy clay soils of limestone ridges, but also on a variety of other soils. Natural regeneration is poor. Seedlings are moderately shade-tolerant but require full sunlight for vigorous growth. Saplings can recover from as much as several decades of suppressed growth under a forest canopy.

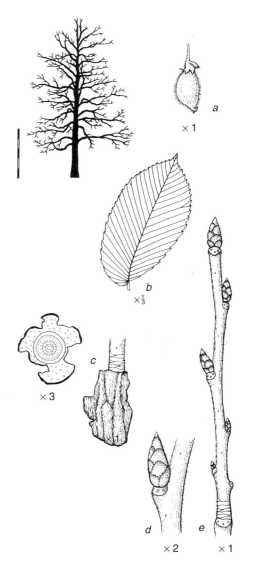

a. Fruit. b. Leaf. c. Corky ridges on 3-year-old twig; cross section (left), lateral view (right). d. Lateral bud and leaf scar. e. Winter twig.

Flowers on slender stalks branching from a central stem.

Fruit entirely covered in hair; wing not distinct from seedcase.

Mature bark has flat-topped ridges with irregular scales.

Young bark gray; branches roughened by corky ridges.

Notes The hardest, heaviest, strongest, and toughest of the elms; wood formerly used where such features are essential, e.g. for making piano frames and hockey sticks, but no longer available in commercial quantities.

Quick Recognition Prominent corky ridges on the branchlets. Leaves shiny dark green; teeth incurved; veins rarely forked. Buds plump, sharp-pointed, dark brown, somewhat hairy, diverging from the twig. Flowers and fruits clustered along a short central stalk; wing hairy, pointed, shallowly notched.

Slippery Elm
Red elm

Ulmus rubra Muhl.

Orme rouge

Not a common species. Noted for its fragrant mucilaginous inner bark.

Leaves 15–20 cm long, fragrant, widest above the middle, tip elongated, base markedly oblique; stalk short, hairy; upper surface dark green, very rough, paler beneath; hairy on both surfaces and on the margin; about 15 veins per side, several forked.

Buds About 6 mm long, tip rounded; dark brown, with reddish-brown hairs, especially at the tip; flower buds almost globular.

Twigs Moderately stout, grayish-brown, hairy; lenticels prominent.

Flowers Each on a separate short stalk, in dense tassel-like clusters (fascicles).

Fruits Almost round, 10–15 mm long; with a line extending from the stalk to the shallow notch at the tip of the wing; reddish-brown hairs cover the seedcase, otherwise hairless.

Bark Brownish, shallowly furrowed with rather irregular, vertical, scaly ridges. Outer bark layers uniform brown; inner bark fragrant, mucilaginous.

Size and Form Medium-sized trees, up to 25 m high, 60 cm in diameter, and 125 years old. Trunk divides into a few wide-spreading upright limbs. Crown broad, flat-topped. Principal branches curve up, then spread out or droop. Root system shallow, wide-spreading.

Habitat Grows best on rich soils along streams and adjacent slopes; sometimes found on rocky ridges. Seedlings thrive in full sunlight or light shade.

11

> **Quick Recognition** Leaves very rough, widest above the middle, several forked veins. Buds blunt, hairy, dark brown. Fruits in tight clusters; seedcase hairy; wing almost round, with a shallow notch. Bark layers uniformly brown.

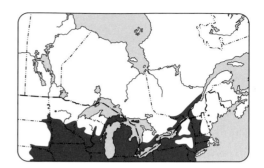

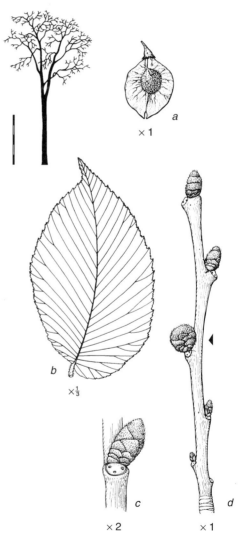

a. Fruit. *b.* Leaf. *c.* Lateral bud and leaf scar. *d.* Winter twig with leaf buds and flower bud (◀).

Flowers on separate short stalks, in dense clusters.

Fruits almost circular, hairless except on seedcase.

Slippery Elm

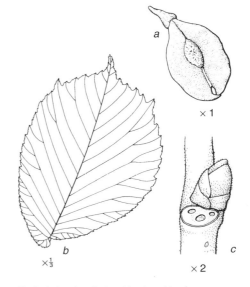

Young bark (inset) dark brown, soon develops irregular ridges. Mature bark shallowly furrowed.

Scotch Elm**

Wych elm

Ulmus glabra Huds.
[syn. *U. campestris* L.]

Orme de montagne

Large trees, native to Europe and western Asia. Resembles slippery elm. Frequently planted in urban areas of eastern North America. Hardy as far north as Zones C4, NA4. **Camperdown elm** is a cultivar of Scotch elm with pendulous branches grafted on a short trunk.

    Leaves 8–16 cm long; tip taper-pointed, often flanked by 2 small lobes; base markedly asymmetrical; very rough above, hairy below, margin hairless. **Buds** triangular, blunt, hairy. **Twigs** grayish-brown, hairy. **Fruits** broadly oval, 20–25 mm long, hairless, with a small notch at the wing tip; almost stalkless; borne in dense, tassel-like clusters. **Bark** smooth compared with other elms, gray.

a. Fruit. *b*. Leaf. *c*. Lateral bud and leaf scar.

Siberian Elm**

Ulmus pumila L.

Orme de Sibérie

Medium-sized trees, native to northeastern Asia. Frequently planted in Canada and elsewhere for hedges; thrive on most soils, often under adverse conditions. Hardy as far north as Zones C3, NA4.

    Leaves narrow, 2–7 cm long, almost symmetrical at the base, margin usually single-toothed; dark green, hairless, smooth, firm; stalk short, tinged dark red. **Buds** small, dark; flower buds very distinctive, large, globular, dark brown; scale margin hairy. **Twigs** slender, brittle, light gray. **Fruits** about 10 mm long, almost round, hairless, with a closed notch at the tip of the wing; short-stalked; borne in small clusters. **Bark** rough, gray.

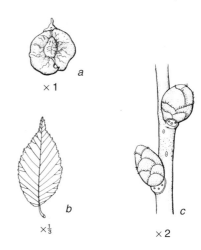

a. Fruit. *b.* Leaf. *c.* Lateral buds; flower bud (above), leaf bud (below).

English Elm**

Ulmus procera Salisb.
[syn. *U. campestris* Mill.]

Orme champêtre

Large trees, common in Great Britain and adjacent Europe. Species status uncertain, possibly a hybrid of *U. glabra* and *U. carpinifolia*. Frequently planted for its graceful form. Hardy as far north as Zones C4, NA4.

    Leaves 6–9 cm long, upper surface rough, hairy beneath; about 12 veins per side, several forked; short-stalked; sometimes remaining on the tree into late autumn. **Buds** ovoid, very dark, mostly hairless. **Twigs** hairy, slender, sometimes with corky ridges. **Fruits** almost round, 10–15 mm across, hairless; wing tip with a closed notch extending inward almost to the seedcase; borne in small tassel-like clusters. **Bark** deeply fissured. Reproduces abundantly by root sprouts; often bears sprouts along the trunk.

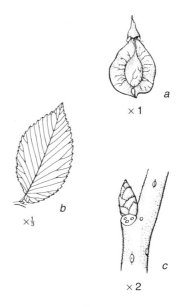

a. Fruit. *b.* Leaf. *c.* Lateral bud and leaf scar.

11

Japanese Zelkova*

Zelkova serrata (Thunb.) Mak.
Ulmaceae: Elm Family

Zelkova du Japon

Medium-sized trees, native to Japan and adjacent parts of Asia. Its vase-shaped form and ascending branches together with its tolerance to Dutch elm disease make Japanese zelkova a candidate to replace the disease-killed elms. Hardy as far north as Zones C5, NA5.

Leaves ovate, 3–5 cm long, upper surface bright green, rough; veins straight, 8–14 per side, each ending in a single sharp tooth. **Buds** ovoid, pointed, with many brown scales; diverging from the twig. **Twigs** brown. **Fruits** 4 mm in diameter, cherry-like but with dry flesh; borne laterally on new shoots. **Bark** smooth, gray, with horizontal pink lenticels.

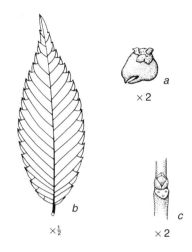

a. Fruit. *b*. Leaf. *c*. Lateral bud and leaf scar.

Hackberry
Northern hackberry
Celtis occidentalis L.
Ulmaceae: Elm Family

Micocoulier occidental

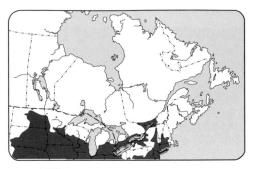

Worldwide there are about 60 species of hackberry, distributed in temperate and in tropical regions; 6 are native to North America and 2 to Canada.

Northern hackberry is sparsely distributed in Ontario and Quebec; also reported from the south end of Lake Manitoba.

Leaves Alternate in 2 rows along the twig, variable in size and shape, 6–9 cm long, widest toward the asymmetrical base, long tapering tip; upper surface deep bluish-green; paler and hairy on the veins beneath; yellow in autumn; short-stalked. Lowest pair of veins originate at an acute angle just below the base of the blade; 4 or 5 veins and 15–40 teeth per side.

Buds Ovoid, 6–8 mm long, flattened, pointed, in 2 rows along the twig, with 5 or more scales in 2 rows. No true terminal bud; end bud originates as a lateral bud, often bent. Leaf scars raised, semi-oval, with 3 vein scars.

Twigs Slender, green, becoming tinged with brown, covered with fine hairs. Pith with cavities separated by transverse partitions, especially at the leaf nodes.

Flowers Small, greenish. Pollen flowers and seed flowers on the same tree. Pollen flowers in small clusters at the base of the new shoot; seed flowers single, in the axils of the new leaves. Wind-pollinated.

Fruits Berry-like drupe, with a pitted stone, reddish-purple, 6–8 mm across, solitary on slender stalk, persisting on the tree in winter; edible. Seed crops abundant most years. Dispersed by fruit-eating birds and mammals.

Vegetative Reproduction By stump sprouts.

Bark Gray to light yellowish-brown; irregular narrow ridges with distinctive wart-like projections; stratified layers in cross section.

Wood Heavy, coarse-textured, hard, weak; brown streaked with yellow; attractive grain-pattern on tangential surfaces. Ring-porous; pores, rays, and annual rings easily visible.

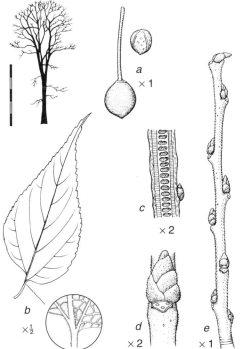

a. Fruit (left); pitted stone (right). b. Leaf with detail of vein division below leaf base. c. Longitudinal section of twig showing chambered pith. d. Lateral bud and leaf scar. e. Winter twig.

Size and Form Small trees, up to 15 m high, 50 cm in diameter, and 150 years old. Crown broad, with ascending, arching branches that droop at tips.

Habitat Grows on a variety of soils. Moderately shade-tolerant.

Notes Used in landscape planting as a substitute for elm; subject to a mite (*Eriophyes* sp.) which causes "witches broom", a cluster of numerous upright small branches.

Common at Point Pelee and other localities on Lake Erie; presumably seed is brought there by birds on their northward migration.

11

Pollen flowers (above) in clusters at base of new shoot. Seed flowers (below) single, in leaf axil.

Hackberry

Fruit solitary on slender stalks.

Young bark (inset) shows ridges and wart-like projections. Mature bark coarsely ridged, scaly.

> **Quick Recognition** Leaves asymmetrical, coarsely toothed; leaf scars raised; fruits berry-like; mature bark with wart-like projections.

Dwarf Hackberry

Georgia hackberry

Celtis tenuifolia Nutt.

Micocoulier rabougri

Shrubs or very small trees of central North America; occurs rarely in the extreme southern part of Ontario. Differs from northern hackberry in its smaller fruit and in leaf features: smaller, broader, more symmetrical; those on fruiting shoots not toothed.

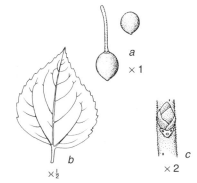

a. Fruit (left); indistinctly pitted stone (right).
b. Leaf . *c.* Lateral bud and leaf scar.

Serviceberries
Juneberry, Saskatoon, shadbush,
Indian pear

Genus *Amelanchier*
Rosaceae: Rose Family

Les amélanchiers

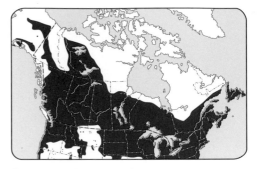

Most of the 16 species of serviceberry are
native to North America; about 10 of these
(5 or 6 of tree size) are native to Canada.
Every province has at least one species of
serviceberry.
    Although it is easy to determine whether
a specimen is a serviceberry, it is often
difficult to identify it to species, especially in
eastern Canada.

Leaves Deciduous, alternate, simple; oval to
almost round, usually less than 8 cm long,
slender-stalked; teeth small, regular, sharp,
1–3 per lateral vein, often toothless toward
the stalk; veins tending to be straight and
parallel, about 10 per side.

Buds Narrowly ovoid, 8–12 mm long,
twisted, tapering to a point; pressed tightly
against the twig; about 5 scales. Terminal
bud much like the lateral buds. Leaf scars
with 3 large vein scars.

Twigs Slender; ridges extend down from
either side of the leaf scar. Pith 5-pointed. A
neoformed shoot usually develops from one
or more leaf axils below a terminal flower
cluster.

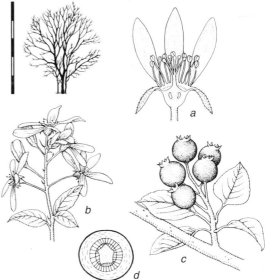

a. Longitudinal section through flower (2 petals
removed) showing enclosed ovary. b. Flower
cluster. c. Fruit cluster. d. Twig cross section.

Flowers White, showy; 5 petals. In elongated clusters (racemes) at the tips of new
leafy shoots; lower stalks longer than the
upper ones. Appear early in spring, before or
with the leaves. Insect-pollinated.

Fruits Berry-like, 6–10 mm across, reddish
or purplish, with 5–10 hard seeds; flesh usually sweet, juicy, edible; ripening in late July
or early August. Dispersed by fruit-eating
birds and mammals.

Seeds Remain viable for some years at
near-freezing temperatures; germinate after
exposure to moist cool conditions.

Seedlings Cotyledons small leafy, raised
above the surface during germination.

Vegetative Reproduction Several species
by root sprouts or underground stems
(stolons), thus forming clumps or thickets.

Bark Smooth, gray, conspicuously marked
by a slightly twisted network of darker
vertical lines; with age becoming rough and
scaly.

Size and Form Some species shrubby;
others become very small trees, up to 10 m
high and 20 cm in diameter. Trunk slender,
with very little taper. Crown narrow, irregular.

Habitat Found in the forest understory, at
forest edges, on sand plains and rock outcrops, and along fencerows.

Notes Twigs, bark, and fruits provide food
for many birds and mammals.

Quick Recognition Leaves oval with prominent straight veins and regular sharp teeth;
buds long, narrow, pointed, twisted; early spring
flower clusters showy; fruits small, berry-like;
bark smooth gray.

11

Serviceberry flowers showy, appear early in spring.

Downy Serviceberry

Downy juneberry

Amelanchier arborea (Michx. f) Fern.

Amélanchier aborescent

Shrubs or small trees up to 12 m high, occurring on dry sites in forested and open areas in southern Ontario and Quebec. **Leaves** oval, 5–9 cm long, tapered to a sharp tip, about 25 teeth per side; less than half as many veins; small, densely hairy, folded at flowering time; hairless or slightly hairy on the undersurface when mature. **Flowers** in dense, erect clusters; petals 8–12 mm long. **Fruits** dry, flavorless; lowermost fruit stalk about 12 mm long.

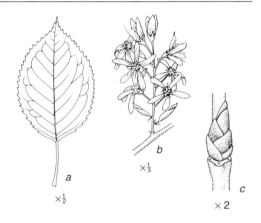

a. Leaf. *b.* Flower cluster. *c.* Lateral bud and leaf scar.

Smooth Serviceberry

Smooth juneberry, Allegheny serviceberry

Amelanchier laevis Wieg.

Amélanchier glabre

Similar to the downy serviceberry and possibly a variety of it; occurs in moist woodlands from Newfoundland to Lake Superior. **Leaves** oval, 3–8 cm long, abruptly tapered to a sharp tip; at flowering time, very distinctively coppery-red, at least half-grown, and hairless; about 25 teeth and 10 veins per side; veins stop short of the teeth. **Flowers** in drooping clusters; petals 10–17 mm long. **Fruits** juicy, sweet; lowermost fruit stalk about 25–45 mm long.

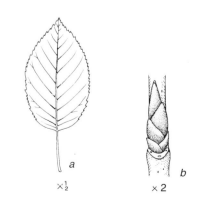

a. Leaf. *b.* Lateral bud and leaf scar.

Saskatoon

Saskatoon-berry, western serviceberry

Amelanchier alnifolia (Nutt.) Nutt.

Amélanchier à feuilles d'aulne

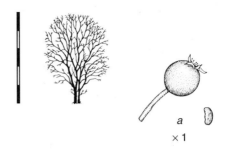

The most common native species of service-berry west of Manitoba.

Leaves Oval to roundish, 2–5 cm long; tip rounded, base rounded or somewhat heart-shaped; upper surface dark green; paler beneath, with fine hairs along the midvein; teeth coarse, somewhat incurved, on the upper two-thirds of the leaf, 1–2 per vein; 8–13 veins per side; stalks 8–18 mm long.

Buds 2–8 mm long, reddish-brown.

Twigs Reddish-brown to grayish-brown, covered with white hairs in spring, becoming hairless.

Flowers White, fragrant, petals 6–10 mm long; in short, dense, erect clusters, 2–3 cm long, with hairy stalks; at flowering time leaves are unfolded but not fully grown.

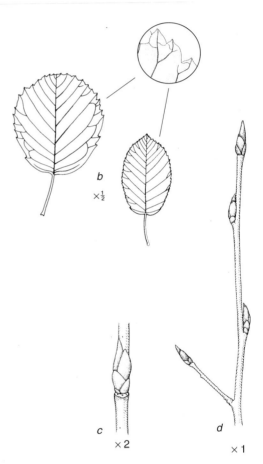

Fruits Globular, 6–10 mm across, dark purple to black, sweet, juicy, with brown seeds; edible.

Bark brown to gray, tinged with red.

Size and Form Shrubs or very small trees up to 4 m, in colonies formed by erect sprouts on underground stems (stolons).

Habitat Occurs on rocky ridges, along streams, and on forest edges.

Quick Recognition Leaf tip rounded; leaves more than half-grown at flowering time; veins 8–13 per side, conspicuous, straight; flower clusters dense, erect; fruits, blackish, juicy; trees in colonies.

a. Fruit (left); seed (right). *b.* Leaf forms and detail of incurved, coarse teeth. *c.* Lateral bud and leaf scar. *d.* Winter twig.

11

Flowers in clusters, appear before leaves are fully grown.

Fruits globular, purple when ripe.

Mature bark smooth, brown or gray.

Saskatoon

Pacific Serviceberry

Western serviceberry

Amelanchier florida Lindl.

Amélanchier de l'Ouest

Shrubs or very small trees of the Pacific coast. **Leaves** elliptic, rounded at both ends, coarsely toothed, thin; upper surface dark green; paler, nearly hairless beneath. **Buds** purplish. **Twigs** reddish-brown, becoming hairless. **Flowers** in clusters, 4–7 cm long. **Fruits** 10–12 mm across, purple, coated with a whitish powder; edible, sweet, juicy. **Bark** gray or brown.

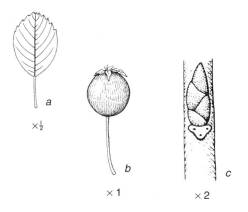

a

$\times \frac{1}{2}$

b

$\times 1$

c

$\times 2$

a. Leaf. *b.* Fruit. *c.* Lateral bud and leaf scar.

Roundleaf Serviceberry

Roundleaf juneberry, New England serviceberry

Amelanchier sanguinea (Pursh) DC.

Amélanchier sanguin

Shrubs or very small trees up to 10 m high, growing in colonies; occurs from Lake Superior eastward. **Leaves** elliptic to roundish, 3–7 cm long, hairless at maturity; coarsely toothed, about 18–20 per side; veins 10–12 per side, each ending in a tooth. **Flowers** in drooping clusters 3–8 cm long, appearing with the unfolding leaves (which are densely hairy at this time); petals 10–15 mm long. **Fruits** 5–10 mm across, dark purple, juicy; stalks 10–30 mm long. **Twigs** reddish when young.

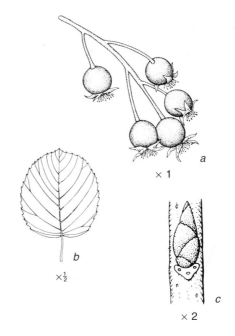

× 1

× ½

× 2

a. Fruit cluster. *b.* Leaf. *c.* Lateral bud and leaf scar.

Mountain Serviceberry

Mountain juneberry

Amelanchier bartramiana (Tausch) M. Roem.

Amélanchier de Bartram

A northern shrubby species occurring on moist, acid sites from Labrador to Lake Superior; a very small, rare tree on good sites. **Leaves** oval, 2.5–5 cm long, usually tapered to both ends; hairless; not folded when emerging in spring; finely toothed. **Flowers** unlike those of the other species: in clusters of 1–3 in the leaf axils. **Fruits** elongated, sweet, dry.

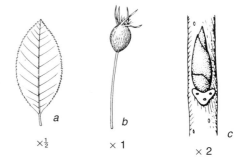

× ½

× 1

× 2

a. Leaf. *b.* Fruit. *c.* Lateral bud and leaf scar.

11

common apple

Siberian crab apple

Flowers and fruits borne on leafy dwarf shoots.

Apples

Crab apples

Genus *Malus*

Rosaceae: Rose Family

Les pommiers

The apple genus comprises about 25 species; 2 are native to Canada. Some authors include both apples and pears in the genus *Pyrus*.

Apples are an important source of food for many birds and mammals. Hundreds of cultivars developed from non-native species are planted in cities and orchards of Canada and elsewhere for their spring blossoms and edible fruits. They are often naturalized.

Characteristic features of the genus include ovate, toothed **leaves**; showy white pink or red **flowers,** which are arranged in small clusters on leafy dwarf shoots and open in spring before the leaves are fully grown. Each flower has 5 petals, many stamens, and a 5-celled ovary; the calyx is green, with 5 lobes (sepals), and is set on a floral cup (hypanthium) that encloses the ovary (compare cherries and plums). The floral cup forms the outer part of the **fruit**; the calyx lobes persist at the apex of the fruit. Seeds dispersed by fruit-eating birds and mammals.

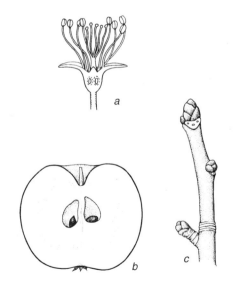

a. Longitudinal section through flower (petals removed) showing enclosed ovary. *b.* Longitudinal section through fruit. *c.* Winter twig with dwarf shoot, terminal bud, and lateral buds.

Wild Crab Apple
Sweet crab apple

Malus coronaria (L.) Mill.
[syn. *Pyrus coronaria* L.]

Pommier odorant

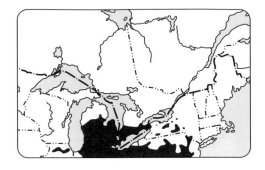

Occurs in southern Ontario near Lakes Erie and Ontario.

Leaves Elongate triangular, 3−8 cm long, base roundish to heart-shaped; almost hairless on the undersurface; teeth increasing in size from the tip downward to become almost lobe-sized near the base.

Buds Terminal bud 3−5 mm long, elongated, blunt, red, hairy, with 5 scales. Lateral buds similar but smaller, 2−4 mm long, pressed against the twig. Vein scars 3 per leaf scar.

Twigs Slender, reddish-brown with a grayish skin that gradually flakes off; hairy during the 1st year. Dwarf shoots often end in a thorn.

Flowers White, streaked with pink, very showy, in small clusters of 2−3, 20−30 mm across; opening after the new leaves have unfolded.

Fruits Almost round, 25−35 mm across, as wide as or wider than long, green, hard, wax-coated, sour; remaining on the tree into winter.

Bark Reddish-brown, scaly.

Size and Form Very small trees, up to 9 m high and 30 cm in diameter. Crown irregular, spreading. Often a thicket-forming shrub.

Habitat Grows in the shade of larger broad-leaf species; also in the open along fences and roadsides.

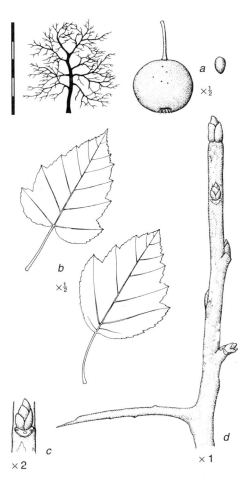

| Quick Recognition | Showy flowers on leafy dwarf shoots, leafy thorns, red buds, apple fruits. |

11

a. Fruit (left); seed (right). *b.* Leaf forms. *c.* Lateral bud and leaf scar. *d.* Winter twig with thorn-tipped dwarf shoot.

Flowers showy, open after leaves have unfolded.

Mature bark scaly, reddish-brown.

Fruit remains green, hard, and waxy.

Young bark shiny, smooth.

Pacific Crab Apple

Oregon crab apple

Malus fusca (Raf.) C.K. Schneid.
[syn. *Pyrus diversifolia* (Bong.) M. Roem.]

Pommier du Pacifique

A western species, similar to the eastern wild crab apple. **Fruits** smaller, 12–20 mm long, longer than wide, red or yellow with red patches; not waxy. Grows in swampy woods and along streams.

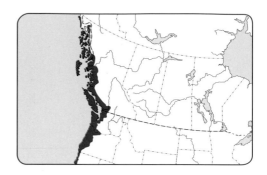

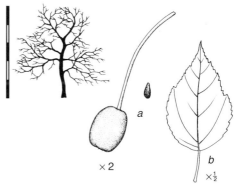

a. Fruit (left), seed (right). *b.* Leaf .

Common Apple**

Malus sylvestris (L.) Mill.
[syn. *Pyrus malus* L.]

Pommier sauvage d'Europe

Numerous cultivars planted in orchards; naturalized in many places. Some hardy as far north as Zones C3, NA3. **Leaves** ovate, 4–10 cm long, blunt-pointed, base rounded, finely toothed, upper surface dark green, densely hairy beneath. **Twigs and buds** hairy. Many dwarf shoots, no thorns. **Flowers** white or pink, opening in spring after the leaves unfold. **Fruits** from seed-grown trees usually inferior; cultivars producing choice fruit have been propagated by grafting.

11

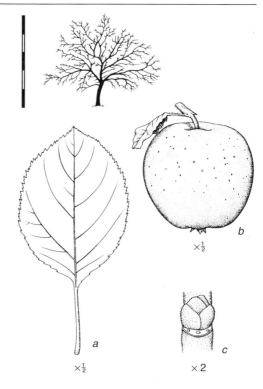

a. Leaf. *b.* Fruit. *c.* Lateral bud and leaf scar.

Pacific Crab Apple

Fruit longer than wide, not waxy.

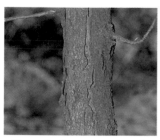

Young bark soon rough and scaly.

Mature bark has large flat scales.

Siberian Crab Apple*
Flowering crab apple

Malus baccata (L.) Borkh.
[syn. *Pyrus baccata* L.]

Pommier de Sibérie

Many crab apple cultivars are planted for their showy blossoms of white, pink, or red; some of them are derived from this species by selection and hybridization. All are characterized by having the calyx attach to the ovary, and apple fruits. Some hardy as far north as Zones C2, NA2.

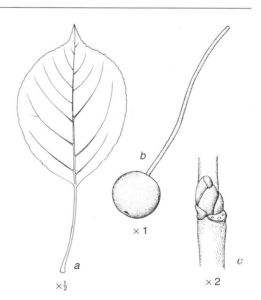

a. Leaf. *b.* Fruit. *c.* Lateral bud and leaf scar.

b

× 1

a

× ½

c

× 2

Cherries and Plums

Genus *Prunus*
Rosaceae: Rose Family

Les cerisiers et les pruniers

A large genus of about 200 species that also includes apricots, peaches, and almonds. About 18 tree-size species of cherry and plum are native to North America; 7 to Canada.

Numerous cultivars of native and introduced species are grown in Canada for their fruits and as ornamentals; propagation is by grafting.

Most cherries and plums are subject to the fungal disease black knot (*Apiosporina morbosa*); it produces large swellings on the branches leading to their death; this renders the tree unsightly and reduces the yield of fruit.

Characteristic features of the genus include showy **flowers** with 5 sepals, 5 petals, and 1 pistil; fleshy **fruits** (drupes) with a single stone; outer flower parts rarely remaining attached to the fruit apex (compare apple); **leaves** toothed, often with a pair of warty glands at the top of the leaf stalk; 10–15 **bud scales**, in several rows; 3 vein scars on each leaf scar; **twigs** with the odor of bitter almond when bruised, covered with a grayish skin that wears off. Wilted leaves may release prussic acid (hydrocyanic acid) and be poisonous to livestock.

The cherries and plums described on the following pages fall into 3 subgenera:

Cerasus (pin cherry, bitter cherry, sweet cherry, sour cherry, and Japanese flowering cherry) are cherries with flowers in few-flowered tassel-like clusters (umbels or umbel-like racemes) on previous year's twigs; fruits smooth, globular with a globular stone; terminal bud present; bark shiny, in thin layers, with horizontal lenticels.

Padus (black cherry, choke cherry, and Amur choke cherry) are cherries with flowers in elongated clusters (racemes) on short leafy shoots; fruits smooth, globular with a globular stone; terminal bud present; bark rough.

Prunophora (Canada plum and wild plum) are plums with flowers in small clusters or solitary, on previous year's twigs; fruits smooth, grooved on one side, elongated, with an elongated flattened stone; no terminal bud; bark rough; often thorny.

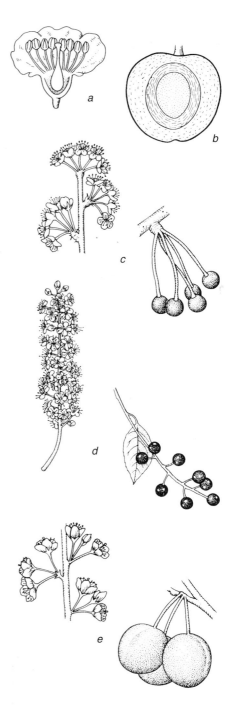

a. Longitudinal section through flower, 2 petals removed. *b.* Longitudinal section throught fruit showing flesh, stone, and seed. *c.* Tassel-like floral clusters and fruit cluster of subgenus *Cerasus*. *d.* Elongated floral cluster and fruit cluster of subgenus *Padus*. *e.* Few-flowered clusters and fruit cluster of subgenus *Prunophora*.

11

pin cherry
Cherries wild or cultivated are showy when in flower.

Sweet Cherry**

Mazzard, gean

Prunus avium (L.) L.
Subgenus *Cerasus*

Cerisier sauvage

A common orchard tree in the milder parts of
Canada; native to Asia Minor. Hardy as far north
as Zones C4, NA3. **Leaves** obovate, double-
toothed, gland-tipped, upper surface dull dark
green, hairy beneath at the vein axils. **Buds** shiny
brown, clustered on dwarf shoots, diverging widely
from the twig. **Flowers** white, in small clusters.
Fruits 25 mm across, deep red, sweet, juicy.

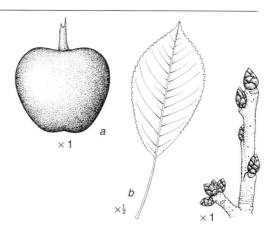

a. Fruit. *b*. Leaf. *c*. Winter twig with dwarf shoot,
lateral buds, and leaf scars.

Sour Cherry**

Pie cherry

Prunus cerasus L.
Subgenus *Cerasus*

Cerisier aigre

A common orchard tree in the milder parts of
Canada (Zones C4, NA3); native to Eurasia.
Leaves elliptic to obovate, blunt-pointed, double-
toothed; leaf stalks without glands; hairless. **Buds**
conical, shiny reddish-brown, clustered on dwarf
shoots, diverging widely from the twig. **Flowers**
white, in small long-stalked clusters on short leafy
shoots. **Fruits** 20 mm across, bright red, tart, juicy.

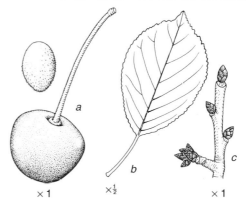

a. Fruit with stone (above). *b*. Leaf. *c*. Winter twig
with dwarf shoot, lateral buds, and leaf scars.

Pin Cherry

Wild red cherry, fire cherry

Prunus pensylvanica L. f.
Subgenus *Cerasus*

Cerisier de Pennsylvanie

Occurs in most of the forested parts of
Canada, from central British Columbia eastward.

Leaves Lance-shaped, 8–15 cm long; gradually tapering to a slender, sharp tip; widest
below or near the middle; thin and fragile;
shiny yellowish-green, hairless, often turning
bright purplish-red in autumn; commonly
curved backward; minute, uneven teeth; margin folded upward along the midvein.

Buds Small, 1–2 mm long, rounded, diverging slightly from the twig; terminal bud
and several similar lateral buds clustered at
the end of the twig.

Twigs Very slender, reddish; neoformed
branches may develop on vigorous shoots.

Flowers White, petals about 5 mm long; in
tassel-like clusters of 4–7, stalks about
20 mm long. Appear when the leaves are
about half-grown.

Fruits Bright red, 6–8 mm across; thin,
sour, edible flesh; slender-stalked. Ripen late
July to early September; produced abundantly.

Seeds May live for some decades on the
forest floor; germination stimulated by exposure to light and by temperature fluctuations
following clearing and disturbance.

Seedlings Cotyledons small, leafy, raised
above the surface.

Vegetative Reproduction Commonly by
root sprouts.

11

Bark Smooth, shiny, dark reddish-brown;
lenticels conspicuous, large, widely spaced,
horizontal, orange powdery; with age, separating into horizontal papery strips.

Size and Form Small trees, up to 12 m high,
25 cm in diameter, and 40 years old; a shrub
on unfavorable sites. Trunk fairly straight.
Crown narrow, round-topped; branches at
first ascending, becoming nearly horizontal
and spreading as the tree ages.

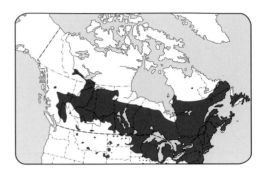

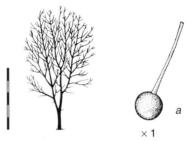

a
× 1

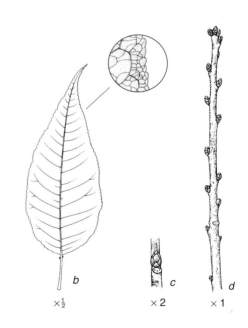

b   c   d
×½   × 2   × 1

a. Fruit . *b.* Leaf with detail of minute uneven
teeth. *c.* Lateral bud and leaf scar. *d.* Winter twig.

Flowers in clusters of 4 – 7.

Fruit slender stalked, bright red.

Young bark shiny, with conspicuous orange lenticels.

Trunk straight; mature bark has horizontal papery strips.

Habitat Usually occurs in areas recently cleared by cutting, windthrow, or burning; also, along rivers in the Prairies; found as scattered individual trees. Intolerant of shade; hence seldom found in mature forests.

Notes Hybridizes with bitter cherry where their ranges overlap.

Quick Recognition Small bright red cherries in small clusters. Shiny reddish-brown bark with conspicuous horizontal lenticels. Lance-shaped leaves with uneven margins. Clusters of small buds at the twig tip.

Bitter Cherry

Prunus emarginata Dougl.
Subgenus *Cerasus*

Cerisier amer

A western species found in moist, sparsely wooded areas along streams and on areas recently cleared; quite similar to the pin cherry; differences are noted.

Leaves Oval or widest above the middle, 3–8 cm long; gradually tapering to the base, tip rounded or wedge-shaped; both surfaces dull yellowish-green, hairless or downy beneath; ascending or spreading stiffly from the twig.

Buds Larger, 3–4 mm long, pointed.

Twigs Downy becoming smooth, darker red.

Flowers In clusters of 5–12.

Fruits Larger, 6–15 mm across, bright red becoming very dark red; bitter, astringent flesh; ripen in July or August.

Bark Bitter taste accentuated.

Size and Form Larger trees, up to 20 m high and 70 cm in diameter. Trunk slender, distinct to the top of a narrow crown of ascending branches; branch-free portion short compared with crown.

Notes Narrow strips of bark ideal for ornamental basketwork.

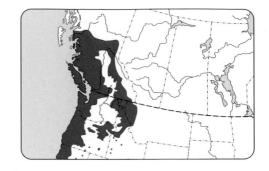
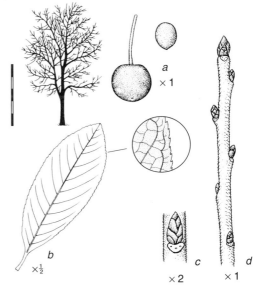

a. Fruit (left); stone (right). *b.* Leaf with detail of minute uneven teeth. *c.* Lateral bud and leaf scar. *d.* Winter twig.

Japanese Flowering Cherry*

Prunus serrulata Lindl.
Subgenus *Cerasus*

Cerisier du Japon

A landscape tree, valued for its abundant showy flowers; native to China and Japan. Hardy only in the mildest parts of Canada and farther south (Zones C6, NA5). Many cultivars have been developed. **Leaves** elliptic, base rounded, tip abruptly long-pointed, sharply toothed. **Buds** of 2 sizes; flowers buds much larger. **Flowers** white or pink; very showy, often double, in large clusters; opening as the leaves develop, or just before. **Bark** grayish-brown, peeling horizontally.

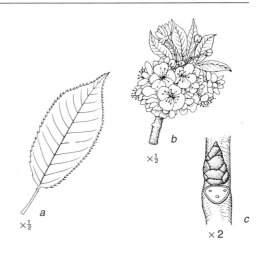

a. Leaf. *b.* Flower cluster. *c.* Lateral bud and leaf scar.

11

Fruit larger than pin cherry; turns dark red.

Flowers similar to pin cherry.

Bitter Cherry

Young bark; orange lenticels prominent.

Mature bark smooth, dark reddish-brown.

Black Cherry

Prunus serotina Ehrh.
Subgenus *Padus*

Cerisier tardif

The largest member of the genus *Prunus* native to Canada; toward its northern limit it may be a shrub.

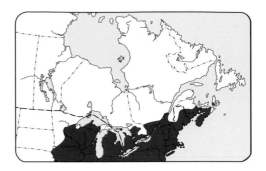

Leaves Lance-shaped, 5–15 cm long, gradually tapering to both ends, sharp-pointed, thick and leathery; upper surface shiny bright green; paler beneath, with a narrow mat of fine brown hairs along each side of the basal part of the midvein; teeth distinctly elongated, with a sharp incurved tip resembling a birds' beak.

Buds 3–4 mm long, blunt, reddish-brown, diverging slightly from the twig; with about 10 scales, brown with darker tips and green bases.

Twigs Slender, reddish-brown.

Flowers In elongated loose clusters, 10–15 cm long, at the end of new short leafy shoots; flower stalk 5 mm long. Open as the leaves reach full size.

Fruits In elongated drooping clusters of 6–12 fruits. Each 8–10 mm across, very dark reddish black, astringent but edible; the lower whorl of flower calyx retained at the base of each fruit. Ripen in August or early September. Abundant seed crops occur every 3–4 years.

Vegetative Reproduction By stump sprouts.

Bark Smooth on young trees; very dark reddish-brown to blackish, with conspicuous, horizontal, dash-like, grayish lenticels; with age, separating into squarish scales, curved outward at their vertical edges, reddish-brown on the inner surface; lenticels still visible.

Wood Moderately heavy, hard, strong, light to dark reddish-brown, decorative, easy to work. Semi-ring-porous; pores visible with a hand lens; rays and annual rings visible to the naked eye.

Size and Form Medium-sized trees, up to 22 m high, 60 cm in diameter, and 150 years old. Trunk sinuous with little taper; branches arching, with drooping tips. Root system shallow, wide-spreading; seedling produces a taproot in the 1st year. Fast-growing when young.

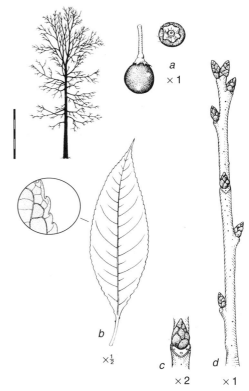

a. Fruit on stalk (left); view of large calyx (right).
b. Leaf with detail of elongated, incurved teeth.
c. Lateral bud and leaf scar. *d.* Winter twig.

Habitat Grows well on a wide variety of soils; mixed with other broadleaf species, such as sugar maple, white ash, basswood, yellow birch, white oak, shagbark hickory, and tulip-tree. Intolerant of shade.

Notes Wood used in furniture-making.

Quick Recognition Undersurface of leaves with a narrow mat of hair on either side of the midvein; buds with touches of green; bark scaly when tree-size; calyx retained at base of each fruit.

11

Black Cherry

Flowers (above) in elongated clusters. Fruit (below) dark reddish-black when ripe.

Young bark (inset) smooth, very dark with gray lenticels. Mature bark conspicuously scaly.

Amur Choke Cherry*

Prunus maackii Rupr.
Subgenus *Padus*

Cerisier de Mandchourie

A landscape tree cultivated for its unusual bark; one of the hardiest cherries; native to Korea and Manchuria. Hardy as far north as Zones C2, NA2. **Leaves** narrowly ovate to elliptic; long-pointed, rounded at the base, finely toothed, glandular dots on the underside. **Twigs** hairy. **Flowers** white, in elongated hairy clusters, on short leafless branches; appearing as the leaves develop. **Fruits** black, 5–6 mm across; ripen in August. **Bark** a rich cinnamon brown, with patches of gold, red and gray; in thin layers, often peeling horizontally in shaggy masses.

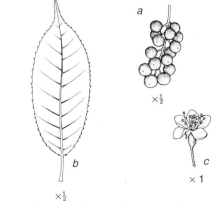

$\times \frac{1}{2}$

a $\times \frac{1}{2}$

b

c $\times 1$

a. Fruit on stalk. *b.* Leaf. *c.* Flower.

Choke Cherry

Eastern choke cherry

Prunus virginiana L. var. *virginiana*
Subgenus *Padus*

Cerisier de Virginie

Transcontinental except for coastal British Columbia.

Leaves Broadly oval to widest above the middle, 5–10 cm long, thin; abruptly tapering to both ends, particularly to the short, sharp tip; upper surface dull green; paler beneath with occasional tufts of hair at the vein axils; teeth sharp, closely spaced, 4–5 per vein, each tooth ending in a straight hair-like point.

Buds 3–4 mm long, sharp-pointed, diverging slightly from the twig; scales dark brown with pale edges.

Twigs Slender to moderately stout, smooth, grayish-brown. Only member of genus without a grayish skin that wears off.

Flowers In elongated, dense, cylindrical clusters, 8–15 cm long, at the end of new short leafy shoots; flower stalks 5–6 mm long. Open before the new leaves are fully developed.

Fruits In elongated drooping clusters of 6–12 fruits. Each 8–10 mm across, varying by tree from yellow to crimson to black, very astringent but edible; minute remnant of the flower calyx at the base of each fruit. Ripen in August or early September.

Vegetative Reproduction Prolifically by root and stump sprouts.

Bark Smooth, or with fine scales, dark grayish-brown when young; becoming almost black with age; lenticels prominent but not extended horizontally as in most species of *Prunus*.

Size and Form Usually a tall shrub, but on favorable sites a very small tree up to 9 m high and 15 cm in diameter. Trunk slender, often inclined, crooked, and twisted; branches slender, upright to slightly spreading.

Habitat Commonly occurs on open sites with rich, moist soils, such as along fence-rows and streams, on cleared land, bordering wooded areas. Relatively intolerant of shade.

11

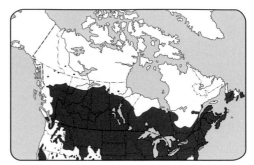

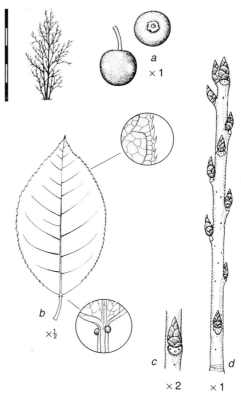

a. Fruit on stalk (left); view of small calyx (right).
b. Leaf with details of closely spaced teeth (above) and glands on leaf stalk (below). *c.* Lateral bud and leaf scar. *d.* Winter twig.

Notes An important source of browse and fruit for a number of mammals and birds.

Quick Recognition Leaves widest above the middle, with small sharp teeth; buds with 2 shades of brown; bitter almond odor of bruised twigs very strong. Flowers earlier than black cherry.

Flowers in dense elongated clusters.

Choke Cherry

Fruits in drooping clusters.

Young bark (inset) dark grayish-brown; lenticels prominent but not horizontally extended. Mature bark almost black.

Western Choke Cherry

Prunus virginiana var. *demissa* (Nutt.) Torr.

Cerisier du Pacifique

Occurs on moist sites in the interior of southern British Columbia. **Leaves** rounded at the base, widest above the middle, dark green, downy beneath; young **twigs** downy. **Fruits** almost black, astringent. Often with several irregular stems.

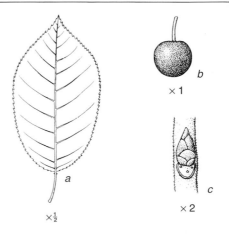

a. Leaf . *b.* Fruit. *c.* Lateral bud and leaf scar.

Canada Plum

Prunus nigra Ait.
Subgenus *Prunophora*

Prunier noir

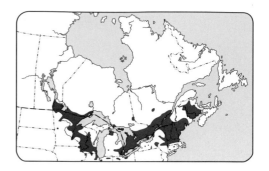

Occurs in the southern parts of eastern Canada. Several cultivars have been developed and planted as ornamentals because of the attractive showy flowers.

Leaves Broadly oval to widest above the middle, 6–12 cm long; narrowing abruptly to a long, slender tip; base widely wedge-shaped, rounded, or heart-shaped; upper surface dull dark green, slightly paler beneath; thin, fragile; double-toothed; teeth prominent, rounded, the smaller ones usually gland-tipped; midvein prominent; lateral veins slender; stalks stout with large, dark glands.

Buds Grayish-brown, 4–8 mm long, appressed against the twig; scale tips thin, pale, frayed.

Twigs Slender, smooth, reddish-brown. Shorter shoots often end in a thorn; may bear leaves and flowers.

Flowers White turning pink, showy, fragrant, 15–25 mm across, on short slender reddish stalk. In small clusters (umbels) of 2–4 flowers, on previous year's twigs. Open at the same time as or just before the new leaves appear.

Fruits Somewhat elongated, 25–30 mm long; skin thick, red, orange-red, or yellow, not powdery; flesh yellow, juicy, sour. Ripen from mid-August to early September.

Bark Black, with slightly elongated grayish lenticels; with age, splitting vertically, the 2 separated parts curling back slightly to reveal the inner bark; gradually becoming scaly.

Size and Form Very small trees, up to 9 m high and 25 cm in diameter; often occurring in clumps. Trunk short, crooked, often divided about a metre above the ground into several stiff, upright, crooked branches. Crown irregular, flat-topped.

Habitat Grows best in river valleys; also found in pastures and along fencerows, especially on lime soils; occurs as widely scattered individual trees.

11

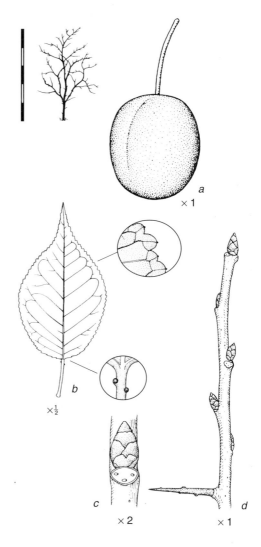

a. Fruit. *b.* Leaf with details of gland-tipped teeth (above) and glands on leaf stalk (below). *c.* Lateral bud and leaf scar. *d.* Winter twig with thorn-tipped dwarf shoot.

Flowers in small clusters on previous year's twigs.

Fruit somewhat elongated, without a powdery coating.

Young bark black, with grayish lenticels.

Mature bark scaly, splits vertically to expose inner bark.

Quick Recognition Twigs with bitter almond flavor, thorny; leaves with blunt, gland-tipped teeth; flowers white, opening before the leaves expand; fruits red to yellow.

American Plum
Wild plum

Prunus americana Marsh.
Subgenus *Prunophora*

Prunier d'Amérique

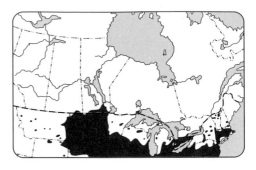

A species of central North America; found in southern Ontario, southern Manitoba, and the extreme southeastern portion of Saskatchewan. Frequently planted in Canada beyond its natural range as an ornamental and for its fruit (Zones NA3, C3). Similar to Canada plum; differences are noted.

Leaves Ovate, usually widest slightly below the middle, 6–10 cm long; fully rounded at the base or with a slight taper toward the stalk; hairless or slightly hairy; double- or single-toothed; teeth sharp, not gland-tipped.

Buds Grayish, pointed, 3–8 mm long; scales 2 shades of pale grayish-brown.

Twigs Grayish- to reddish-brown.

Flowers White, about 25 mm across, on slender greenish-brown stalks. In small clusters (umbels). Opening as the leaves expand.

Fruits About 20–25 mm long, orange or reddish, surface slightly powdery; flesh sweet or sour.

Bark Reddish-brown or dark gray to nearly black.

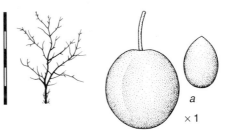

a

×1

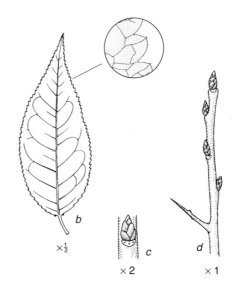

b

×½

c

×2

d

×1

Quick Recognition Twigs with bitter almond flavor, thorny; leaves with sharp, non-glandular teeth; flowers white, opening as the leaves expand; fruits red to yellow.

a. Fruit (left); stone (right). *b.* Leaf with detail of sharp non-glandular teeth. *c.* Lateral bud and leaf scar. *d.* Winter twig with thorn-tipped dwarf shoot.

11

Flowers on slender greenish-brown stalks.

Fruit orange or reddish, surface slightly powdery.

Young stems bear thorn-tipped dwarf shoots.

Mature bark similar to Canada plum.

Hawthorns

Genus *Crataegus*
Rosaceae: Rose Family

Les aubépines

Hawthorns comprise a very large genus of small trees and coarse shrubs distributed throughout the North Temperate Zone; over 100 species are native to North America; about 30 to Canada, most occurring from southern Ontario eastward to Nova Scotia. The 10 native species described below are wide ranging or at least locally common; 2 frequently planted introduced species are also described.

Leaves Deciduous, simple, alternate; small, 4–10 cm long; with sharp teeth of 2 sizes, sometimes lobed especially on vigorous shoots; base tapering in most species, but sometimes broadly wedge-shaped or rounded; often with 2 toothed leaf-like stipules at the base of the leaf stalk. Larger leaves on short shoots most useful in species identification.

Buds Terminal bud broad, rounded, smooth, shiny reddish-brown, with 5–10 scales. Lateral buds similar, somewhat smaller, often 2 or 3 side by side, one developing into a thorn, the other into a new shoot bearing leaves or flowers or both. Leaf scars narrow, somewhat elevated, with 3 vein scars.

Twigs Shoots of 2 kinds; long shoots straight or zigzag, lustrous, pale gray to orange-brown, with smooth, shiny, rigid, very sharp thorns that are sometimes branched; dwarf shoots often bearing terminal flower clusters and fruit.

Flowers Arranged in flat-topped clusters on the ends of dwarf shoots, each flower with 5 greenish sepals, 5 white or occasionally pink petals, 5–25 stamens, and 1–5 pistils; 1–2 cm across; often very showy like apple blossoms but blooming a few weeks later, with the leaves. Odor sweet, fetid.

Fruits Similar to a small apple (a pome) with a thin flesh and 1–5 seeds, often called a haw; usually red, but may be orange, yellow, blue, or black; edible; often remaining on the tree during winter. Dispersed by fruit-eating birds and mammals.

Seeds Germinate after cool moist stratification.

Seedlings Bear 2 small leaf-like cotyledons that are raised above the ground.

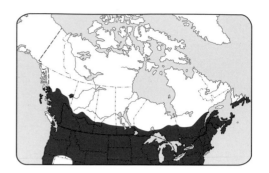

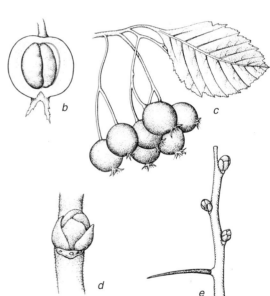

a. Flower; side view (left), radial view (right).
b. Longitudinal section through fruit. *c.* Fruit cluster and typical leaf. *d.* Lateral bud and leaf scar. *d.* Winter twig and thorn.

Bark Evenly separated into firm shreds that become loose at both ends.

Wood Hard, heavy.

Size and Form Shrubs to small trees up to 12 m high and 30 cm in diameter. Often with a distinct, crooked trunk; sometimes multistemmed and shrubby. Crown low, wide-spreading, somewhat rounded or flat topped.

11

Flowers in clusters on dwarf twigs.

Young bark (inset) smooth and gray. Mature bark separates into loose shreds.

Fruits usually red, may be orange, yellow, blue, or black.

Pear Hawthorn

Habitat Occurring on abandoned farmland, along streams, and in forest openings, especially on soils high in calcium. Moderately shade-tolerant. Often forming thickets of several different species.

Notes Wood is suitable for carving and turnery. Hawthorn thickets provide protected habitats and a food source in edible fruit, bark, and twigs for many small birds and mammals. Frequently planted for landscape purposes.

Hawthorns can reproduce by apomixis, a process by which a cell in the gametophyte divides and forms an embryo in the absence of fertilization by pollen. The resulting progeny are exact copies of the parent tree. In addition, hawthorn species hybridize freely. Many features of the progeny are interme-diate between the parent trees. As a result, while it is easy to identify a tree as a hawthorn, determining the species is often difficult. When hybridization is followed by apomixis, populations with intermediate characters may persist for generations. The clearing and abandonment of land has provided an increased habitat for the proliferation of hawthorns, with a presumably enhanced opportunity for hybridization and evolution. There may be more hawthorn species extant now than before European settlement.

Quick Recognition Smooth shiny sharp thorns; showy 5-petaled flowers in flat-topped clusters; broad rounded terminal buds; small apple-like fruits, often persisting in winter.

Pear Hawthorn

Crataegus calpodendron (Ehrh.) Medic.

Aubépine à fruits piriformes

Many-stemmed shrubs or very small trees, 3–4 m tall, common in southern Ontario. Similar to fleshy hawthorn. Produces leaves and flowers much later in spring than most other hawthorns. **Leaves** 5–8 cm long, hairy, dull yellowish-green (occasionally darker green) on the upper surface. **Thorns** few or none, 3–4 cm long, stout, glossy deep brown. **Fruits** very small, 4–8 mm across, pear-shaped, shiny orange-red.

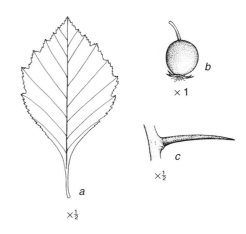

a. Leaf. *b.* Fruit. *c.* Thorn.

Columbia Hawthorn

Crataegus columbiana T.J. Howell

Aubépine occidentale

Shrubs or very small trees to 6 m tall, occurring on dry sites in interior British Columbia, on Vancouver Island, and in the Cypress Hills, and south into the United States. **Leaves** 3–7 cm long, distinctly lobed, with irregular, gland-tipped teeth; hairy on both surfaces. **Thorns** stout, 4–6 cm long. **Fruits** globular, 8–11 mm across, dark red.

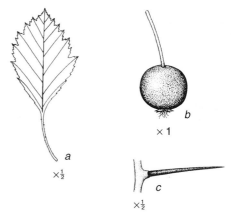

a. Leaf. *b.* Fruit. *c.* Thorn.

Black Hawthorn

Crataegus douglasii Lindl.

Aubépine noire

11

The only hawthorn that regularly grows as a tree in British Columbia; up to 11 m tall. Its range extends through Alberta into Saskatchewan and south to California, with a disjunct occurrence around Lake Superior. **Leaves** small, 2–8 cm long, coarsely double-toothed to shallowly lobed, almost hairless. **Thorns** short, usually less than 3 cm long. **Fruits** ovoid, 8–10 mm across, dark reddish-purple to black.

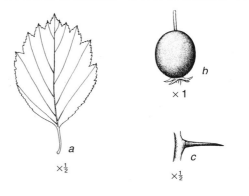

a. Leaf. *b.* Fruit. *c.* Thorn.

Fireberry Hawthorn
Crataegus chrysocarpa Ashe

Aubépine dorée

Shrubs or very small trees up to 6 m tall, ranging from Newfoundland to the Rocky Mountains, north to the Peace River in Alberta, and south into the United States. **Leaves** small, about 4 cm long, almost circular, lobed and toothed, almost hairless, glandular on the teeth, leaf stalk glandular and hairy. **Thorns** about 6 cm long, blackish, slender, straight. **Fruits** hairy, globular, 10-15 mm across, deep red (seldom yellow, despite the Latin name), fruit stalk hairy.

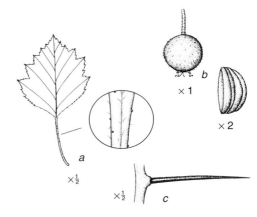

a. Leaf with detail of glandular hairy leaf stalk.
b. Fruit (above); nutlet (below). c. Thorn.

Dotted Hawthorn
Whitehaw

Crataegus punctata Jacq.

Aubépine ponctuée

Very small trees up to 8 m tall, usually single-stemmed, with layered, horizontal branches and pale gray twigs conspicuous in winter, occurring in southern Quebec and Ontario, and south to Georgia. **Leaves** 5−7 cm long, widest near the top, gradually tapering to the base, toothed to slightly lobed, with many veins ascending obliquely. **Thorns** slender, 2−8 cm long, straight to slightly curved, sometimes branched. **Fruits** globular, conspicuously dotted, 10−15 mm across, dull to bright red or sometimes yellow.

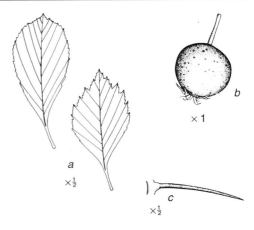

a. Leaf forms. b. Fruit. c. Thorn.

Downy Hawthorn
Crataegus mollis Scheele

Aubépine duveteuse

Very small trees up to 10 m high, common to the southern parts of eastern Canada and south to Texas. **Leaves** 4−8 cm long, widest below the middle, abruptly tapered to the base, coarsely double-toothed to shallowly lobed, densely hairy. **Thorns** 2−6 cm long, straight, relatively slender, not numerous. **Fruits** nearly round, 10−12 mm across, scarlet to dull crimson.

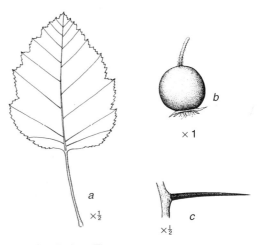

a. Leaf. b. Fruit. c. Thorn.

Cockspur Hawthorn
Crataegus crus-galli L.

Aubépine ergot-de-coq

Very small trees up to 10 m high with wide-spreading, horizontal branches (similar to those of dotted hawthorn, though not so pale); occurring in southern Quebec and Ontario, and south to Massachusetts, Iowa, and Florida. **Leaves** 3–5 cm long, mostly unlobed, usually with a rounded wide tip, sharply toothed, with secondary veins branching before reaching margin, hairless, leathery, shiny on the upper surface, more than twice as long as wide. **Thorns** slender, numerous, 5–7 cm long. **Fruits** 6–10 mm across, deep orange-red when ripe.

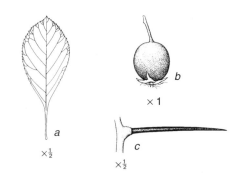

a. Leaf. *b*. Fruit. *c*. Thorn.

Scarlet Hawthorn
Crataegus coccinea L.

Aubépine écarlate

Coarse shrubs or very small variable trees, to 10 m tall; occurring in southern Ontario and Quebec and south to Maine, Iowa, and North Carolina. **Leaves** large, 6–9 cm long, widest below the middle, sharp-pointed, double-toothed to shallowly lobed. **Thorns** stout, to 6 cm long, usually curved. **Fruits** globular, 10–14 mm across, bright red.

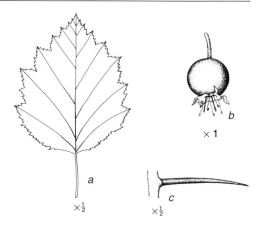

a. Leaf. *b*. Fruit. *c*. Thorn.

Fleshy Hawthorn
Succulent hawthorn

Crataegus succulenta Schrad. ex Link

Aubépine succulente

Shrubs, or occasionally multistemmed shrubby trees up to 8 m high, with conspicuously dark twigs and thorns, occurring from southern Manitoba to Nova Scotia and south to Maine and Colorado. **Leaves** 3–8 cm long, shallowly lobed about the middle, with small sharp teeth; leaf stalks grooved above, with wings on the sides. **Thorns** stout, up to 8 cm long, blackish and glossy. **Fruits** glossy deep red, 6–8 mm across, usually in erect or drooping clusters.

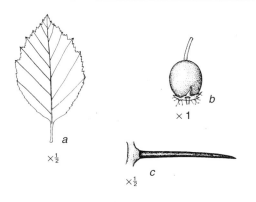

a. Leaf. *b*. Fruit. *c*. Thorn.

11

Fanleaf Hawthorn

New England hawthorn

Crataegus flabellata (Bosc) K. Koch

Aubépine flabelliforme

Shrubs or very small trees up to 5 m tall, very thorny, occurring from southern Ontario to Nova Scotia, south to Louisiana and Georgia. **Leaves** to 5 cm long, with 7–13 sharp lobes, often reflexed; lobes with several small sharp teeth; leaf stalks grooved above, with wings on the sides. **Thorns** slender, numerous, slightly curved, 5–6 cm long. **Fruits** 8–10 mm across, crimson, with thick flesh.

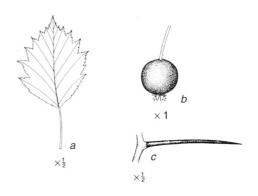

a. Leaf. *b.* Fruit. *c.* Thorn.

One-Seeded Hawthorn**

Maythorn, English hawthorn

Crataegus monogyna Jacq.

Aubépine monogyne

Very small trees to 8 m tall, introduced from Europe; naturalized in Canada; frequently planted as ornamentals. Hardy as far north as Zones C3, NA4. **Leaves** 2–3 cm long, deeply 3–7 lobed; cut nearly to the midvein; lobes narrow, obscurely toothed; veins running to the notches as well as to the lobes. **Thorns** short, 1–2 cm long, gray, straight. **Flowers** rose-colored; dark red, bright red, pink, or white. **Fruits** scarlet to deep red, glossy, 6–8 mm across, with 1 seed.

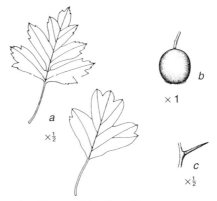

a. Leaf forms. *b.* Fruit. *c.* Thorn.

Washington Hawthorn*

Crataegus phaenopyrum (L. f.) Medic.

Aubépine de Washington

Very small trees up to 10 m high; native to the eastern United States; planted in Canada as ornamentals. Hardy as far north as Zones C6, NA5. **Leaves** 3–8 cm long, somewhat triangular, often 3–5 lobed, not strongly toothed; notches wide; veins running to notches and lobes; hairless; upper surface dark glossy green; bright orange to bright red in autumn. **Thorns** numerous, slender, 3–8 cm long. **Flowers** small, on hairless stalks. **Fruits** very small, 4–6 mm across, shiny orange-red.

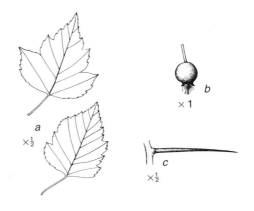

a. Leaf forms. *b.* Fruit. *c.* Thorn.

Group 12.

Leaves alternate, simple; edges smooth; deciduous (or evergreen)

The alternate arrangement of leaves is by far the most common; alternate leaves are usually arranged spirally, but in a few cases they are in 2 ranks (rows) on opposite sides of the stem. Most trees in Group 12 are deciduous with simple leaves (blade not divided into leaflets); arbutus and rhododendron have evergreen leaves.

The leaves of species in Group 12 have smooth edges, which are usually easy to recognize; black-gum has leaves that are also wavy.

Some species key out in Group 12, but are described in other groups because most of the species in their respective genera have opposite, lobed, or toothed leaves: alternate-leave dogwood (Group 8); sassafras (Group 10); European beech, several willows, the hollies, mountain-holly, and glossy buckthorn (Group 11).

The following species have leaves with smooth edges but are in Group 8 (and its key) because the leaves are opposite, subopposite, or whorled: silver buffalo-berry, the catalpas, katsura-tree, the dogwoods, common lilac, European buckthorn, and the euonymus.

Rhododendrons

Genus *Rhododendron*
Ericaceae: Heath Family

Les rhododendrons

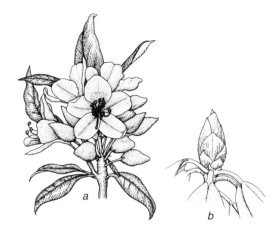

Well-known as cultivated shrubs and small trees bearing spectacular clusters of flowers. Several species of rhododendron have been introduced from Asia and innumerable cultivars developed from them. Abundantly planted in southwestern British Columbia, where they are favored by the mild climate and acid soils. Several hardy cultivars can be grown as far north as Zones C3 and NA3.
  Leaves evergreen (or deciduous), clustered toward the end of the shoot. Terminal **bud** large, with many overlapping scales. **Flowers** in showy terminal or lateral clusters; appear in spring before or with the new leaves. **Fruit** a woody capsule. **Bark** becoming thin, reddish-brown, scaly, with age.

a. Flower cluster. *b.* Scaly terminal bud.

Pacific Rhododendron

Rhododendron macrophyllum
D. Don

Rhododendron à grandes feuilles

A rare shrub in British Columbia; a very small tree in the western United States. **Leaves** evergreen, oblong, 6–25 cm long, leathery, shiny on the upper surface, whitened beneath; margin rolled under. **Flowers** bell-shaped; rose-purple, pink, or white; appear in June.

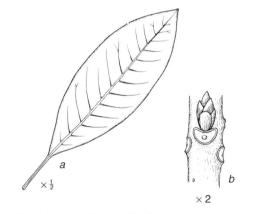

$\times \frac{1}{2}$

$\times 2$

a. Leaf. *b.* Lateral bud and leaf scar.

Rosebay Rhododendron*

Great-laurel

Rhododendron maximum L.

Rhododendron géant

A small eastern tree up to 12 m high, resembling Pacific rhododendron. Cultivated in the Atlantic provinces. Hardy as far north as Zones NA3, C3. Reports of its occurrence in Ontario and Nova Scotia as a native species are unconfirmed. Forms large dense thickets in the Appalachian Mountains of the United States.

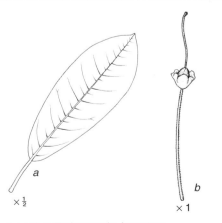

$\times \frac{1}{2}$

$\times 1$

a. Leaf. *b.* Fruit capsule, immature.

Arbutus

Arbutus menziesii Pursh
Ericaceae: Heath Family

Arbousier d'Amérique

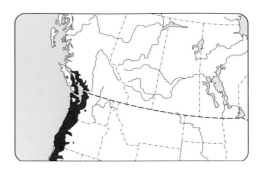

A Pacific coast species; the only native broadleaf evergreen that regularly reaches tree size; frequently planted as an ornamental (Zones C8, NA8). Worldwide, there are about 12 species of arbutus; only *Arbutus menziesii* is native to Canada.

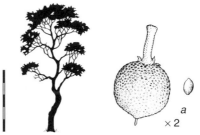
a
×2

Leaves Evergreen, alternate, simple; oval, 7–12 cm long, thick, leathery; upper surface dark glossy green, paler beneath and hairless; margin usually smooth, but may be finely toothed on leaves of young plants and vigorous shoots; principal veins branched, pinnately arranged along a thick pale midvein, 20–30 veins per side. No stipules. Older leaves turn reddish, shed during the 2nd growing season.

Buds Terminal bud about 8 mm long, green, pointed, with many overlapping scales, lower scales ridged. Lateral buds much smaller. Leaf scars with 1 vein scar.

Twigs Stout, pale green in spring becoming reddish-brown, hairless. Pith round, thick, uniform.

Flowers White, with a strong honey odor, urn-shaped, about 1 cm long; in drooping, branched clusters, at the shoot tip. Appear with the new leaves.

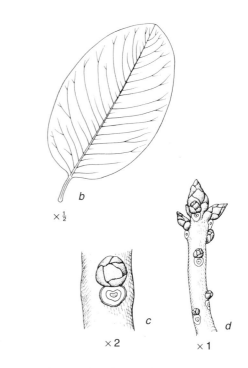
b
×½

Fruits Berry-like, less than 1 cm across, orange-red, granular surface, many seeded; in clusters. Dispersed by birds.

Vegetative Reproduction Sometimes by stump sprouts.

Bark Thin, smooth, reddish-brown, peeling in papery flakes and strips; newly exposed surfaces yellowish-green, soon becoming reddish-brown; thicker with age, broken into many small flakes.

c   *d*
×2   ×1

a. Fruit (left); seed (right). *b.* Leaf. *c.* Lateral bud and leaf scar. *d.* Twig terminal.

Wood Heavy, hard, reddish-brown, close-grained; warps and checks on drying. Diffuse-porous.

12

Size and Form Small trees, up to 20 m high, 60 cm in diameter, and 200 years old. Trunk leaning, crooked, extending into the irregular crown.

Flowers urn-shaped; appear with the new leaves.

Mature bark becomes flaky and scaly.

Fruits in many branched clusters; fruit surface glandular.

Young bark reddish-brown, peels in papery strips.

Habitat Restricted to a narrow belt along the Pacific coast; occurs on rocky shores and forest edges, and as an understory tree. Often invades newly cleared land. Intolerant of shade.

Quick Recognition Leaves evergreen, thick, shiny. Bark smooth, blotched, reddish-brown, peeling; a distinctive feature.

Pawpaw

Asimina triloba (L.) Dunal
Annonaceae: Custard-apple Family

Asiminier trilobé

Rare in Canada; occurs in Ontario north of Lake Erie. Sometimes planted for its unusual fruit. The pawpaw genus occurs only in North America; *Asimina triloba* is the sole tree-size species.

Leaves　Deciduous, alternate, simple; 15–30 cm long, widest above the middle, thin; upper surface green, paler beneath; with reddish-brown hairs when young; pendulous; unpleasant odor when bruised. Principal veins prominent, looped together near the margin, about 15 per side, reddish-brown on undersurface. Stalks short, grooved.

Buds　Terminal bud elongated, flattened, without scales; the exposed immature leaves covered with reddish-brown hairs. Lateral buds similar, smaller, pressed against the twig; often 2 together, the lowermost very small. Flower buds larger, globular, on a stalk. Leaf scars crescent-shaped, almost encircling the bud; covered with a membrane for a short time after leaf fall; 5–7 vein scars.

Twigs　Slender, zigzag, becoming hairless; brownish, streaked with fine, whitish, shallow grooves. Pith whitish, solid, banded.

Flowers　Solitary or in small clusters, 4 cm across, reddish-purple, showy, with an unpleasant odor; drooping on previous year's twigs. Appear as the leaves begin to unfold. Insect-pollinated.

Fruits　Fleshy, with edible pulp; irregular cylindrical, up to 12 cm long; pale greenish-yellow becoming yellow to brownish when ripe; solitary or in clusters.

Seeds　Dark brown, flattened, several per fruit. Cotyledons raised above the surface.

Vegetative Reproduction　By root sprouts; can be propagated by root cuttings.

Bark　Thin, smooth, shiny dark brown with grayish blotches when young; becoming rough with age.

12

Size and Form　Very small trees, up to 10 m high, often in thickets. Crown broad, with straight, spreading branches.

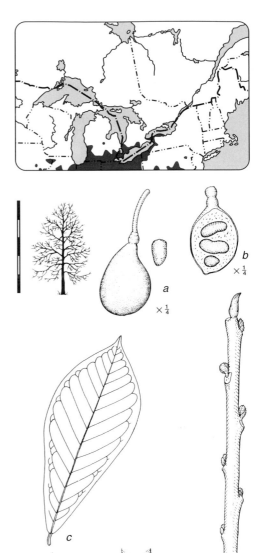

a. Fruit (left); seed (right). *b.* Longitudinal section through fruit showing seeds. *c.* Leaf. *d.* Lateral bud and leaf scar. *e.* Winter twig.

Flowers showy; appear with leaves.

Fruit irregularly shaped; turns yellow to brown when ripe.

Young bark smooth, grayish-brown.

Tree usually small, single-stemmed; often grows in thickets.

Habitat Occurs on the rich moist soils of floodplains and wet woods; in colonies as an understory tree. Shade-tolerant.

Quick Recognition Leaves large, pendulous, widest above the middle, tapered toward the base; with looped veins and short stalks. Terminal bud flattened, flower buds globular, all with reddish hairs. No scales cover the immature leaves that form the bud.

Redbud

Judas-tree

Cercis canadensis L.
Caesalpiniaceae: Cassia Family

Gainier rouge

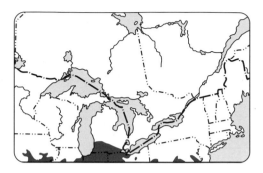

Native to eastern North America from Mexico north; probably native to Canada on Pelee Island in Lake Erie, the most southerly part of Ontario; natural range uncertain because it has become naturalized from planting. Frequently planted as an ornamental north and west of its range (Zones C6, NA4). The redbud genus is small and has only 2 species native to North America.

Leaves Deciduous, alternate, simple; heart-shaped to rounded, 8–13 cm wide, tip short, blunt; upper surface pale bluish-green, lighter beneath; smooth-margined; principal veins 5–9, prominent, radiating from the base of the blade. Stalk noticeably swollen at the top.

Buds No terminal bud. Lateral buds dark reddish-brown, slender, flat, pressed against the twig, with 5 or 6 scales; a second very small bud may be present, partially covered by the leaf scar. Flower buds larger, rounded, in small clusters, stalked, on branches even on trunk. Leaf scars raised, inversely triangular, with 3 vein scars.

Twigs Slender, zigzag, reddish-brown, with 3 ridges extending down from each leaf scar. Pith pinkish, with darker streaks.

Flowers Pea-flower shape, about 10 mm long, on stalks about as long; deep pink in the bud, becoming pale pink as the petals open; abundant, in small clusters along the trunk and branches. Appear early in spring before the leaves. Flowering begins at about 5 years and continues almost every year thereafter.

Fruits Flat, reddish-brown pods, 5–10 cm long, pointed at both ends, short-stalked, hanging in small lateral clusters from the branches.

Seeds Shiny, brown, flat, round, hard; 10–12 per pod.

Bark Smooth, gray with reddish streaks when young; becoming reddish-brown, scaly with age.

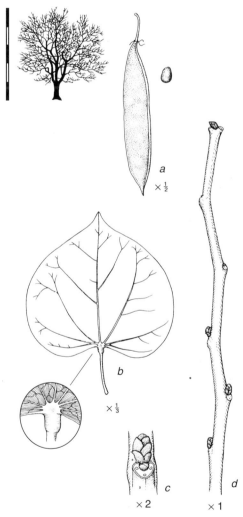

a. Fruit (left), seed (right). *b.* Leaf with detail of swollen leaf stalk. *c.* Lateral buds and leaf scar. *d.* Winter twig.

12

Flowers in small clusters; appear before the leaves.

Fruit a flat pod, tapered at both ends.

Young bark (above) smooth, gray, with reddish streaks. Mature bark reddish-brown, scaly.

Size and Form   Very small trees, up to 8 m high. Trunk low, branched. Crown spreading, with horizontal and ascending branches.

Habitat   A tree of the understory and forest edges, often along streams. Shade-tolerant.

Quick Recognition   Leaves heart-shaped. Principal veins radiate from the leaf base. Leaf stalk swollen where it meets the blade. Flowers pea-like, pink, conspicuous, occurring in small clusters along the branches. Bark smooth, gray.

Black-Gum

Black tupelo, sourgum, pepperidge

Nyssa sylvatica Marsh.
Cornaceae: Dogwood Family

Nyssa sylvestre

Rare in Canada; occurs in Ontario north of Lake Erie; planted as an ornamental for its attractive form and scarlet autumn leaves (Zones C5, NA4). Black-gum is the only species of *Nyssa* native to Canada.

Leaves Deciduous, alternate, simple; 5–12 cm long, variable in shape but usually widest above the middle; in clusters on dwarf branches and at the end of major shoots; upper surface shiny dark green, whitened beneath; sparsely hairy on the veins; margin smooth, wavy. Stalk reddish.

Buds Terminal bud 7 mm long, curved, pointed; 5 scales, yellowish-brown to dark reddish-brown, hairy at tips. Lateral buds somewhat smaller, diverging widely from the twig. Leaf scars broadly crescent-shaped, with 3 sunken whitish vein scars.

Twigs Moderately slender, reddish-brown with a grayish skin. Pith with hard, greenish crossbars. Dwarf branches present.

Flowers Small, inconspicuous, greenish-white; on long hairy stalks in small clusters in the leaf axils, especially on the dwarf branches. Pollen flowers and seed flowers on separate trees; some flowers may bear both pollen and ovules. Appear in late spring after the leaves are full size.

Fruits Plum-like, 1–3 cm long, blue-black; flesh thin, oily, sour; stone indistinctly ribbed; solitary or in clusters at the ends of long stalks.

Vegetative Reproduction By root sprouts.

Bark Gray, flaky when young; becoming dark gray, with thick irregular ridges broken into block-like segments.

Wood Brownish-gray, close-grained, moderately heavy, hard, and strong; resistant to abrasion. Diffuse-porous.

Size and Form Small trees, up to 20 m high, larger farther south. Trunk distinct into the upper part of the crown. Crown broad, flat-topped, with crooked, horizontal branches.

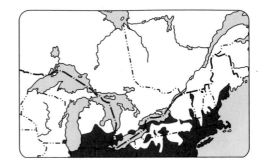

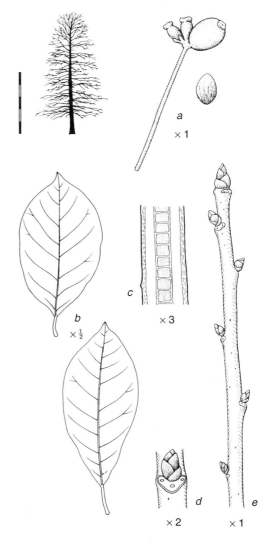

a. Fruit cluster (left); seed (right). *b.* Leaf forms. *c.* Longitudinal section through twig showing banded pith. *d.* Lateral bud and leaf scar. *e.* Winter twig.

12

Pollen flowers (above) and seed flowers (below) inconspicuous; occur on separate trees.

Fruit plum-like, bluish-black when mature.

Young bark (inset) gray and flaky. Mature bark gray, irregularly ridged.

Habitat Occurs as an understory tree on low, wet ground along streams or in swamps. Moderately shade-tolerant.

Notes A variety exists that has gray twigs and thin, almost translucent leaves with many clear dots on the undersurface.

Quick Recognition Leaves shiny dark green, clustered on dwarf branches and at the end of the major shoots; margin smooth, wavy. Fruits in long-stalked clusters in leaf axils. Terminal bud pointed; lateral buds diverging from the twig.

Magnolias

Genus *Magnolia*
Magnoliaceae: Magnolia Family

Les magnolias

The magnolia genus comprises about 80 species; 8 are native to North America; only cucumber magnolia is native to Canada. A number of native and introduced species and their hybrids have been cultivated in North America for landscape purposes.

Magnolias have large showy flowers, which are solitary on the ends of shoots;

large leaves; large terminal buds with a single bud scale formed by the fusion of 2 stipules; twigs encircled with a stipule scar at each bud; fruits in cone-like aggregates; and aromatic bark.

Cucumber-Tree

Cucumber magnolia

Magnolia acuminata L.

Magnolia acuminé

Rare in Canada; occurs only in southern parts of Ontario north of Lake Erie. An endangered species, threatened with imminent extirpation throughout its Canadian range.

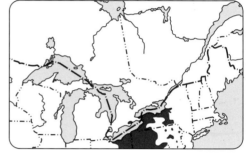

Leaves Deciduous, alternate, simple; 10–25 cm long; widest near the middle, abruptly tapered to a short, sharp tip; rounded to slightly tapered at the base; slightly hairy on the undersurface; very thin; smooth-margined; principal veins prominent, branched, widely spaced, about 10 per side. Stalk flattened and split toward the base. Stipules shed soon after bud burst.

Buds Terminal bud 15–20 mm long; the remnant of an aborted leaf can be seen at the base of the solitary hairy bud scale. Lateral buds much smaller, with a tuft of hair at the tip. Leaf scars horseshoe-shaped, with 5–9 vein scars.

Twigs Stout, reddish-brown to grayish, aromatic. Pith white, solid, round in cross section. Dwarf shoots present on the branchlets.

Flowers Large, bell-shaped with petals 5–8 cm long, greenish-yellow; less conspicuous than other magnolias. Appear as the leaves reach full size.

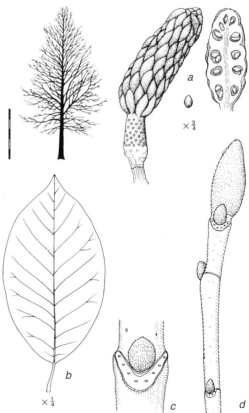

12

a. Fruit (left); seed (center); longitudinal section through fruit (right). b. Leaf. c. Lateral bud and leaf scar. d. Winter twig.

Flowers large, greenish-yellow, not showy.

Fruit conelike, turns red when ripe.

Young bark (inset) smooth with whitish streaks. Mature bark has flat scaly ridges.

Fruits Cone-like, composed of small fleshy pod-like follicles spirally arranged on a stout stalk; green until mature and somewhat resembling a cucumber; 5–8 cm long. Each follicle turns reddish at maturity and splits along one side releasing 1 or 2 large shiny orange or scarlet seeds which hang for a time on slender white threads. Dispersed by seed-eating birds and animals. Some fruit each year, with good crops every 3–5 years.

Bark Grayish-brown, furrowed into long, narrow, flattish, scaly ridges.

Wood Soft, weak, close-grained; durable and easily worked. Diffuse-porous.

Size and Form Medium-sized trees, up to 25 m high and 75 cm in diameter; larger farther south. Crown broadly pyramidal with upper branches curved upward; lower branches droop. Root system deep.

Habitat Occurs on rich soils in moist to wet sites; singly or in small groups mixed with other broadleaf trees such as oak, ash, tuliptree, maple and beech. Intolerant of shade.

Quick Recognition Leaves large, pointed, smooth-margined. Flowers large, bell-shaped, at the end of the shoot. Fruits cucumber-like when immature; later splitting to expose large orange seeds. Terminal bud large, with 1 hairy bud scale.

Saucer Magnolia*

Magnolia ×soulangiana Soul.-Boud.

Magnolia de Soulange

Spreading shrubby trees; the commonest of the
cultivated magnolias in Canada. Widely planted as
an ornamental in the milder parts of Canada, and
in the United States. Hardy as far north as Zones
C5, NA4. **Leaves** large, widest toward the tip, hairy
beneath. Terminal **bud** large, hairy. **Flowers** up to
15 cm across, in various shades of pink.
Sometimes blooming a second time in late
summer. Spectacular in early spring when the
leafless tree is covered with a multitude of large
pink flowers.

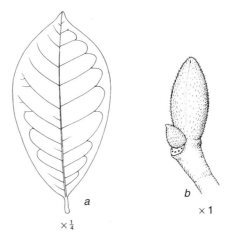

a. Leaf. *b.* Terminal floral bud with lateral leaf bud
and leaf scar.

Sea-Buckthorn*

Hippophae rhamnoides L.
Elaeagnaceae: Oleaster Family

Argousier faux-nerprun

Shrubby trees native to Europe and Asia. Planted
in North America for the gray foliage, abundant
bright orange fruits, and tolerance to salt spray
along highways. Hardy as far north as Zone C2,
NA3.

   Leaves deciduous, alternate, simple; narrow,
stalkless, silvery-scaly especially below. No terminal bud. Lateral **buds** small, golden brown, with
2 loose scales. **Twigs** often become thorns.
Flowers small, yellow, appearing before the
leaves; seed flowers and pollen flowers on separate trees. **Fruits** small, 6–10 mm long, plum-like,
globular, bright orange; clustered along the branches; persisting through the winter.

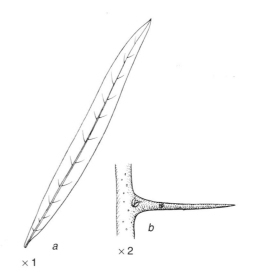

a. Leaf. *b.* Lateral bud and thorny twig.

12

Elaeagnus

Oleasters

Genus *Elaeagnus*

Elaeagnaceae: Oleaster Family

Les chalefs

The elaeagnus genus comprises about 40 species; only silverberry is native to Canada. The introduced species, Russian-olive, is more commonly found.

   Species of the elaeagnus genus are characterized by a dense coating of silvery (or rusty brown) scales on the leaves, twigs, flowers, and other parts. **Leaves** deciduous, alternate, simple; leathery, smooth-margined. **Fruits** single-seeded. Shrubs or small trees frequently

Russian-olive

Fruit silvery; leaves willow-like.

planted for landscape purposes; tolerant of harsh conditions.

Russian-Olive**

Russian elaeagnus

Elaeagnus angustifolia L.

Olivier de Bohême

A very small tree native to Europe and western Asia, planted in North America for its gray foliage and tolerance to salty soil. Naturalized in some parts of western North America. Hardy as far north as Zones C3, NA2.

   Leaves narrowly oblong, 4–8 cm long, dull green on the upper surface, covered with silvery scales beneath. **Buds** small, ovoid, with several bud scales. **Leaf scars** with a single extended vein scar. **Twigs** silvery-scaly and of 3 kinds: normal twigs with a small terminal bud; weak slender twigs that rarely survive the winter; and short stunted twigs that form thorns. **Flowers** small, yellow inside, silvery outside, very fragrant, in small clusters in leaf axils near the base of new shoots. **Fruits** olive-like but dry, silvery, about 2 cm long.

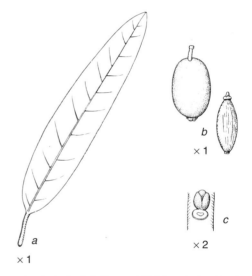

a. Leaf. *b.* Fruit (left); seed (right). *c.* Lateral bud and leaf scar.

Silverberry

Silver elaeagnus, wolf-willow

Elaeagnus commutata Bernh.

Chalef argenté

Shrubs or shrubby trees, occurring in eastern Canada; resemble Russian-olive. **Leaves** broader, scaly on both surfaces; **twigs** reddish-brown; no thorns; **fruits** smaller, with a hard outer coat. Spreads by root sprouts, usually forming clumps.

a. Leaf. *b.* Fruit. *c.* Lateral bud and leaf scar.

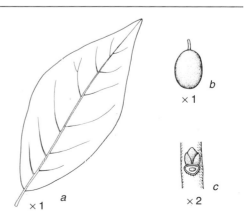

Common Smoke-Tree*
Smokebush

Cotinus coggygria Scop.
Anacardiaceae: Cashew Family

Sumac fustet

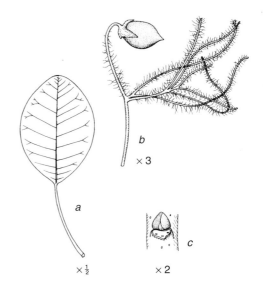

Broad shrubby trees, native to Europe and Asia.
Only 2 species in the genus; the other is native to
the southern United States. Frequently planted;
cultivars with purple leaves are especially common. Hardy as far north as Zones C5, NA4.
　　Leaves deciduous, alternate, simple; broadly
oval, bluish-green, long-stalked. **Buds** 2 mm long,
with several reddish-brown scales. Cut **twigs**
exude a gummy, strong-smelling sap. Plumose, or
"smokey", effect is due to numerous long hairs on
the fruit stalks (which are usually without fruits).

a. Leaf. *b.* Fruit. *c.* Lateral bud and leaf scar.

Osage-Orange*
Hedge-apple, bodark, bowwood

Maclura pomifera (Raf.) C.K Schneid.
Moraceae: Mulberry Family

Oranger des Osages

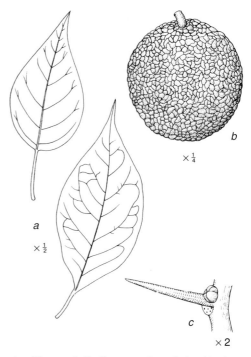

Small trees, native to Texas and adjacent states.
Only species in the genus. Planted for hedges and
windbreaks. Thrives under a variety of difficult
conditions. Intolerant of shade. Hardy as far north
as Zone C6, NA4.
　　Leaves small, 7–12 cm long, oval, shiny, narrowing abruptly into long, slender tips; smooth-
margined. **Buds** small, globular, 5–scaled, partly
embedded in the bark. **Twigs** orange-brown, zigzag, commonly bearing a stiff thorn at each leaf
node; in the 2nd year, a dwarf shoot may develop
beside the thorn. A milky juice exudes from cut
twigs. **Flowers** wind-pollinated; pollen flowers and
seed flowers on separate trees. **Fruits** 10–14 cm
across, resembling a green orange, but not edible.
Bark orange-brown. **Wood** orange-yellow, very
hard, flexible, tough, heavy, durable in contact with
soil; ring-porous. Reproduces by root sprouts.

a. Leaf forms. *b.* Fruit aggregate. *c.* Lateral bud,
leaf scar, and thorn.

12

Keys to Groups and Selected Genera

Key to Group 1
Short needles or scales; evergreen; closely spaced in opposite pairs or whorls of 3, often overlapping and obscuring the stem; seeds in cones, some cones berry-like

1 Seed cones oval, scales in opposite pairs; twigs flattened to somewhat rounded; leaf sprays flattened; leaves scale-like ....*Thuja*, THUJAS   22

1 Seed cones globular, scales fused; twigs 4-sided; leaf sprays erect to horizontal with drooping tips; leaves scale-like or needle-like

  2 Leaves exclusively scale-like; seed cones greenish with prominent triangular projections; medium-sized trees to 25 m with drooping leading shoot and branch tips ....*Chamaecyparis*, FALSE CYPRESSES   28

  2 Leaves exclusively needle-like, or needle-like on vigorous shoots and elsewhere scale-like; seed cones bluish-black, often with a powdery coating; very small trees or shrubs, upright to sprawling .... *Juniperus*, JUNIPERS   16

Key to the Junipers (Genus *Juniperus*)

1 Leaves needle-like, in whorls of 3; cones in leaf axils .... *J. communis,* COMMON JUNIPER   19

1 Leaves mostly scale-like, but needle-like on young, vigorous shoots, mostly in opposite pairs; cones terminal on branchlets

  2 Branches slender and ascending; small trees

    3 Needle-shaped leaves blunt-tipped; mature seed cones violet or brownish ....*J. chinensis*, CHINESE JUNIPER   21

    3 Needle-shaped leaves sharp-tipped; mature seed cones green to dark blue

      4 Scale leaves pale yellowish-green, successive pairs barely overlapping; seed cones maturing in 2 years ....*J. scopulorum*, ROCKY MOUNTAIN JUNIPER   18

      4 Scale leaves dark blue-green, successive pairs strongly overlapping; seed cones maturing in 1 year ....*J. virginiana,* EASTERN REDCEDAR   20

2 Branches spreading, trailing or prostrate; low shrubs

> **5** Needle-shaped leaves numerous, dark green, upper surface with a prominent midvein; foliage with a strong, unpleasant odor when bruised; seed cones maturing in 1 year .... *J. sabina,* SAVIN JUNIPER   21

> **5** Needle-shaped leaves few, bluish-green, upper surface lacking a prominent midvein; foliage aromatic; seed cones maturing in 2 years .... *J. horizontalis,* CREEPING JUNIPER   19

Key to the Thujas (Genus *Thuja*)

1 Twig complex vertical; cone scales thick, sharply pointed; seeds not winged ....*T. orientalis,* ORIENTAL-CEDAR   27

1 Twig complex horizontal; cone scales thin, leathery; seeds laterally winged

> **2** Leaves dull yellowish-green, turning bronze in winter, outer surface with conspicuous glandular dot; twigs flattened, yellowish-green above and below; occurring eastward from Manitoba ....*T. occidentalis,* EASTERN WHITE-CEDAR   26

> **2** Leaves shiny yellowish-green year-round, outer surface with inconspicuous glandular dot; twigs rounded, yellowish-green above, whitish below; occurring westward from Alberta ....*T. plicata,* WESTERN REDCEDAR   24

Key to the False Cypresses (Genus *Chamaecyparis*)

1 Leaves dull blue-green on both surfaces; twigs somewhat 4-sided, branch tips drooping; native to the Pacific coast .... *C. nootkatensis,* YELLOW-CEDAR   28

1 Leaves bright green and shiny above, whitish and glandular below; twigs flattened, branch tips horizontal; introduced, planted as ornamentals

> **2** Leaves marked below with indistinct white lines; pollen cones bright red; bark thick .... *C. lawsoniana,* LAWSON-CYPRESS   31

> **2** Leaves marked below with distinct white lines; pollen cones yellow; bark thin, smooth, peeling in strips

>> **3** Leaves sharp-pointed at the tip, in lateral and vertical pairs of equal size and length; seed cones about 5 mm in diameter, scales 10, rarely 12 .... *C. pisifera,* SAWARA-CYPRESS   31

>> **3** Leaves blunt-pointed at the tip, lateral pairs larger and longer than the vertical pairs; seed cones about 7 mm in diameter, scales 8, rarely 10 .... *C. obtusa,* HINOKI-CYPRESS   31

Key to Group 2
Introduced species hardy in southwest British Columbia; leaves various shapes; seeds in cones

1 Leaves to 15 cm long, in united pairs, on microshoots growing in the axils of small scale-leaves, in whorls of 10–30 along major shoots .... *Sciadopitys verticillata,* UMBRELLA-PINE   38

1 Leaves to 5 cm long, solitary, not in whorls

  2 Leaves opposite, deciduous, to 3 cm long on major shoots, to 2 cm long on deciduous shoots .... *Metasequoia glyptostroboides,* DAWN REDWOOD   35

  2 Leaves alternate or arranged spirally

    3 Leaves deciduous, 1.5–2 cm long, appearing 2-ranked on deciduous shoots, spiral on persistent shoots, yellowish-green above, whitened below; seed cones reddish-purple, disintegrating at maturity .... *Taxodium distichum,* BALD-CYPRESS   36

    3 Leaves and shoots persistent

      4 Principal branches not in whorls; seed cones egg-shaped to oblong

        5 Leaves on horizontal shoots linear, to 2 cm long, 2-ranked, on vigorous upright shoots and young trees broadly linear, to 0.5 cm long, diverging in all directions; winter buds scaly; seed cones 2–3 cm long, maturing in 1 year .... *Sequoia sempervirens,* COAST REDWOOD   34

        5 Leaves on horizontal shoots scale-like, to 0.5 cm long, spirally arranged, appressed and overlapping, on vigorous upright shoots and young trees lance-shaped, to 1.2 cm long, spreading; winter buds without scales; seed cones 5–10 cm long, maturing in 2 years and persisting .... *Sequoiadendron gigateum,* SIERRA REDWOOD   33

      4 Principal branches in whorls; seed cones globular

        6 Leaves triangular, to 5 cm long, clasping the stem but diverging and overlapping, tip a sharp prickle; seed cones 10–18 cm across, maturing in 2–3 years .... *Araucaria araucana,* MONKEY-PUZZLE   39

        6 Leaves narrowly awl-shaped, to 1.2 cm long, lower part clasping the stem, upper part pointing forward and curving inward, tip blunt-pointed; seed cones 1.2–1.8 cm across, cone scales toothed at the tip, maturing in 1 year .... *Cryptomeria japonica,* JAPANESE-CEDAR   37

Key to Group 3
Needles evergreen; in bundles of 2, 3, or 5; seeds in cones

Key to the Pines (Genus *Pinus*)

1 Leaves in bundles of 5

    2 Leaf margins toothed, rough to the touch

8 Leaves shorter than 7 cm, twisted or divergent

    10 Leaves scarcely spread apart, spirally twisted

        11 Leaves bluish-green; bark on branchlets and upper stem becoming flaky, bright yellowish-orange; seed cones stalked, falling at maturity .... *P. sylvestris,* SCOTS PINE   62

        11 Leaves yellowish-green; bark on branchlets and upper stem slightly scaly, orange-brown; seed cones stalkless, persisting on the tree after maturity

            12 Trunk straight, little tapered, to 30 m tall; bark thin, with fine scales .... *Pinus contorta* var. *latifolia,* LODGEPOLE PINE   60

            12 Trunk crooked, tapered, to 15 m tall; bark thick, deeply furrowed .... *Pinus contorta* var. *contorta,* SHORE PINE   61

    10 Leaves spread apart to form a V, only slightly twisted

        13 Leaves yellowish-green; seed cones asymmetrical, mostly in pairs; trees of the boreal forest .... *P. banksiana,* JACK PINE   58

        13 Leaves dark green; seed cones ovoid, mostly in groups of more than 2; shrubs or small trees, introduced, widely planted as ornamentals .... *P. mugo,* MUGHO PINE   65

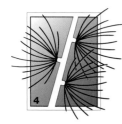

Key to Group 4
Needles deciduous (or evergreen); in tufts of 10 or more on dwarf shoots; also single on long shoots; seeds in cones

1 Leaves soft to somewhat stiff, turning yellow and shed annually in autumn; scales of seed cones persistent .... *Larix*, LARCHES 67

1 Leaves rigid, evergreen, persistent; scales of seed cones deciduous, cone axes persist for many years .... *Cedrus*, CEDARS 78

Key to the Larches (Genus *Larix*)

1 Leaves 4-sided in cross section; twigs densely hairy, becoming black with age; a western species, occurring on high slopes above elevations of 1200 m .... *L. lyallii*, SUBALPINE LARCH 72

1 Leaves mostly flattened or triangular in cross section; twigs hairless to sparsely hairy

  2 Mature seed cones 1–2 cm long, scales 20 or fewer .... *L. laricina*, TAMARACK 74

  2 Mature seed cones more than 2 cm long, scales more than 20

    3 Mature seed cones with bracts extending well beyond the cone scales; native to southern British Columbia, Alberta, and the United States northwest .... *L. occidentalis*, WESTERN LARCH 70

    3 Mature seed cones with bracts shorter than the cone scales; introduced species, in plantations and landscape plantings

      4 Leaves broad, with conspicuous white dots below; cone scales curving outward at the tips .... *L. kaempferi*, JAPANESE LARCH 76

      4 Leaves narrow, with inconspicuous white dots below; cone scales straight or curving inward at the tips

        5 Cone scales mostly hairless, mature seed cones to 3 cm long; twigs somewhat hairy, yellowish-brown, turning reddish in winter .... *L. gmelinii*, DAHURIAN LARCH 77

        5 Cone scales finely hairy, mature seed cones to 4 cm long

          6 Leaves bright green, to 5 cm long; buds only slightly resinous; twigs hairless, pale yellow; cone scales more than 30, seed wings extending to rounded scale margins .... *L. decidua*, EUROPEAN LARCH 76

          6 Leaves green to grayish-blue, to 4 cm long; buds resinous; twigs hairy in spring, becoming smooth, yellowish-gray; cone scales about 30, seed wings not extending to wavy scale margins .... *L. sibirica*, SIBERIAN LARCH 77

Key to the Cedars (Genus *Cedrus*)

1 Leaves less than 26 mm long, thicker than wide; mature seed cones less than 8 cm long, scales less than 4 cm wide .... *C. atlantica,* ATLAS CEDAR   79

1 Leaves more than 26 mm long, as wide as or wider than thick; mature seed cones more than 8 cm long, scales more than 4 cm wide

    2 Leaves as wide as thick, dark blue-green; mature seed cones rounded at the tip .... *C. deodara,* DEODAR CEDAR   78

    2 Leaves wider than thick, light green to silvery-blue; mature seed cones flattened or concave at the tip .... *C. libani,* CEDAR-OF-LEBANON   79

Key to Group 5
Needles evergreen; single; flat or 4-sided; seeds in cones

1 Twigs ridged and grooved, roughened by raised pegs where leaf bases attach

    2 Leaves flat in cross section, tip rounded or indented, base narrowing abruptly to a thread-like stalk; buds rounded; seed cones terminal on twigs; leading shoot oblique .... *Tsuga*, **HEMLOCKS**   114

    2 Leaves mostly 4-sided in cross section, tip bluntly to sharply pointed, base jointed, stalkless; buds blunt-pointed; seed cones both terminal and lateral on twigs; leading shoot upright .... *Picea*, **SPRUCES**   95

1 Twigs not grooved (may have undulating ridges), with flat, round or oval leaf scars where leaf bases attach

    3 Leaves 1.5–7.5 cm long, stalkless, often twisted at the base; twigs stout, somewhat brittle; buds rounded, mostly resinous; mature seed cones upright on branches, scales shed as seeds dispersed, the central axis persisting .... *Abies*, **FIRS**   81

    3 Leaves 2–3 cm long, stalked; twigs moderately stout, flexible; buds sharply pointed, not resinous; mature seed cones hanging from branches, falling intact after seeds dispersed

        4 Leaves bright yellowish-green, slightly apple-scented when crushed; seed cones with extended, 3–pronged, straight bracts; crown narrow, conical; very large Pacific coastal trees, commonly to 60 m tall .... *Pseudotsuga menziesii* var. *menziesii,* **DOUGLAS-FIR**   110

        4 Leaves bluish-green, slightly orange-scented when crushed; seed cones with extended, 3–pronged bracts bent backward; crown long, limby; Rocky Mountain trees, elsewhere planted ornamentally, seldom taller than 40 m .... *Pseudotsuga menziesii* var. *glauca,* **ROCKY MOUNTAIN DOUGLAS-FIR**   113

Key to the Firs (Genus *Abies*)

1 Leaves marked with white stomatal lines on upper and lower surfaces

    2 Twigs and young branches hairy

        3 Leaves not grooved along the upper surface .... *A. magnifica,* **RED FIR**   92

        3 Leaves grooved along the upper surface

            4 Leaves along the sides of the twig curving upward and standing erect, those below the twig appressed; mature seed cones less than 10 cm long, with bracts shorter than the scales .... *A. lasiocarpa,* **SUBALPINE FIR**   86

            4 Leaves along upper surface of twig pressed down and pointing forward; mature seed cones greater than 11 cm long, with bracts longer than the scales .... *A. procera,* **NOBLE FIR**   92

2 Twigs and branches hairless

    5 Leaves greater than 40 mm long, whitish-green, 2- ranked, rubbery, fragrant when crushed .... *A. concolor,* WHITE FIR   94

    5 Leaves less than 15 mm long, dark green, not 2-ranked, stiff ...... *A. pinsapo,* SPANISH FIR   93

1 Leaves marked with white stomate lines only on lower surface

    6 Twigs on lower and less vigorous branches with all leaves spreading horizontally, appearing 2-ranked

        7 Twigs greenish-gray; mature seed cones barrel-shaped, bracts usually shorter than the scales, with a long needle-like tip and rounded shoulders, scales 1.0 to 1.5 cm wide; ranging eastward from central Alberta .... *A. balsamea,* BALSAM FIR   84

        7 Twigs olive-green to dark orange-brown; mature seed cones cylindrical, bracts shorter than the scales, with a short tooth-like tip and broad shoulders, scales 2.5–3 cm wide; ranging westward from south-central British Columbia .... *A. grandis,* GRAND FIR   90

    6 Twigs on lower and less vigorous branches with lateral leaves spreading horizontally, the upper leaves pressed down and pointing forward

        8 Twigs dark yellowish-brown to grayish-brown; buds rounded, resinous; seed cones with bracts shorter than scales, scales 2–3 cm wide; trees of the Pacific coastal forest .... *A. amabilis,* AMABILIS FIR   88

        8 Twigs greenish-brown; buds ovoid, non-resinous; seed cones with recurved bracts longer than scales, scales 3–4 cm wide; introduced, planted as ornamentals .... *A. nordmanniana,* NORDMANN FIR   93

Key to the Spruces (Genus *Picea*)

(For proper identification, examine 2-year-old twigs.)

1 Leaves flattened in cross section; bands of white dots on lower surface very prominent

    2 Leaves very sharp-tipped, to 30 mm long, yellowish-green above; twigs hairless; seed cones broadly cylindrical, 5–9 cm long, scales papery thin, outer scale margins wavy and irregularly toothed; very large trees of the Pacific coast .... *P. sitchensis,* SITKA SPRUCE   98

    2 Leaves blunt-tipped, to 20 mm long, dark green above; twigs hairy; seed cones egg-shaped, 4–6 cm long, scales stiff and woody, outer scale margins rounded, weakly toothed; medium sized trees, introduced, planted as ornamentals .... *P. omorika,* SERBIAN SPRUCE   108

1 Leaves 4-sided in cross section, bands of white dots on lower surface may be visible, but not prominent

    3 Twigs mostly hairless, smooth

        4 Secondary branches often strongly and distinctly pendulous; leaves shiny dark green; cones large, to 18 cm long .... *Picea abies,* NORWAY SPRUCE   108

4 Secondary branches mostly horizontal; leaves dull green to bluish-green; cones to 12 cm long

    5 Leaves blunt-tipped, to 22 mm long, green to bluish-green, pungently aromatic when crushed; twigs greenish-gray, often tinged orange or purple, leaf-cushions laterally rounded; seed cones 3–6 cm long, scales close-fitting, outer margins smooth; transcontinental in range .... *P. glauca,* WHITE SPRUCE   102

    5 Leaves very sharp-tipped, to 30 mm long, green to striking blue, smelling resinous when crushed; twigs orange-brown to grayish-brown, leaf-cushions laterally rounded but often rising to a central ridge; seed cones 5–12 cm long, scales loose-fitting, outer margins irregularly toothed; introduced, widely planted as ornamentals .... *P. pungens,* COLORADO SPRUCE   109

3 Twigs mostly with fine, soft hairs

    6 Leaves to 20 mm long, pungently aromatic when crushed; seed cones 3–7 cm long, cylindrical to narrowly ovoid, scales flexible, loose-fitting; a western species, occurring on mountain slopes and along streams at lower elevations .... *P. engelmannii,* ENGLEMANN SPRUCE   100

    6 Leaves to 16 mm long, pleasantly aromatic when crushed; seed cones to 5 cm long, ovoid to broadly ovoid, scales stiff, close-fitting

        7 Twigs with leaf-cushions laterally rounded, the grooves between V-shaped; leaves shiny yellowish-green, often pressed to the twig; seed cones 3–5 cm long, falling after 1 season .... *P. rubens,* RED SPRUCE   104

        7 Twigs with leaf-cushions flattened, the groves between closed; leaves waxy, dull grayish-green, often projecting away from the twig; seed cones 2–3 cm long, persisting for several seasons .... *P. mariana,* BLACK SPRUCE   106

Key to the Hemlocks (Genus *Tsuga*)

1 Leaves rounded in cross section, of uniform length, extending outward in all directions from around the twig; seed cones 3–8 cm long .... *T. mertensiana,* MOUNTAIN HEMLOCK   118

1 Leaves flat in cross section, of varying lengths, appearing 2-ranked along the sides of the twig; seed cones to 2.5 cm long

    2 Leaves slightly tapered towards the tip; seed cones stalked, cone scales rounded; trees of eastern North America .... *T. canadensis,* EASTERN HEMLOCK   120

    2 Leaves not tapering towards the tip, except on vigorous twigs; seed cones without stalks, cone scales rectangular; trees of western North America .... *T. heterophylla,* WESTERN HEMLOCK   116

Key to Group 6
Needles evergreen, single, flat; twigs and buds green; single seed in a fleshy cup

Key to the Yews (Genus *Taxus*)

1 Branches low, trailing, upturned at the tips; leaves pale green, abruptly pointed, turning reddish in winter; seeds wider than long, equal in width to the aril opening; a many-stemmed shrub, ranging through eastern Canada and the adjacent United States ....*T. canadensis,* CANADA YEW   125

1 Branches upright and spreading; seeds longer than wide; small to medium-sized trees

   2 Leaves tapering gradually to a point; twigs greenish for several growth seasons; seeds elliptical; trunk often short and massive; medium-sized trees, to 25 m tall ....*T. baccata,* ENGLISH YEW   124

   2 Leaves abruptly pointed; twigs turning brownish after 1 growth season; shrubs or small trees

      3 Leaves 1.5−2 mm wide, with a slightly elevated midvein, dark yellowish-green above, pale green below, distinctly 2-ranked; aril about 8 mm long, seed oval; an understory species, occurring westward from Alberta ....*T. brevifolia,* WESTERN YEW   124

      3 Leaves 2−3 mm wide, with a prominent midvein, pale green above, with 2 broad, yellow bands below, indistinctly 2-ranked and upright along the twig, forming a V-shaped trough; aril about 6 mm long, seed narrow and compressed; introduced, planted in ornamental settings ....*T. cuspidata,* JAPANESE YEW   125

Key to Group 8
Leaves in opposite pairs (or subopposite or whorled); blade simple or compound; edges lobed, toothed, or smooth

1 Leaves simple, composed of only a single blade

  2 Leaves narrow, needle-like, on major shoots individual, on deciduous lateral shoots in pairs of 25–30, borne in 2 flattened rows; fruit a globular, greenish, long-stalked cone; a conifer, introduced, planted as ornamentals ....(Group 2) *Metasequoia glyptostroboides*, DAWN REDWOOD   35

  2 Leaves broad, not needle-like; fruit not a cone

    3 Leaves with lateral veins curving forward and following the margin to the leaf tip

      4 Leaf margins smooth; leaves always in opposite pairs; twigs with prominent lenticels, shoots often ending in a swollen floral bud; flowers terminal in compact clusters; fruits with 1 or 2 hard stones .... *Cornus*, DOGWOODS   180

      4 Leaf margins sharply double-toothed, leaves often subopposite along vigorous shoots; twigs with small, dash-like lenticels, shoots often ending in a thorn-like prickle; flowers lateral in lower leaf axils; fruits with 3 or 4 seeds ....(Group 11) *Rhamnus cathartica*, EUROPEAN BUCKTHORN   278

    3 Leaves with lateral veins extending directly towards the leaf margin

      5 Leaves heart-shaped

        6 Leaves with weakly toothed margins; flowers small, without petals; fruits small pods .... *Cercidiphyllum japonicum*, KATSURA-TREE   193

        6 Leaves with smooth margins, rarely with a few large teeth or shallow lobes; flowers large, with petals fused, forming a tube

          7 Leaves often in whorls of 3, large, 10–30 cm long, softly hairy below, stalks 10–16 cm long; fruits long, cylindrical, many-seeded .... *Catalpa*, CATALPAS   188

          7 Leaves always in opposite pairs, small, 5–12 cm long, hairless below except along the midvein, stalks 2–3 cm long; fruits small, flattened, 2 seeded ....*Syringa vulgaris*, COMMON LILAC   187

      5 Leaves lance-shaped, elliptical or lobed, not heart-shaped

        8 Twigs roughened by thorns or ridges

          9 Twigs thorny, covered in silver-brown scales; leaves scaly on one or both surfaces, leaf margins without teeth; fruits berry-like, red and sour ....*Shepherdia argentea*, SILVER BUFFALO-BERRY   194

          9 Twigs 4-sided or with longitudinal corky wings; leaves smooth, margins toothed; fruits 4-lobed capsules, becoming pink at maturity .... *Euonymus*, EUONYMUS   190

8 Twigs smooth, without thorns or ridges

10 Leaves broad; buds in leaf axils with several scales or naked; fruits not borne on catkins

11 Leaves always opposite; flowers in clusters; fruits not borne in globular clusters

1 Leaves compound, with 3 to several leaflets

13 Leaves pinnately compound with 3–13 leaflets; flowers small, mostly inconspicuous; fruits winged or berry-like

14 Leaflets regular in shape, leaflet margins regularly toothed or smooth, not lobed

15 Lateral buds visible in axils of leaf stalks; leaflet margins toothed

Key to the Maples (Genus *Acer*)

1 Leaves pinnately compound, with 3–9 stalked, coarsely toothed or lobed, asymmetrical leaflets; seedcases elongated, wrinkled, wings incurved, forming an angle of less than 45° .... *Acer negundo*, MANITOBA MAPLE   144

1 Leaves simple, lobed, or rarely with 3 stalkless leaflets

    2 Leaves typically 7–9 lobed, broadly oval to circular

        3 Lobes triangular, sides parallel or angling inward from base to tip; leaf margins with short-pointed single or double teeth; seed wings forming an angle of 180° .... *A. circinatum*, VINE MAPLE   152

        3 Lobes long, taper-pointed, sides curved, widest in the middle and tapering inward at both ends, leaf margins with long-pointed double teeth; seed wings forming an angle of 100° .... *A. palmatum*, JAPANESE MAPLE   155

    2 Leaves typically 3–7 lobed, square or rectangular

        4 Central lobe narrowing progressively from base to tip

            5 Leaves longer than wide, margins coarsely and irregularly single-toothed; seed wings forming an angle of less than 90° .... *A. spicatum*, MOUNTAIN MAPLE   146

            5 Leaves as wide as or wider than their length, margins double-toothed

                6 Leaves shallowly 3-lobed at the tip, often without lobes on vigorous shoots, teeth fine and regular; bark conspicuously striped; seed wings forming an angle of 90° .... *A. pensylvanicum*, STRIPED MAPLE   148

                6 Leaves 3- to indistinctly 5-lobed, occasionally with lobes divided to the leaf stalk, forming 3 stalkless leaflets, teeth coarse and irregular; bark not striped; seed wings forming an angle of less than 45° .... *A. glabrum* var. *douglasii*, DOUGLAS MAPLE   150

        4 Central lobe with sides parallel or diverging above the base

            7 Leaves more than 15 cm wide, central and lateral lobes often overlapping; seedcases hairy .... *A. macrophyllum*, BIGLEAF MAPLE   138

            7 Leaves less than 15 cm wide, central and lateral lobes rarely overlapping; seedcases hairless

                8 Leaves with numerous double teeth along the margin; fruit maturing and falling before midsummer

                    9 Leaves deeply notched between central and lateral lobes, central lobe with sides diverging above the base; seedcases ribbed, often only 1 seed developed, seed wings more than 4 cm long, forming an angle of 90° .... *A. saccharinum*, SILVER MAPLE   142

                    9 Leaves shallowly notched between the central and lateral lobes, central lobe with sides parallel to the midrib; seedcases swollen, both seeds developing, seed wings less than 2.5 cm long, forming an angle of 60° .... *A. rubrum*, RED MAPLE   140

8 Leaves with coarse, single teeth along the margin; fruit maturing and falling in autumn

 10 Central lobe not less than twice the length of the lateral lobes .... *A. ginnala,* Amur maple 154

 10 Central lobe approximately equal in length to the lateral lobes

  11 Leaves thick, wrinkled, lobes with numerous coarse teeth and white hair on each side of the main vein below; mature bark scaly, shedding; seeds occasionally in groups of 3 .... *A. pseudoplatanus,* sycamore maple 153

  11 Leaves thin, smooth, lobes with a few irregular, wavy teeth, hairless or entirely covered with hair below; mature bark ridged, not shedding; seeds always in pairs

   12 Leaves 5−7 lobed, teeth and lobes bristle-tipped, leaf stalks exuding milky sap when cut; flowers 10 mm across, appearing with the leaves in spring; seedcases flattened, seed wings 3.5−5 cm long, forming an angle of 180° .... *A. platanoides,* Norway maple 154

   12 Leaves 3−5 lobed, teeth and lobes blunt-tipped, leaf stalks not exuding milky sap when cut; flowers 5 mm across, appearing before the leaves in spring; seedcases plump, seed wings 3−3.5 cm long, almost parallel

    13 Leaves typically 5-lobed, pale green and hairless below, central lobe square, separated from lateral lobes by wide, rounded notches, brilliant orange or scarlet in autumn .... *A. saccharum,* sugar maple 134

    13 Leaves typically 3-lobed, appearing wilted and droopy, yellowish-green and covered with dense, velvety bown hair below, central lobe slightly tapered, separated from the lateral lobes by shallow, open notches, yellow in autumn .... *A. nigrum,* black maple 136

Winter Key to the Maples (Genus *Acer*)

1 Twigs lacking a terminal bud, ending in a withered stub or pair of lateral buds

 2 Buds showing many pairs of overlapping scales; twigs brownish, many-sided in cross section ....*A. ginnala,* Amur maple 154

 2 Buds showing 1 or 2 pairs of scales; twigs reddish, round in cross section

  3 Leaf scars elevated, forming flared collars around the buds; twig often covered in waxy powder; introduced, widely planted as ornamentals .... *A. palmatum,* Japanese maple 155

  3 Leaf scars not greatly elevated; twigs not powder-coated; native to the Pacific coastal forest region, often vine-like and coiling around other trees ...... *A. circinatum,* vine maple 152

1 Twigs ending with a terminal bud

 4 Terminal buds sharply pointed, with 4−8 pairs of bud scales

  5 Bud scales only faintly hairy, medium to dark grayish-brown; twigs shiny reddish-brown; mature bark divided into firm, irregular strips, curling outward along one side, becoming scaly .... *A. saccharum,* sugar maple 134

424   **Leaves in opposite pairs; simple or compound**

5 Bud scales hairy, dark brown; twigs dull reddish-brown; mature bark divided into deep furrows with long, irregular ridges, occasionally scaly .... *A. nigrum,* BLACK MAPLE   136

4 Terminal buds blunt or rounded, with 1–4 pairs of bud scales

   6 Buds with 1 pair of scales visible

      7 Buds and twigs appear dull, covered in short grayish hairs, terminal buds 2–3 times longer than wide .... *A. spicatum,* MOUNTAIN MAPLE   146

      7 Buds and twigs shiny, hairless, terminal buds 2 times longer than wide

         8 Terminal buds about 10 mm long; bark smooth, green, becoming conspicuously marked with vertical white stripes; trees of eastern North America .... *A. pensylvanicum,* STRIPED MAPLE   148

         8 Terminal buds 2–5 mm long; bark rough, dark reddish-brown; trees of western North America .... *A. glabrum* var. *douglasii,* DOUGLAS MAPLE   150

   6 Buds with more than 1 pair of scales visible

      9 Bud scales covered by fine whitish hairs; twigs often coated in an easily removed, waxy powder; terminal and lateral buds of similar size, lateral buds pressed against the twig .... *A. negundo,* MANITOBA MAPLE   144

      9 Bud scales hairless; twigs without a powdery coating; terminal buds larger than lateral buds, lateral buds diverging from the twig

         10 Twigs with dwarf shoots that bear plump, reddish floral buds clustered at the tips; more than 1 lateral bud at some leaf scars

            11 Twigs with an unpleasant odor when bruised; young bark gray, with age becoming grayish-brown and shaggy, with ridges loose at the ends .... *A. saccharinum,* SILVER MAPLE   142

            11 Twigs lacking an unpleasant odor when bruised; young bark light gray, with age becoming very dark and developing scaly ridges .... *A. rubrum,* RED MAPLE   140

         10 Twigs lacking dwarf shoots; 1 lateral bud at each leaf scar

            12 Leaf scars with 5–9 vein scars arranged in a V- or U-shape; native to the Pacific coastal forest region .... *A. macrophyllum,* BIGLEAF MAPLE   138

            12 Leaf scars with 3 or rarely 5 vein scars arranged in a line; introduced European species, widely planted

               13 Bud scales reddish or greenish-purple; opposite leaf scars meeting around the twig; mature bark very dark gray, with low, intersecting ridges .. .. *A. platanoides,* NORWAY MAPLE   154

               13 Bud scales greenish with dark margins; opposite leaf scars not meeting around the twig; mature bark pinkish-brown, conspicuously mottled, scaly and shedding .... *A. pseudoplatanus,* SYCAMORE MAPLE   153

Key to the Ashes (Genus *Fraxinus*)

1 Leaflets with smooth margins (occasionally with a few faint, rounded teeth)

    2 Leaflets hairless below except along the veins; twigs very stout, hairless, often tinged purplish; seeds cylindrical, seed wings tapered, enclosing only the tip of the seedcase .... *F. americana,* WHITE ASH   162

    2 Leaflets densely hairy below, twigs stout, densely hairy

        3 Leaflets 4–10 cm long; fruits 2.5–5 cm long, seed wing uniformly broad; a tree of the Pacific coastal forest .... *F. latifolia,* OREGON ASH   169

        3 Leaflets 12–25 cm long; fruits 5–8 cm long, seed wing widest at or above the middle; a tree of southeastern North America, ranging north to the west end of Lake Erie .... *F. profunda,* PUMPKIN ASH   170

1 Leaflets with margins faintly to strongly toothed above the middle

    4 Leaflets stalked

        5 Twigs conspicuously ridged and 4-sided; leaflets asymmetrical at the base; seeds flattened, seed wings twisted and entirely enclosing the seedcase .... *F. quadrangulata,* BLUE ASH   168

        5 Twigs not ridged, cylindrical; leaflets symmetrical at the base; seeds cylindrical, seed wings flat, often notched at the tip, enclosing about one-half of the seedcase

            6 Leaflet lower surface, leaf stalk, and twigs hairless or nearly so .... *F. pennsylvanica* var. *subintegerrima,* GREEN ASH   165

            6 Leaflet lower surface, leaf stalk, and twigs densely hairy

                7 Leaflet margins prominently toothed; seed wing broad, spatula-shaped .... *F. pennsylvanica* var. *austini,* NORTHERN RED ASH   165

                7 Leaflet margins toothed above the middle; seed wing slender .... *F. pennsylvanica* var. *pennsylvanica,* RED ASH   164

    4 Leaflets stalkless

        8 Leaflets with dense tufts of rust-coloured hair at the point of attachment to the central leaf stalk, elsewhere hairless, 7–11 per leaf; seed wings often twisted; trees of eastern North America .... *F. nigra,* BLACK ASH   166

        8 Leaflets with dense hair along the sides of the midvein below, hairless at point of attachment to the central leaf stalk, 9–15 per leaf; seed wings flat; introduced, widely planted .... *F. excelsior,* EUROPEAN ASH   170

Winter Key to the Ashes (Genus *Fraxinus*)

1 Twigs conspicuously ridged, 4-sided in cross section .... *F. quadrangulata,* BLUE ASH   168

1 Twigs not ridged, round in cross section

   2 Buds conspicuously deep inky-black .... *F. excelsior,* EUROPEAN ASH   170

   2 Buds reddish to dark brown

     3 Leaf scars with a deep, V-shaped notch partially enclosing the lateral bud; uppermost lateral buds set close to the terminal bud, without bark showing in between .... *F. americana,* WHITE ASH   162

     3 Leaf scars without a deep, V-shaped notch, the lateral bud set entirely above the leaf scar or in a curved indentation; uppermost lateral buds off-set from the terminal bud, some or much bark showing in between

       4 Buds dark brown, uppermost lateral pair on twig set distinctly below the terminal bud, bark clearly visible in between, lenticels on young twigs purple; leaf scars round; bark on young trees soft and corky, easily indented, becoming scaly .... *F. nigra,* BLACK ASH   166

       4 Buds reddish-brown, uppermost lateral pair on twig set slightly below the terminal bud, bark only slightly visible in between, lenticels on young twigs white; leaf scars crescent-shaped or semicircular; bark on young trees firm, not corky

         5 Leaf scars crescent-shaped; trees of the Pacific coastal forest; twigs densely hairy, reddish-brown .... *F. latifolia,* OREGON ASH   169

         5 Leaf scars semicircular; trees of central and eastern North America

           6 Twigs hairless or nearly so .... *Fraxinus pennsylvanica* var. *subintegerrima,* GREEN ASH   165

           6 Twigs densely hairy

             7 Lateral buds shallowly set in the upper margin of the leaf scar; trunk not swollen or buttressed at the base; young branches often tinged with red; mature bark with an irregular diamond pattern, ridges slightly raised .... *F. pennsylvanica,* RED ASH   164

             7 Lateral buds deeply set in the upper margin of the leaf scar; trunk often swollen and buttressed at the base; young branches light gray; mature bark fissured into broad, scaly ridges .... *Fraxinus profunda,* PUMPKIN ASH   170

Key to Group 9
Leaves alternate, compound
(divided into 3 or more leaflets)

1 Leaves with 3 leaflets

    2 Leaf stalk 10–15 cm long, leaflets 10–15 cm long, with many translucent dots; flowers small, greenish-white; fruit 1 or 2-seeded, surrounded by a veined wing .... *Ptelea trifoliata*, COMMON HOPTREE   230

    2 Leaf stalk about 5 cm long, leaflets about 5 cm long, without dots; flowers showy, bright yellow, in elongated, drooping clusters; fruit a flattened pod 3–5 cm long .... *Laburnum ×watereri*, LABURNUM   218

1 Leaves with 5 or more leaflets (rarely, some leaves with 3 leaflets)

    3 Twigs with thorns or spines

        4 Leaf stalks with prickles, fruit not a pod

            5 Leaves 50–150 cm long, many-branched; twigs grayish, with many prickles; flowers at shoot tip; white; fruit black, plum-like, 6 mm across.......... .... *Aralia elata*, JAPANESE ANGELICA-TREE   220

            5 Leaves 10–20 cm long, unbranched; twigs dark brown, with a pair of spines flanking the leaf scar; flowers on previous year's twigs, greenish; fruit a reddish-brown, round capsule, 4–5 mm across .... *Zanthoxylum americanum*, COMMON PRICKLY-ASH   220

        4 Leaf stalks without prickles, fruit a pod

            6 Twigs with single, mostly 3-branched thorns, major branches and trunk with many thorns in groups; preformed leaves singly compound, neoformed leaves doubly compound; fruit pods 15–40 cm long .... *Gleditsia triacanthos*, HONEY-LOCUST   214

            6 Twigs with paired spines, set at the base of each leaf; leaves exclusively singly compound; fruit pods less than 15 cm long

                7 Spines broad; leaflets bristle-tipped, on stalks; lateral buds without scales, sunken; flowers white, in loose, hanging clusters; fruit pod flat, 7–10 cm long .... *Robinia pseudoacacia*, BLACK LOCUST   216

                7 Spines needle-shaped; leaflets short-pointed, stalkless; lateral buds with chaff-like scales, not sunken; flowers bright yellow, in small clusters; fruit pod twisted, 4–5 cm long .... *Caragana arborescens*, SIBERIAN PEA-TREE   218

    3 Twigs without thorns or spines

        8 Leaflets with smooth margins (rarely with a few minute teeth)

            9 Lateral buds enclosed within the leaf stalks

                10 Leaflets widest in the middle, blunt-pointed, base rounded; flowers large, white, in clusters at the shoot tip; fruit a thin pod, 7–8 cm long; small trees, to 18 m tall .... *Cladrastis lutea*, YELLOW-WOOD   219

10 Leaflets with parallel sides, long-pointed, base wedge-shaped; flowers small, yellowish, in clusters in the leaf axils; fruit plum-like, glossy white, 10–13 mm across; shrubs or very small trees, to 6 m tall ....*Toxicodendron vernix,* POISON-SUMAC   228

 9 Lateral buds visible in the axils of the leaf stalks

    11 Leaves singly or doubly compound, 15–30 cm long, pale green, leaflets elliptical, occasionally minutely toothed; twigs slender, ending in a withered stump, leaf scars small, U-shaped; fruit a 15–40 cm long, flat, leathery pod ....*Gleditsia triacanthos,* HONEY-LOCUST   214

    11 Leaves exclusively doubly compound, 30–90 cm long, bluish-green, leaflets ovate; twigs very stout, tapering to a blunt point, leaf scars large, inversely heart-shaped; fruit a 12–20 cm long, thick, woody pod ....*Gymnocladus dioicus,* Kᴇɴᴛᴜᴄᴋʏ ᴄᴏꜰꜰᴇᴇᴛʀᴇᴇ   212

 8 Leaflets with toothed margins

    12 Leaflets with coarse, irregular teeth becoming lobe-like at the leaflet base, with a warty gland beneath, often only on one side; fruit a rounded seedcase set in the center of a long, twisted wing .... *Ailanthus altissima,* AILANTHUS   232

    12 Leaflets with sharp, regular teeth; fruits not winged

        13 Lateral buds enclosed beneath the leaf stalks; leaf stalks and twigs exuding milky sap when leaves removed; fruits small, plum-like, in erect, reddish, conical masses .... *Rhus,* SUMACS   225

        13 Lateral buds visible in the axils of the leaf stalks; leaf stalks and twigs not exuding milky sap when leaves removed; fruits nuts in husks, or berry-like, in hanging clusters

            14 Leaflets of similar size and shape; flowers 5-petaled, white, in showy terminal clusters, appearing after the leaves have fully developed; fruits berry-like, in persistent clusters, 6–12 mm across, orange or red ....*Sorbus,* MOUNTAIN-ASH   221

            14 Leaflets dissimilar in size and shape; flowers without petals, green, small and inconspicuous, appearing in spring with the leaves; fruits hard-shelled nuts in husks, green or brown

                15 Leaves 12–25 cm long; leaf scars with many scattered vein scars; pollen catkins 3-branched; nuts with a husk splitting into 4 sections, surface of nuts smooth, with 4 lines or ridges ....*Carya,* HICKORIES   202

                15 Leaves 20–60 cm long; leaf scars with 3 conspicuous vein scars (or with vein scars in 3 conspicuous clusters); pollen catkins unbranched; nuts with nonsplitting husk, surface of nuts furrowed ....*Juglans,* WALNUTS   196

Key to Group 10
Leaves alternate, simple; edges lobed

Key to the Oaks (Genus *Quercus*)

3 Largest lateral lobes of leaves growing in direct sunlight much longer than the leaf is wide between opposite notches

    4 Buds very pale gray-brown, hairless; acorns 15–30 mm long ....
.... *Q. shumardii*, SHUMARD OAK   254

    4 Buds brown, may have hairs at the tip; acorns 9–20 mm long

        5 Acorn cup saucer-shaped, much wider than deep; acorns 9–13 mm long ....
.... *Q. palustris*, PIN OAK   252

        5 Acorn cup bowl-shaped, about as wide as deep; acorns 12–20 mm long

            6 Leaf stalks not more than 1 mm in diameter; acorn cups 10–15 mm wide; acorns without rings at the tip .... *Q. ellipsoidalis*, NORTHERN PIN OAK   254

            6 Leaf stalks more than 1 mm in diameter; acorn cups 15–20 mm wide; acorns commonly with rings at the tip .... *Q. coccinea*, SCARLET OAK   255

1 Leaves with lobes lacking bristle-tips, or with coarsely toothed margins, the teeth may bear short points less than 0.5 mm in length; acorns mature in one year

    7 Leaves with coarsely toothed margins

        8 Leaf lower surface with 2 types of hairs – small, flat stellate hairs and longer hairs diverging from the leaf surface; acorns on stalks 2–10 cm long ....
.... *Q. bicolor*, SWAMP WHITE OAK   260

        8 Leaf lower surface with only small stellate hairs; acorns sessile or short-stalked

            9 Stellate hairs with some rays spreading away from leaf surface; acorn cup enclosing 1/2 of the nut .... *Q. montana* (syn. *Q. prinus*), CHESTNUT OAK   266

            9 Stellate hairs with rays flat and parallel to leaf surface; acorn enclosing 1/3 to 1/2 of the nut

                10 Leaves 5–15 cm long with 4–9 principal veins per side; small to large shrubs, rarely very small trees .... *Q. prinoides*, DWARF CHINQUAPIN OAK   263

                10 Leaves 10–18 cm long with 10–15 principal veins per side; shrubs to medium-sized trees .... *Q. muehlenbergii*, CHINQUAPIN OAK   262

    7 Leaves with lobes

        11 Mature leaves hairless or slightly hairy below

            12 Leaves 10–20 cm long, shiny green above, with 7–9 narrow lobes; acorn 12–20 mm long, stalk less than 2 cm long .... *Q. alba*, WHITE OAK   256

            12 Leaves 5–12 cm long, dark green above, with 3–7 lobes; acorn 15–40 mm long, stalk 3–8 cm long .... *Q. robur*, ENGLISH OAK   266

        11 Leaves hairy below

            13 Acorns on stalks 2–10 cm long; twigs usually hairless; leaves usually shallowly and irregularly lobed .... *Q. bicolor*, SWAMP WHITE OAK   260

13 Acorns on stalks less than 2 cm long; first-year twigs hairy; leaves deeply lobed, at least towards the base

14 Leaves pliant, variably lobed, with an expanded terminal lobe, deep notches below the middle, 2–3 pairs of lower lobes, pale green to whitish below; acorn cup scales knobby, forming a marginal fringe .... *Q. macrocarpa*, BUR OAK   258

14 Leaves thick, stiff, with 5–7 lobes separated by deep notches, dull yellowish-green below; acorn cup scales hairy, not forming a marginal fringe .... *Q. garryana*, GARRY OAK   264

Winter Key to the Oaks (Genus *Quercus*)

1 Buds sharply conical, narrow; twigs sometimes bearing immature acorns; mature bark firmly furrowed and ridged

2 Terminal buds on vigorous twigs average greater than 6 mm long

3 Buds hairless or very slightly downy, very pale gray-brown; twigs gray to grayish-brown, slender .... *Q. shumardii*, SHUMARD OAK   254

3 Buds with few to many long hairs; twigs reddish-brown, moderately stout

4 Buds entirely covered with dense, grayish-white hairs; mature bark with long, irregular, rounded ridges broken horizontally into squarish patches, inner bark yellowish .... *Q. velutina*, BLACK OAK   250

4 Buds hairy only above the middle; mature bark with long, mostly unbroken vertical ridges, inner bark pinkish-red

5 Twigs shiny reddish-brown; buds with a few hairs at the tip; bark on young trunks slate-gray, older trunks developing flattish, pale gray ridges ..... *Q. rubra*, RED OAK   248

5 Twigs light brown; buds more or less covered with silky-white hair above the middle; bark on young trunks brown, older trunks developing irregular, dark brown to nearly black ridges .... *Q. coccinea*, SCARLET OAK   255

2 Terminal buds on vigorous twigs average 6 mm long or less

6 Terminal bud 2–4 mm long; twigs hairless; inner bark reddish .... *Q. palustris*, PIN OAK   252

6 Terminal bud 4–6 mm long; young twigs hairy; inner bark light yellow .... *Q. ellipsoidalis*, NORTHERN PIN OAK   254

1 Buds oval to broadly conical; twigs not bearing immature acorns; mature bark rough, ridged, breaking into plates or scales

7 Terminal bud on vigorous twigs pointed, elongated

8 Buds densely hairy; twigs dark reddish brown, densely hairy in 1st year; occurring along the Pacific coast... *Q. garryana*, GARRY OAK   264

8 Buds hairless; twigs grayish to orange-brown, hairless; trees of eastern North America

    9 Bark pale gray to whitish with thin narrow scales .... *Q. muehlenbergii*, CHINQUAPIN OAK[1]   262

    9 Bark dark brown to reddish brown with broad, rounded ridges .... *Q. montana* (syn. *Q. prinus*), CHESTNUT OAK   266

7 Terminal bud on vigorous twigs blunt, ovoid or round

    10 Terminal bud 7–9 mm long; twigs slender, slightly powdery on the surface; leaves often persisting long into winter; cultivars with strongly upright branches common .... *Q.robur*, ENGLISH OAK   266

    10 Terminal bud less than 6 mm long; twigs stout, not powdery on the surface

        11 Buds hairy, laterals pressed against the twig; twigs somewhat hairy, yellowish-brown, developing conspicuous corky ridges .... *Q. macrocarpa*, BUR OAK   258

        11 Buds mostly hairless, laterals diverging from the twig; twig hairless, without corky ridges

            12 Terminal bud 3–5 mm long, reddish-brown; twigs reddish-green when young, becoming gray; bark smooth on upper branches, mature bark scaly, deeply fissured, with rounded ridges .... *Q. alba*, WHITE OAK   256

            12 Terminal bud 2–4 mm long, chestnut-brown; twigs reddish-brown; bark ragged and peeling on upper branches, mature bark scaly, fissured, with flat ridges .... *Q. bicolor*, SWAMP WHITE OAK   260

[1] Chinquapin oak and dwarf chinquapin oak (*Q. prinoides*) are not reliably separable on the basis of winter characteristics. Chinquapin oak is more likely to attain tree size.

Key to Group 11
Leaves alternate, simple; edges toothed

12 Mature leaves hairless below; buds not clustered at twig tips

    13 Leaves leathery, dark bluish green above; terminal and most lateral buds long and slender; fruit in a reddish-brown, bristly husk; bark smooth, light bluish-gray, often mottled .... *Fagus*, BEECHES   268

    13 Leaves rough, bright green above; buds ovoid; fruit a cherry-like drupe; bark smooth, gray, with horizontal, pinkish lenticels .... *Zelkova serrata*, JAPANESE ZELKOVA   361

10 Leaf with lateral veins not extending to each marginal tooth

    14 Leaves narrow, more than 3 times as long as wide

        15 Buds covered by a single, cap-like scale .... *Salix*, WILLOWS   312

        15 Buds covered by more than 1 scale

            16 Leaf with a mat of fine brown hairs along the midrib below .... *Prunus serotina*, BLACK CHERRY   380

            16 Leaf lacking a mat of fine brown hairs along midrib below

                17 Leaves coarsely toothed, broadest above the middle, glandular below; rare along the coast of southern British Columbia .... *Myrica californica*, PACIFIC BAYBERRY   275

                17 Leaves finely toothed, broadest near or below the middle, resin-stained below; scarce in southwest Saskatchewan and adjacent Alberta .... *Populus angustifolia*, NARROWLEAF COTTONWOOD   342

    14 Leaves oval, less than 3 times as long as wide

        18 Fruit borne in persistent, woody, cone-like catkins .... *Alnus*, ALDERS   296

        18 Fruit not in persistent, woody, cone-like catkins

            19 Leaf scars with 1 vein scar .... *Ilex*, HOLLIES   274

            19 Leaf scars with 3 vein scars

                20 Pith round in cross section

                    21 Leaf stalk hairless, with warty glands at the top; twigs with an odor of bitter almond when bruised .... *Prunus*, PLUMS AND CHERRIES   374

                    21 Leaf stalk hairy, without warty glands; twigs with a licorice-like flavor .... *Malus*, APPLES   369

                20 Pith 5-pointed in cross section

                    22 Leaves oval to nearly round, with straight lateral veins; young bark smooth, gray, often marked with vertical lines .... *Amelanchier*, SERVICEBERRIES   364

                    22 Leaves 2 times as wide as long, with irregular, branching lateral veins; young bark smooth, whitish-, yellowish- or grayish-green .... *Populus*, POPLARS   335

6 Leaf margin double-toothed

23 Leaf base symmetrical

24 Leaf margin with large, pointed teeth; leaf stalk without warty glands at the leaf blade

25 Pith round or oval in cross section; fruit not in woody, cone-like catkins

26 Leaves and branchlets not gland-dotted; all buds with similar number of visible scales

27 Leaves elliptical and tapered towards the tip, regular in shape; fruits arranged along drooping, flexible stems

Key to the Birches (Genus *Betula*)

1 Twigs fragrant, smelling and tasting of wintergreen

1 Twigs lacking fragrance or taste of wintergreen

   3 Mature bark dark brown to gray, never white

      4 Leaves yellowish-green and shiny above, teeth sharp; fruit scale lateral lobes shorter than middle lobe; mature bark purplish-brown; shrubs or small trees of western North America, excluding the coastal forest .... *B. occidentalis,* WATER BIRCH   290

      4 Leaves dull dark green and hairy above, teeth coarse; fruit scale lateral lobes equal in length to middle lobe; mature bark varying from gray to reddish- to dark-brown; small trees of Yukon and Alaska .... *B. kenaica,* KENAI BIRCH   293

   3 Mature bark white, may be tinged copper or pink

      5 Leaves conspicuously dotted with resin glands

         6 Leaves 6–12 cm long with tiny resin glands on upper surface, veins 7–9 per side, base variable, rounded or straight, usually heart-shaped; an eastern North American species .... *B. cordifolia,* MOUNTAIN PAPER BIRCH   286

         6 Leaves 4–7 cm long with tiny resin dots on lower surface, veins 4–5 per side, base broadly wedge-shaped; a western boreal forest species .... *B. neoalaskana,* ALASKA PAPER BIRCH   292

      5 Leaves without conspicuous resin glands

         7 Leaf tip short-pointed, wedge-shaped

            8 Leaves 5–10 cm long, neoformed leaves hairy below in vein axils; mature bark peeling in large sheets; branches ascending; transcontinental in range .... *B. papyrifera,* WHITE BIRCH   284

            8 Leaves 3–7 cm long, hairless; mature bark peeling in long strands; branches slender, drooping; introduced from Europe, widely planted, becoming naturalized .... *B. pendula,* EUROPEAN WHITE BIRCH   289

         7 Leaf tip long-pointed and tapered

            9 Leaves 5–9 cm long, bluish-green above, base variable; mature seed catkins 2.5–3.5 cm long; mature bark peeling easily in sheets .... *B. ×caerulea,* BLUELEAF BIRCH   287

            9 Leaves 4–7 cm long, shiny dark green and rough above, base almost straight; mature seed catkins 1.5–2 cm long; mature bark peeling with difficulty into thin, rectangular plates .... *B. populifolia,* GRAY BIRCH   288

Winter Key to the Birches (Genus *Betula*)

1 Twigs fragrant, smelling and tasting of wintergreen

    2 Wintergreen fragrance moderate; buds mostly hairy, pressed against the twig; twigs uniformly brown; bark shiny reddish-brown, becoming dull yellow, darkening to bronze with age .... *B. alleghaniensis,* YELLOW BIRCH   294

    2 Wintergreen fragrance strong; buds mostly hairless, diverging from the twig; twigs brown to black; bark dark cherry-red to almost black, becoming gray with age .... *B. lenta,* CHERRY BIRCH   295

1 Twigs lacking fragrance or taste of wintergreen

    3 Bark easily separating into papery layers

        4 Twigs abundantly hairy, entirely covered with dense, often crystalline, resin glands; a western boreal forest species .... *B. neoalaskana,* ALASKA PAPER BIRCH   292

        4 Twigs hairless to slightly hairy, smooth to somewhat glandular

            5 Twigs slightly hairy

                6 Twigs glandular, lenticels average about 19 mm long; peeling bark separating into thin, ragged layers, tinged copper on inner surfaces; an eastern North American species .... *B. cordifolia,* MOUNTAIN PAPER BIRCH   286

                6 Twigs smooth or very sparsely glandular, lenticels average about 13 mm long; peeling bark separating in large sheets, lacking a copper tinge on the inner surface; transcontinental in range .... *B. papyrifera,* WHITE BIRCH   284

            5 Twigs always hairless

                7 Twigs long, drooping, sparsely glandular; buds blunt, shiny green, to 4 mm long; bark bright chalky-white, with vertical black fissures at base of mature trees; introduced from Europe, widely planted, becoming naturalized .... *B. pendula,* EUROPEAN WHITE BIRCH   289

                7 Twigs upright, glandular; buds sharp-pointed, brownish, to 6 mm long; bark white; ranging from Nova Scotia and northern Quebec to northern New England and eastern New York state .... *B.* ×*caerulea,* BLUELEAF BIRCH   287

    3 Bark not easily separating into papery layers

        8 Bark dull chalky white, often with black, triangular patches below the branches; a small tree of eastern North America .... *B. populifolia,* GRAY BIRCH   288

        8 Bark not white; lacking black triangular patches below the branches; shrubs or small trees of western North America

            9 Bark variably reddish or dark brown, may be tinged pinkish or light gray; ranging through Yukon and Alaska .... *B. kenaica,* KENAI BIRCH   293

            9 Bark very dark brown, becoming purplish-brown with age; ranging throughout forested regions of western North America, excluding the coastal forest .... *B. occidentalis,* WATER BIRCH   290

Key to the Alders (Genus *Alnus*)

1 Buds stalkless; immature fruit catkins enclosed within buds in winter, opening as leaves reach full size; fruit with wings as wide as the nutlet

> **2** Leaf margin with teeth distinctly 2-sized, appearing shallowly lobed, leaf base linear, asymmetrical at stalk; tall shrubs or small trees of the Pacific coast and Rocky Mountains .... *A. viridis* ssp. *sinuata,* SITKA ALDER   302

> **2** Leaf margin with many sharp teeth of equal size, leaf base rounded or heart-shaped; large shrubs of the boreal forest .... *A. viridis* ssp. *crispa,* GREEN ALDER   304

1 Buds stalked; immature fruit catkins naked in winter, open in spring before leaves appear; fruit with wings narrower than the nutlet

> **3** Leaf margin wavy, with a single row of fine teeth; shrubs or very small trees of the coastal region of eastern North America .... *A. serrulata,* HAZEL ALDER   305

> **3** Leaf margin with coarse teeth of 2 sizes

>> **4** Leaves widest at or above the middle

>>> **5** Leaves oval, tapered at both ends, 8–15 veins per side, not sticky, leaf margin rolled under; a native species of the Pacific Coast forest .... *A. rubra,* RED ALDER   298

>>> **5** Leaves circular, blunt at tip, wedge-shaped at base, 5–6 veins per side, sticky when young, leaf margin flat; introduced from Europe, frequently planted .... *A. glutinosa,* EUROPEAN BLACK ALDER   305

>> **4** Leaves widest below the middle

>>> **6** Leaves thick, wrinkled, with veins impressed above and projecting conspicuously below; mature fruit catkins 13–16 mm long; very small trees to 8 m tall, common through eastern and central North America, ranging north into the Yukon .... *A. incana* ssp. *rugosa,* SPECKLED ALDER   300

>>> **6** Leaves thin, smooth, the veins not impressed, mature fruit catkins 9–13 mm long; very small trees to 10 m tall, common through western Canada inland from the Pacific coast .... *A. incana* ssp. *tenuifolia,* MOUNTAIN ALDER   301

Winter Key to the Alders (Genus *Alnus*)

1 Terminal bud stalkless, sharp-pointed, with 3–5 overlapping bud scales; fruit catkins on stalks equal to their length

> **2** Bud scales purplish-brown; fruit catkins to 20 mm long; tall shrubs or small trees of the Pacific coast and Rocky Mountains .... *A. viridis* ssp. *sinuata,* SITKA ALDER   302

> **2** Bud scales reddish or greenish; fruit catkins to 15 mm long; large shrubs of the boreal forest .... *A. viridis* ssp. *crispa,* GREEN ALDER   304

1 Terminal bud stalked, blunt, with 2–3 bud-scales meeting along the edges; fruit catkins longer than the stalks

Key to the Willows (Genus *Salix*)

7 Leaves without a fragrance of resin

    8 Leaves light yellowish-green; twigs flexible, arching near the tip, yellowish-brown, becoming gray .... *S. amygdaloides,* PEACHLEAF WILLOW   316

    8 Leaves dark green; twigs stout, rigid, upright

        9 Twigs dark shining purple, coated with bluish-white waxy powder .... *S. daphnoides,* VIOLET WILLOW   334

        9 Twigs yellowish to reddish-brown, become shiny

            10 Leaves shiny, hairless below; ranging through eastern North America, west to Saskatchewan .... *S. lucida* ssp. *lucida,* SHINING WILLOW   324

            10 Leaves whitish, slightly hairy below; ranging through western North America, east to Saskatchewan .... *S. lucida* ssp. *lasiandra,* PACIFIC WILLOW   322

3 Leaves more than 5 times as long as wide, widest below the middle

    11 Leaves rounded at the base

        12 Stipules to 10 mm long, semicircular, shallowly toothed; leaves uniformly dark green above and below; small trees, to 12 m tall .... *S. nigra,* BLACK WILLOW   326

        12 Stipules to 20 mm long, round-tipped, deeply toothed; leaves tinged reddish-purple, lower surface paler than upper; shrubs or very small trees, not more than 6 m tall

            13 Twigs brownish, sometimes with long shaggy hairs; a western species, to 5 m tall .... *S. prolixa,* MACKENZIE WILLOW   331

            13 Twigs reddish-brown to yellowish-green, hairy in spring, becoming smooth; an eastern species, to 6 m tall .... *S. eriocephala,* HEARTLEAF WILLOW   332

    11 Leaves tapered to the base

        14 Mature leaves glossy green above

            15 Leaf margin irregularly, coarsely toothed, teeth and leaf stalk glandular; twigs brittle, easily detached at the base, shiny yellowish-green to dark red; medium-sized trees, introduced from Europe, naturalized in eastern North America .... *S. fragilis,* CRACK WILLOW   327

            15 Leaf margin finely toothed, teeth glandular towards the leaf tip; twigs flexible, slender, yellow-green to olive-brown, become very dark brown; very small trees or shrubs, ranging from New Brunswick to northern British Columbia .... *S. petiolaris,* MEADOW WILLOW   332

    14 Mature leaves dull green or faintly hairy above

        16 Twigs dull olive-brown; leaves at first finely hairy below but soon hairless .... *S. alba* ×*fragilis,* HYBRID WHITE WILLOW   329

16 Twigs light yellow; leaves with fine, silky hair below

    17 Branchlets pendulous ....
        .... *S. alba* var. *vitellina* cv. Pendula, GOLDEN WEEPING WILLOW   328

    17 Branches upright, not weeping .... *S. alba* var. *vitellina,* GOLDEN WILLOW   328

2 Mature leaves densely hairy below, may be hairy above

  18 Leaves less than 5 times as long as wide, widest at or above the middle

    19 Leaf margin with many regular, fine teeth .... *S. arbusculoides,* LITTLETREE WILLOW   319

    19 Leaf margin with a few sparse, wavy teeth, or without teeth

      20 Leaves smooth above, densely hairy below

        21 Leaves bluish-green below, with silky hairs, especially along the midvein;
        twigs remaining densely hairy for at least 2 years; seed capsules hairless
        .... *S. hookeriana,* HOOKER WILLOW   323

        21 Leaves white below with felt-like woolly hairs; twigs hairless after 1 year; seed
        capsules hairy .... *S. alaxensis,* FELTLEAF WILLOW   331

      20 Leaves hairy above, densely hairy below

        22 Leaves shiny below with silky hairs; stipules small, 1 mm long; twigs brittle,
        dark reddish-brown .... *S. sitchensis,* SITKA WILLOW   330

        22 Leaves whitish below with rust-colored hairs; stipules 1–3 mm long, longer
        on vigorous shoots; twigs stiff, yellowish to greenish-brown ....
        .... *S. scouleriana,* SCOULER WILLOW   330

  18 Leaves more than 5 times as long as wide, widest at the base

    23 Leaf margin with small glandular teeth; fruit capsules mostly hairless ....
    .... *S. exigua,* SANDBAR WILLOW   317

    23 Leaf margin rolled under, smooth or slightly wavy; fruit capsules densely hairy

      24 Leaves very narrow, to 15 mm wide; twigs flexible, greenish-yellow to reddish-
      brown, hairy when young, becoming smooth and shiny; an introduced, European
      species .... *S. viminalis,* BASKET WILLOW   333

      24 Leaves narrow, to 25 mm wide; twigs brittle, yellowish to dark reddish-brown,
      often coated with a waxy powder; a native species, ranging from Labrador to
      Saskatchewan .... *S. pellita,* SATINY WILLOW   333

Key to the Poplars (Genus *Populus*)

1 Leaf stalk laterally flattened near junction with the leaf base

  2 Leaves somewhat triangular or 4-sided, taper-pointed; buds resinous

    3 Leaves 4-sided, diamond-shaped, mostly shorter than 10 cm; a clonal cultivar, planted
    as ornamentals, often columnar .... *P. nigra* cv. Italica, LOMBARDY POPLAR   350

3 Leaves somewhat triangular, mostly longer than 10 cm; branches spreading, crown broad and open

    4 Leaves similar in color below and above, petiole where flattened more than twice as deep as wide

        5 Margin of early leaves with 5–15 round teeth per side; buds covered with minute hairs; leaf base somewhat rounded .... *P. deltoides* ssp. *monilifera*, PLAINS COTTONWOOD   345

        5 Margin of early leaves with 20–25 round teeth per side, buds hairless; leaf base linear or wedge-shaped

            6 Leaf base tending to straight, mostly with 3–5 warty glands where the leaf and stalk meet; trunk often divided into wide-spreading limbs, annual nodes often irregularly spaced .... *P. deltoides* ssp. *deltoides*, EASTERN COTTONWOOD   344

            6 Leaf base tending to wedge-shaped, mostly lacking glands; trunk generally distinct to top of tree, annual nodes regularly spaced .... *P.* ×*canadensis*, CAROLINA POPLAR   351

    4 Leaves paler below than above, petiole where flattened less than twice as deep as wide

        7 Leaves mostly lance-shaped to broadly lance-shaped, at least 1.5 times as long as wide, margins toothed along the middle portion; small trees, to 20 m tall, common along streams in the Rocky Mountain region, southern Alberta to Texas .... *P.* ×*acuminata*, LANCELEAF COTTONWOOD   343

        7 Leaves mostly broadly ovate triangular, less than 1.5 times as long as wide, margins toothed along most of their length; medium-sized trees, to 30 m tall

            8 Leaves to 10 cm long, leaf stalk hairless; ranging across North America where *P. deltoides* and *P. balsamifera* grow together .... *P.* ×*jackii*, JACK'S HYBRID POPLAR   341

            8 Leaves to 17 cm long, leaf stalk densely hairy; a clonal cultivar, commonly planted as ornamentals .... *P.* ×*jackii* cv. Gileadensis, BALM-OF-GILEAD   341

2 Leaves oval to elliptical or lobed, short-pointed; buds non-resinous

    9 Margin of leaves with few rounded teeth, leaves palmately 3–5 lobed; leaves, buds and twigs conspicuously covered in white, woolly hairs; introduced, planted as ornamentals .... *P. alba*, EUROPEAN WHITE POPLAR   351

    9 Margin of leaves with at least 7 teeth per side, leaves never lobed; leaves, buds and twigs hairless to somewhat hairy

        10 Margin of early leaves with 20–30 fine, irregular teeth per side; leaf stalk longer than leaf blade .... *P. tremuloides*, TREMBLING ASPEN   346

        10 Margin of early leaves with 7–15 coarse, wavy teeth per side; leaf stalk shorter than leaf blade .... *P. grandidentata*, LARGETOOTH ASPEN   348

1 Leaf stalk round or dorsally compressed at junction with leaf blade, often channeled above

    11 Leaves very narrow, lance-shaped; a western species, range in Canada southwest Saskatchewan and southern Alberta .... *P. angustifolia*, NARROWLEAF COTTONWOOD   342

11 Leaves ovate to somewhat 4-sided

    12 Leaf stalk to 2 cm long, leaves somewhat 4-sided, widest above the middle, whitish beneath; introduced, planted as ornamentals, often columnar .. *P. simonii,* SIMON POPLAR   350

    12 Leaf stalk 3 cm or longer, leaves ovate, widest at or below the middle, silvery-green and stained with resin blotches beneath

        13 Leaf stalk 7–10 cm long, leaf base rounded, occasionally with warty glands; fruit capsules hairless, splitting in 2 parts when mature; medium-sized trees to 25 m tall, transcontinental in range .... *P. balsamifera,* BALSAM POPLAR   340

        13 Leaf stalk 3–4 cm long, leaf base round, wedge- or heart-shaped, lacking warty glands; fruit capsules with short hairs, splitting in 3 parts when mature; large trees to 35 m tall, ranging through most of British Columbia and western Alberta .... *P. trichocarpa,* BLACK COTTONWOOD   338

Winter Key to the Poplars (Genus *Populus*)

1 Buds resinous, sticky when squeezed

    2 Resin on buds fragrant, smelling of balsam

        3 Twigs angular in cross section; introduced, planted as ornamentals, often columnar in form .... *P. simonii,* SIMON POPLAR   350

        3 Twigs round to slightly angular in cross section

            4 Terminal buds with 5 visible bud scales

                5 Terminal bud 6–12 mm long, brownish; young twigs yellowish-brown, becoming ivory-white with age; range in Canada southwest Saskatchewan and southern Alberta .... *P. angustifolia,* NARROWLEAF COTTONWOOD   342

                5 Terminal bud to 25 mm long, orange-brown; young twigs bright reddish-brown, becoming dark orange, then gray with age; range transcontinental .... *P. balsamifera,* BALSAM POPLAR   340

            4 Terminal buds with 6–7 visible bud scales

                6 Terminal bud 9–12 mm long, buds orange; small trees, to 20 m tall when mature, common along streams in southern Alberta .... *P.* ×*acuminata,* LANCELEAF COTTONWOOD   343

                6 Terminal bud 17–20 mm long, buds red; large trees, to 35 m tall when mature, ranging through British Columbia and southern Alberta .... *P. trichocarpa,* BLACK COTTONWOOD   338

    2 Resin on buds not strongly fragrant

        7 Terminal bud to 8 mm long; lateral buds pressed tightly against the twig; mature bark shallowly ridged; introduced species, crown narrow, columnar .... *P. nigra* cv. Italica, LOMBARDY POPLAR   350

        7 Terminal bud to 20 mm long; lateral buds diverging from the twig; mature bark deeply furrowed; crown broad, spreading

Key to the Elms (Genus *Ulmus*)

2 Leaves average more than 15 pairs of lateral veins; smooth to somewhat rough above

    5 Leaves to 10 cm long, shiny and smooth above, lateral veins rarely forking, leaf base nearly symmetrical; fruits hairy, seedcase indistinct .... *U. thomasii,* ROCK ELM   356

    5 Leaves to 15 cm long, mostly rough above, 2 or 3 lateral veins at the leaf base forking, leaf base asymmetrical; fruits hairy only along wing margin, seedcase distinct .... *U. americana,* WHITE ELM   354

Winter Key to the Elms (Genus *Ulmus*)

1 Buds sharply pointed; young twigs finely hairy, older twigs hairless, developing conspicuous, corky ridges .... *Ulmus thomasii,* ROCK ELM   356

1 Buds blunt-pointed, twigs without corky ridges

    2 Twigs hairless, or only sparsely hairy

        3 Twigs grayish-brown, noticeably zig-zaged; buds roughly equal in size, end bud strongly oblique; large trees of central and eastern North America .... *U. americana,* WHITE ELM   354

        3 Twigs light gray, brittle, nearly straight; flower buds globular, much larger than the leaf buds, end bud weakly oblique; introduced, medium-sized trees, often planted and trimmed in hedge-form .... *Ulmus pumila,* SIBERIAN ELM   360

    2 Twigs noticeably hairy

        4 Twigs slender, pale brown, densely hairy, lacking a slippery, mucilaginous, fragrant inner bark; buds ovoid, very dark, mostly hairless, leaf buds small, flower buds larger, rounded; bark becoming ridged, showing deep fissures .... *U. procera,* ENGLISH ELM   360

        4 Twigs moderately stout, grayish brown, hairy, with a slippery, mucilaginous and fragrant inner bark

            5 Twigs with prominent lenticels; buds, especially at the tips, coated in reddish-brown hairs; mature bark brown, shallowly furrowed; native to eastern North America .... *Ulmus rubra,* SLIPPERY ELM   358

            5 Twigs without prominent lenticels; buds covered in brown hairs, leaf buds almost black, flower buds dark brown; mature bark gray, remaining smooth for many years; introduced, planted as ornamentals .... *U. glabra,* SCOTCH ELM   359

Key to the Serviceberries (Genus *Amelanchier*)

1 Leaves unfolded when emerging in spring; flowers in groups of 1–3, located in the leaf axils .... *A. bartramiana,* MOUNTAIN SERVICEBERRY   368

1 Leaves folded when emerging in spring; flowers in many-flowered clusters located at the tips of new leafy shoots

     2 Leaf margins coarsely toothed, 3–5 teeth/cm, leaves unfold soon after emerging from winter buds

         3 Leaves to 7 cm long, margin toothed to near the leaf base; occurring eastward from Lake Superior .... *A. sanguinea,* ROUNDLEAF SERVICEBERRY   368

         3 Leaves to 5 cm long, margin not toothed on lower one-third of the leaf; occurring westward from Manitoba

             4 Leaves with fine hair below along the midvein; flower clusters 2–3 cm long; fruit 6–10 mm across, purple; ranging widely through the northwest of Canada and the United States .... *A. alnifolia,* SASKATOON   366

             4 Leaves nearly hairless below; flower clusters 4–7 cm long; fruit 10–12 mm across, purple with a whitish powdery coat; restricted to the Pacific coastal forest .... *A. florida,* PACIFIC SERVICEBERRY   367

     2 Leaf margin finely toothed, 6–12 teeth/cm, leaves remain folded at flowering time

         5 Leaves at flowering time small, densely hairy; flower clusters erect, petals 8–12 mm long; fruit dry, tasteless, lowermost fruit stalk about 12 mm long .... *A. arborea,* DOWNY SERVICEBERRY   365

         5 Leaves at flowering time distinctly copper-red, at least half-grown and hairless below; flower clusters drooping, petals 10–17 mm long; fruit juicy, sweet, lowermost fruit stalk 25–45 mm long .... *A. laevis,* SMOOTH SERVICEBERRY   365

Key to the Cherries and Plums (Genus *Prunus*)

1 Twigs with a terminal bud, without thorns; fruit spherical

     2 Leaves oval or elliptical

         3 Leaf margin finely and singly toothed

             4 Leaves with glandular dots below; bark rich cinnamon-brown; fruit black, about 5 mm across; an introduced, ornamental species ..... *P. maackii,* AMUR CHOKE CHERRY   381

             4 Leaves lack glandular dots below; bark dark, grayish, becoming black with age; fruit dark crimson to black, to 15 mm across; native species

                 5 Leaves dull green, pale below and hairless except at vein axils, leaf base tapered; ranging across North America .... *P. virginiana* var. *virginiana,* CHOKE CHERRY   382

                 5 Leaves dark green, downy below, leaf base rounded; occurring in the interior of British Columbia .... *P. virginiana* var. *demissa,* WESTERN CHOKE CHERRY   383

3 Leaf margin coarsely double-toothed

    6 Leaves long-pointed, teeth bristle-tipped, very sharp; flowers pink or white; fruit black when mature .... *P. serrulata,* JAPANESE FLOWERING CHERRY   378

    6 Leaves short-pointed, teeth sharp, lacking bristle tips; flowers white; fruit bright red when mature

        7 Leaves hairless below, leaf stalk lacking glands; fruit 20 mm across, tart tasting .... *P. cerasus,* SOUR CHERRY   375

        7 Leaves hairy below at vein axils, leaf stalk with glands; fruit 25 mm across, sweet tasting .... *P. avium,* SWEET CHERRY   375

2 Leaves long, narrowly lance-shaped

    8 Midvein on lower leaf surface covered at the base with narrow mat of brown hair; teeth elongated, incurved; mature bark reddish-brown to blackish, distinctly scaly .... *P. serotina,* BLACK CHERRY   380

    8 Midvein on lower leaf surface lacking mat of brown hair; teeth uneven, minute, often gland-tipped; mature bark reddish, peeling

        9 Leaves 8–15 cm long, widest below the middle, shiny yellowish-green, hairless, tip slender and gradually tapered; widely occurring , ranging from Newfoundland to central British Columbia .... *P. pensylvanica,* PIN CHERRY   376

        9 Leaves 3–8 cm long, widest above the middle, dull yellowish-green, downy below, tip rounded or wedge-shaped; occurring in the southern part of central and coastal British Columbia .... *P. emarginata,* BITTER CHERRY   378

1 Twigs lacking a terminal bud, thorny; fruit elongated

    10 Leaves widest above the middle, teeth rounded, the smaller teeth usually with glands at the tips; fruit 25–30 mm long, not powdery on surface .... *P. nigra,* CANADA PLUM   384

    10 Leaves widest below the middle, teeth sharp, lacking glands at the tips; fruit 20–25 mm long, slightly powdery on surface .... *P. americana,* WILD PLUM   386

Winter Key to the Cherries and Plums (Genus *Prunus*)

1 Twigs without thorns, terminal buds present

    2 Buds clustered at the ends of all twigs; young twigs very slender, less than 2 mm thick

        3 Buds 3–4 mm long; twigs dark red; narrow-crowned trees, to 20 m tall; occurring in the southern part of central and coastal British Columbia .... *P. emarginata,* BITTER CHERRY   378

        3 Buds 1–2 mm long; twigs red; round-topped trees, to 12 m, shrubby on poor sites; ranging from Newfoundland to central British Columbia .... *P. pensylvanica,* PIN CHERRY   376

2 Buds clustered at the ends of dwarf shoots but not at the end of vigorous twigs; young twigs mostly more than 2.5 mm thick

    4 Dwarf shoots rare or absent; native species

        5 Buds 6–7 mm long, sharp-tipped, scales chocolate-brown, with pale gray margins; twigs gray-brown, lenticels prominent but not horizontally extending; mature bark smooth or with fine scales; very small trees, often only shrubs .... *P. virginiana,* CHOKE CHERRY   382

        5 Buds 3–4 mm long, blunt-tipped, scales brown with green bases and dark tips; twigs reddish-brown, with conspicuous, horizontal dash-like lenticels; mature bark distinctly scaly; medium-sized trees, to 22 m tall .... *P. serotina,* BLACK CHERRY   380

    4 Dwarf shoots with terminal bud clusters numerous; introduced European species, cultivated or escaped

        6 Bark rich cinnamon-brown, peeling horizontally in shaggy masses; cultivated trees to 15 m tall, crown irregularly low-domed .... *P. maackii,* AMUR CHOKE CHERRY   381

        6 Bark darker, shiny red-brown to dull grayish-brown, becoming broken and scaly with age

            7 Buds sharply pointed, of 2 sizes; bark along the trunk often frost-cracked and issuing sap; small landscape trees with sparse, widely arching branches .... *P. serrulata,* JAPANESE FLOWERING CHERRY   378

            7 Buds blunt-pointed, more-or-less all equal in size

                8 Buds conical, shiny reddish-brown, inner scales upright; bark grayish, scaly; very small trees with flattened, rounded crowns, often lacking a central trunk .... *P. cerasus,* SOUR CHERRY   375

                8 Buds slender, shiny light brown, inner scales bent back from tip; bark reddish-brown, becoming purplish and fissured; small trees with regular, pyramidal crowns and slender, single trunks .... *P. avium,* SWEET CHERRY   375

1 Twigs thorny, lacking terminal buds

    9 Buds grayish-brown, bud scales thin, frayed and pale at the edges; twigs reddish-brown; bark black .... *P. nigra,* CANADA PLUM   384

    9 Buds grayish, bud scales 2 shades of pale grayish-brown; twigs grayish to reddish brown; bark reddish-brown or dark gray to nearly black .... *P. americana,* WILD PLUM   386

Key to Group 12
Leaves alternate, simple; edges smooth; deciduous (or evergreen)

1 Leaves with more than 1 prominent vein beginning at or just beyond the leaf base

1 Leaves with a single prominent vein beginning at the leaf stalk

   3 Leaves covered below with silvery-brown scales

   3 Leaf lower surface not scaly

      5 Twigs without thorns

         6 Leaves thick, leathery, persisting for at least 1 winter, changing color and falling individually

            7 Leaves 6–25 cm long, occasionally with margins finely toothed on vigorous shoots; fruit in terminal clusters

6 Leaves thin, flexible, collectively changing color and shed before winter

9 Buds without scales, covered by small, immature leaves

10 Leaves 3–8 cm long, with 5–10 pairs of conspicuous, straight veins, margins often faintly wavy; flowers greenish-yellow; fruit round, 2 or 3-seeded, purplish-black when ripe ....(Group 11) *Rhamnus frangula,* GLOSSY BUCKTHORN   278

10 Leaves 15–30 cm long, with lateral veins looping at the margin; flowers reddish-purple; fruit irregularly pear-shaped, many-seeded, pale greenish-yellow to brownish when ripe .... *Asimina triloba,* PAWPAW   398

9 Buds covered by scales

11 Lateral buds covered by a single, cap-like scale ....
....(Group 11) *Salix,* WILLOWS   312

11 Lateral buds covered by more than 1 scale

12 Leaves with lateral veins curving forward and following the leaf margin towards the leaf tip, appearing opposite or whorled on short twigs, margins often slightly wavy ....
.... (Group 8) *Cornus alternifolia,* ALTERNATE-LEAF DOGWOOD   184

12 Leaves with lateral veins extending directly towards the leaf margin

13 Leaves almost circular; twigs exuding a strong-smelling, gummy sap when cut; fruits in conspicuous, long-haired clusters ....
....*Cotinus coggygria,* COMMON SMOKE-TREE   408

13 Leaves longer than wide; twigs not exuding strong-smelling sap when cut

14 Twigs slightly zig-zag, dwarf shoots rare; leaves with prominent lateral veins, margin wavy; buds slender, pointed ....
....................................(Group 11) *Fagus sylvatica,* EUROPEAN BEECH   269

14 Twigs straight, dwarf shoots numerous on branches

15 Twigs encircled by a thin line at the upper edge of each leaf scar; lateral veins branching, not entirely extending to leaf margin; flowers bell-shaped, showy .... *Magnolia,* MAGNOLIAS   404

15 Twigs not encircled by a thin line at the upper edge of each leaf scar

16 Leaves of varying shape, to 12 cm long, leaf stalk reddish; twigs reddish-brown, with a grayish skin; fruit plum-like, blue-black, containing a single, indistinctly ribbed stone ....
.... *Nyssa sylvatica,* BLACK-GUM   402

16 Leaves oval, to 7 cm long, leaf stalk purple; twigs purplish, smooth; fruit round, purplish-red, containing 4 or 5 bony nutlets ....
...........(Group 11) *Nemopanthus mucronata,* MOUNTAIN-HOLLY   275

Winter Keys to the Genera
(excluding the evergreens in Groups 1–6)

Note: When identifying trees in winter, a hand lens of 10× or greater power is essential for examining leaf and vein scars. A sharp knife or small razor blade for cutting twigs is also useful and will allow observation of inner features such as pith shape, size, and color.

452

Winter Key A
Leafless conifers: crown conical, central stem distinct; cone-bearing

1 Buds arranged oppositely along major shoots ... *Metasequoia glyptostroboides*, DAWN REDWOOD   35

1 Buds arranged alternately along major shoots

   2 Dwarf shoots lacking, shoots bear small circular scars lacking vein scars; winter buds with few scales; seed cones not persistent ....*Taxodium distichum*, BALD-CYPRESS   36

   2 Dwarf shoots abundant, shoots bear minute, triangular scars with a single vein scar; winter buds with many pointed bud scales; seed cones persistent on twig .... *Larix*, LARCHES   67

Winter Key B
Broadleaf evergreens: green leaves present on branches; leaves broad, not needle or scale-like

1 Leaves oppositely arranged; twigs slightly 4-sided in cross-section ....
.... *Euonymus fortunei*, WINTER-CREEPER EUONYMUS   192

1 Leaves alternately arranged; twigs round in cross section

   2 Leaves with margins coarsely toothed

      3 Leaves (at least in lower crown) spiny along margins, oval; fruits bright red, berry-like ....
.... *Ilex*, HOLLIES   274

      3 Leaves not spiny along margins, narrow; fruits dark purple, in globular clusters ....
.... *Myrica californica*, PACIFIC BAYBERRY   275

   2 Leaves with margins smooth or finely toothed

      4 Leaves oval, leaf margin may be finely toothed on vigorous shoots; fruit berry-like; mature bark peeling in thin, papery flakes or strips .... *Arbutus menziesii*, ARBUTUS   396

      4 Leaves oblong, leaf margin rolled under; fruit an elongated, sticky capsule; mature bark thin, scaly .... *Rhododendron*, RHODODENDRONS   395

Winter Key C
Leafless trees with buds opposite
(or subopposite or whorled)

1 Terminal bud present on most twigs

    2 Terminal bud large, conspicuous; leaf scars large, shield-shaped .... *Aesculus*, HORSECHESTNUTS   158

    2 Terminal bud smaller; leaf scars crescent-shaped or semicircular

        3 Twigs with raised ridges, appearing winged or 4-sided

            4 Lateral buds stalked, often clustered; crown regular and pyramidal .... *Metasequoia glyptostroboides*, DAWN REDWOOD   35

            4 Lateral buds not stalked, mostly single; crown irregular, spreading

                5 Vein scars single; twigs slender, laterally lined or conspicuously winged; shrubs or very small trees .... *Euonymus*, EUONYMUS   190

                5 Vein scars many; twigs stout and conspicuously 4-sided; small trees .... *Fraxinus quadrangulata*, BLUE ASH   168

        3 Twigs without raised ridges, circular or faintly many-sided in cross section

            6 Twigs thorny, covered in silver-brown scales .... *Shepherdia argentea*, SILVER BUFFALO-BERRY   194

            6 Twigs without thorns, not covered in silver-brown scales

                7 Mature bark conspicuously corky and deeply fissured; young twigs yellowish with prominent lenticels; buds small, naked .... *Phellodendron amurense*, AMUR CORKTREE   171

                7 Mature bark smooth to scaly or ridged, not deeply fissured

                    8 Vein scars 5 to many

                        9 Terminal bud pyramidal, with 1–3 pairs of bud scales .... *Fraxinus*, ASHES   160
                        (See winter key to ash species, p. 426)

                        9 Terminal bud blunt, with 3–4 pairs of bud scales .... *Acer macrophyllum*, BIGLEAF MAPLE   138

                  8 Vein scars 3

                      10 Leaf scars raised above the twig .... *Cornus*, DOGWOODS   180

                    10 Leaf scars flat to the twig

                        11 Buds with 2 or more pairs of scales .... *Acer*, MAPLES   423
                        (See winter key to maple species, p. 423)

11 Buds naked or with 1 pair of scales

 12 Twigs often with an unpleasant odor when bruised; fruits berry-like, with a single flat seed, born in clusters .... *Viburnum*, VIBURNUMS   176

 12 Twigs lacking an unpleasant odor when bruised; fruits winged seeds, joined in pairs .... *Acer*, MAPLES   132
 (See winter key to maple species, p. 423)

1 Terminal bud absent, twig tip may be withered, thorn-like, or have an oblique end bud

 13 Buds on current, vigorous shoots subopposite

  14 Buds with a single cap-like bud scale .... *Salix purpurea*, PURPLE-OSIER WILLOW   334

  14 Buds with 2 or more bud scales

   15 Buds with 2 scales; twigs swollen below the buds; dwarf shoots not ending in a thorn-like spike; medium-sized trees .... *Cercidiphyllum japonicum*, KATSURA-TREE   193

   15 Buds with many scales; twigs not swollen below the buds; dwarf shoots ending in a thorn-like spike; shrubs or very small trees .... *Rhamnus cathartica*, EUROPEAN BUCKTHORN   278

 13 Buds on current, vigorous shoots not subopposite

  16 Buds arranged in whorls of 3 and in pairs

   17 Twigs stout, lateral buds 2–5 mm long, rounded; fruits cylindrical capsules, 15–60 cm long, persisting in clusters; small or medium-sized trees .... *Catalpa*, CATALPAS   189

   17 Twigs slender, lateral buds tiny, almost submerged; fruits cone-shaped nutlets, persisting in a globular cluster 2 cm across; shrubs or very small trees .... *Cephalanthus occidentalis*, BUTTON-BUSH   193

  16 Buds arranged exclusively in pairs

   18 Twigs stout, with conspicuous warty lenticels; leaf scars with 5–7 vein scars; fruits berry-like, 3–7 mm across; coarse shrubs or very small trees .... *Sambucus*, ELDERS   172

   18 Twigs slender and smooth; leaf scars with 1–3 vein scars

    19 Leaf scars with single vein scars; buds large, stout, with 3–5 pairs of green and brown bud scales; fruit a flattened 2-valved capsule .... *Syringa vulgaris*, COMMON LILAC   187

    19 Leaf scars with 3 vein scars; lateral buds small, with 1 or 2 pairs of green or red bud scales; fruit a joined pair of winged seeds .... *Acer*, MAPLES   132
    (See winter key to maple species, p. 423)

Winter Key D
Leafless trees with buds alternate; distinct terminal buds present; leaf scars with 3 or fewer vein scars

1 Leaf scars with 1 vein scar

  2 Twigs spiny or thorny

    3 Twigs silvery scaly, with short stunted twigs forming thorns .... *Elaeagnus angustifolia*, RUSSIAN-OLIVE   407

    3 Twigs green, with needle-like spines in pairs flanking the leaf scar .... *Caragana arborescens*, SIBERIAN PEA-TREE   218

  2 Twigs lacking spines or thorns

    4 Twigs scaly, reddish brown or silvery .... *Elaeagnus commutata*, SILVERBERRY   407

    4 Twigs not scaly

      5 Twigs yellow-green, with a spicy fragrance when bruised .... *Sassafras albidum*, SASSAFRAS   242

      5 Twigs brownish or purplish, lacking a spicy fragrance when bruised

        6 Twigs slender, purplish, smooth, with numerous, pale lenticels; buds with upper bud scale thickened; fruits not persisting into winter .... *Nemopanthus mucronata*, MOUNTAIN-HOLLY   275

        6 Twigs stout, brownish, finely ridged, with warty lenticels; buds occasionally flanked by small, black stipules; fruits berry-like, red or orange, persisting long into winter .... *Ilex verticillata*, COMMON WINTERBERRY   274

1 Leaf scars with more than 1 vein scar

  7 Leaf scars with 2 vein scars; dwarf branches prominent, twig stout, smooth, with a pale, yellowish pith .... *Ginkgo biloba*, GINKGO   126

  7 Leaf scars with 3 vein scars

    8 Twigs spiny or thorny

      9 Twigs with paired, broad, flat-based spines flanking the leaf scar .... *Zanthoxylum americanum*, COMMON PRICKLY-ASH   220

      9 Twigs with single thorns or thorn-like dwarf shoots

        10 Twigs with smooth, very sharp, axillary thorns 1–8 cm long; buds spherical; mature bark pale gray-brown, separating in loose-ended shreds .... *Crataegus*, HAWTHORNS   388

        10 Twig with dwarf shoots ending in a sharp thorn; buds elongated; mature bark reddish brown, scaly .... *Malus*, APPLES   370

8 Twigs lacking spines or thorns

11 Buds lacking scales

12 Buds stalked, densely covered with yellowish-brown hairs; twig with persistent, orange-yellow fruit capsules .... *Hamamelis virginiana*, WITCH-HAZEL 272

12 Buds stalkless, covered with rusty-brown hairs; twig lacking persistent fruit capsules .... *Rhamnus*, BUCKTHORNS 276

11 Buds with scales

13 Buds with lowest bud scale set directly above leaf scar .... *Populus*, POPLARS 335
(See winter key to poplar species, p. 443)

13 Buds with lowest bud scale set to side of leaf scar

14 Twigs stout, pith chambered; leaf scars large, conspicuous, 3-lobed .... *Juglans*, WALNUTS 196

14 Twigs slender to moderately stout, pith solid; leaf scars small or narrow, triangular, elliptical or crescent-shaped

15 Vein scars whitish, prominent against darker leaf scar

16 Branches developing corky ridges; pith star-shaped; fruits spherical, persisting into winter .... *Liquidambar styraciflua*, SWEETGUM 240

16 Branches lacking corky ridges; pith with hard, greenish cross-bars; fruits plum-like, not persisting into winter ... *Nyssa sylvatica,* BLACK-GUM 402

15 Vein scars not whitish

17 Twigs exuding gummy, strong-smelling juice when cut; wood mustard yellow .... *Cotinus coggygria*, COMMON SMOKETREE 408

17 Twigs not exuding gummy juice when cut; wood not mustard yellow

18 Leaf scars mostly triangular; buds mostly stalked; fruit persistent in woody, cone-like catkins .... *Alnus*, ALDERS 296
(See winter key to alder species, p. 438)

18 Leaf scars rounded; buds stalkless; fruit not in woody cone-like catkins

19 Terminal buds with 4 or fewer visible bud scales

20 Buds rounded, silvery, with 2–4 scales.................................. .... *Laburnum* x*watereri*, LABURNUM 218

20 Buds pointed, reddish brown, with 2 or 3 scales .... *Cornus alternifolia,* ALTERNATE-LEAVED DOGWOOD 184

19 Terminal buds with more than 4 visible bud scales

21 Bud scales 10–15 in several rows; twigs with a strong odor and taste of bitter almond when bruised .... *Prunus*, CHERRIES 447
(See winter key to cherry and plum species, p. 447)

21 Bud scales 5–7; twigs without strong odor or taste of bitter almond when bruised

Winter Key E
Leafless trees with buds alternate;
distinct terminal buds present;
leaf scars with 4 or more vein scars

Winter Key F
Leafless trees with buds alternate; end bud a lateral resembling adjacent laterals; leaf scars with 3 or fewer vein scars

8 Twigs zigzag, without a citrus odor when bruised; fruit a pod 15–40 cm long, persisting into winter .... *Gleditsia triacanthos*, HONEY-LOCUST   214

6 Buds not sunken in the twig

    9 Lateral buds on long shoots with 1–3 visible bud scales

        10 Buds with a single, cap-like bud scale .... *Salix*, WILLOWS   312

        10 Buds with more than 1 bud scale

            11 Dwarf shoots frequent on long shoots; lateral buds on long shoots with 3 visible scales, solitary buds on dwarf shoots with 5–7 visible scales...............................
.... *Betula*, BIRCHES   282
(See winter key to birch species, p. 437)

            11 Dwarf shoots infrequent on long shoots; buds all with 2 or 3 visible scales ....
.... *Castanea*, CHESTNUTS   270

    9 Lateral buds on long shoots with 4 or more visible bud scales

        12 Leaf scar upper edge with a fringe of hair; twigs marked by 3 ridges extending from leaf scars; pith pinkish, with darker streaks ... *Cercis canadensis*, REDBUD   400

        12 Leaf scar upper edge lacking a fringe of hair; twigs without ridges extending from leaf scars; pith not pinkish

            13 Bud scales in regular, vertical rows; pollen catkins not evident on twigs

                14 Bud scales in 2 rows

                    15 Buds diverging from twig; pith solid; mature bark with rough, irregular ridges; fruit not persisting into winter .... *Ulmus*, ELMS   352
(See winter key to elm species, p. 445)

                    15 Buds pressed against the twig; pith often chambered, especially at leaf nodes; mature bark with corky thickenings and wart-like projections; fruits berry-like, often persisting into winter ....
.... *Celtis occidentalis*, HACKBERRY   362

                14 Bud scales in 4 rows

                    16 Buds pointed, diverging from the twig, bud scales brown, without white margins; mature bark smooth, gray, with horizontal pink lenticiels .... *Zelkova serrata*, JAPANESE ZELKOVA   361

                    16 Buds blunt, pressed to the twig, bud scales reddish-brown with white margins; mature bark smooth, dark purplish-gray, with muscle-like, wavy longitudinal ridges ....
.... *Carpinus caroliniana*, BLUE-BEECH   306

            13 Bud scales not in regular, vertical rows; pollen catkins usually evident on twigs

                17 Buds pointed, bud scales finely grooved; mature bark broken into short, longitudinal strips, loose at both ends, grayish brown ....
...., *Ostrya virginiana*, IRONWOOD   308

                17 Buds blunt, bud scales not grooved; mature bark smooth, gray ....
.... *Corylus*, HAZELNUTS   310

Winter Key G
Leafless trees with buds alternate;
end bud a lateral resembling adjacent laterals;
leaf scars with 4 or more vein scars

Bibliography

ALAM, M.T., AND W.F. GRANT. 1972. Interspecific hybridization in birch (*Betula*). *Le Naturaliste canadien* 99: 33–40

AMBROSE, J., ED. 1987. *Honour Roll of Ontario Trees*. Ontario Forestry Association, Toronto.

ARGUS, G.W. 1964. Salicaceae; the genus *Salix* — the willows. *Preliminary Reports on the Flora of Wisconsin*. Transactions of the Wisconsin Academy of Science, Arts and Letters 53: 217–272.

ARGUS, G.W. 1973. *The Genus* Salix *in Alaska and the Yukon*. National Museum of Natural Sciences (now Canadian Museum of Nature), Publications in Botany, No. 2. Ottawa.

ARGUS, G.W. 1986. *The Genus* Salix *(Salicaceae) in the Southeastern United States*. Systematic Botany Monograph 9, American Society of Plant Taxonomists. Ann Arbor, MI.

ARGUS G.W., K.M. PRYER, D.J. WHITE, AND C.J. KEDDY. 1982–87. *Atlas of the Rare Vascular Plants of Ontario*. Parts 1–4. National Museum of Natural Sciences (now Canadian Museum of Nature), Ottawa.

BAILEY, L.H., AND E.Z. BAILEY. 1976. *Hortus Third*. Macmillan Co., New York.

BARBOUR, M.G., AND W.D. BILLINGS, ED. 1988. *North American Terrestrial Vegetation*. Cambridge University Press, New York.

BARNES B.V., AND W.H. WAGNER JR. 1981. *Michigan Trees*. University of Michigan Press, Ann Arbor.

BARNHART, J.H., ED. 1965. *The New York Botanical Garden Biographical Notes upon Botanists*. 3 vols. G.K. Hall and Co., Boston.

BEARNS, E.R. 1973. *Native Trees of Newfoundland and Labrador*. 5th ed. Newfoundland Ministry of Forestry and Agriculture, St. John's.

BELL, R. 1897. The geographical distribution of forest trees in Canada. *The Scottish Geographical Magazine* 13 (June): 281–296.

BENNETT, K.D. 1987. Holocene history of forest trees in Southern Ontario. *Canadian Journal of Botany* 65: 1792–1801.

BENSON, L. 1959. *Plant Classification*. Heath and Co., Lexington, MA.

BLAKESLEE, A.F., AND C.D. JARVIS. 1913. *Trees in Winter*. Macmillan Co., New York.

BOIVIN, B. 1966. Énumération des plantes du Canada. *Provancheria* no 6, Université Laval et Ministère de l'agriculture, Ottawa.

BRAYSHAW, T.C. 1960. *Key to the Native Trees of Canada*. Canada Dep. Forestry, Bulletin 125. Ottawa.

BRITTAIN, W.H., AND W.F. GRANT. 1967. Observations on Canadian birch (*Betula*) collections at the Morgan Arboretum. V. *B. papyrifera* and *B. cordifolia* from eastern Canada. *Canadian Field-naturalist* 81 (4): 251–262.

BRUMMITT, R.K., AND C.E. POWELL, EDS. 1992. *Authors of Plant Names*. Royal Botanic Gardens, Kew.

BUCKLEY, A.R. 1980. *Trees and Shrubs of the Dominion Arboretum*. Agriculture Canada, Publication 1697. Ottawa.

BURNHAM, C.R. 1988. The restoration of the American chestnut. *American Scientist* 76: 478–487.

BURNS, R.M., AND B.H. HONKALA. 1990. *Silvics of North America.* 2 vols. USDA Forest Service, Agriculture Handbook 654. Washington, D.C.

CATLING, P.M., AND K.W. SPICER. l988. The separation of *Betula populifolia* and *Betula pendula* and their status in Ontario. *Canadian Journal of Forest Research* 18: 1017–1026.

CHELIAK W.M., J. WANG., AND J.A. PITEL. 1988. Population structure and genetic diversity in tamarack. *Canadian Journal of Forest Research* 18: 1318–1324.

COOMBES, A.J. 1985. *Dictionary of Plant Names.* Collingridge Books, Twickenham, Middlesex, UK.

COPE, E.A. 1986. *Native and Cultivated Conifers of Northeastern North America.* Cornell University Press, Ithaca, NY.

CORE, E.L., AND N.P. AMMONA. 1973. *Woody Plants in Winter.* Boxwood Press, Pacific Grove, CA.

CORMACK, R.G.H. 1964. *Trees and Shrubs of Alberta.* Dep. Lands and Forests, Edmonton.

CRONQUIST, A.J. 1981. *An Integrated System of Classification of Flowering Plants.* Columbia University Press, New York.

CRONQUIST, A.J. 1988. *The Evolution and Classification of Flowering Plants.* 2nd ed. New York Botanical Garden, New York.

DALLIMORE, W., AND A.B. JACKSON. 1966. *A Handbook of Coniferae and Ginkgoaceae.* 4th ed. Revised by S.G. Harrison. Edward Arnold, London.

DESMOND, R. 1977. *Dictionary of British and Irish Botanists and Horticulturists.* Taylor and Francis, London.

DIRR, M.A. 1983. *Manual of Woody Landscape Plants: Their Identification, Ornamental Characteristics, Culture, Propagation and Uses.* Stipes Publishing Co., Champaign, IL.

DONLY, J.F. 1960. *Identification of Nova Scotia Woody Plants in Winter.* Nova Scotia, Dep. Lands and Forests, Bulletin 19.

DORN, R.D. 1976. A synopsis of American *Salix. Canadian Journal of Botany* 54: 2769–2789.

ECKENWALDER, J.E. 1977. North American cottonwoods *Populus* (Salicaceae) of sections *Abasco* and *Aigeros. Journal of the Arnold Arboretum* 58: 193–207.

ECKENWALDER, J.E. 1984. Natural intersectional hybridization between North American species of *Populus* (Salicaceae) in sections *Aigeros* and *Tacamahaca.* II. Taxonomy. *Canadian Journal of Botany* 62 (2): 325–335.

ELIAS, T.S. 1989. *Field Guide to North America Trees.* Rev. ed. Grolier Books Clubs Inc.; Danbury, CT.

FERNALD, M.L. 1950. *Gray's Manual of Botany.* D. Van Nostrand Co., New York.

FURLOW, J.J. 1979. The systematics of the American species of *Alnus* (Betulaceae). *Rhodora* 81: 1–121, 151–248.

GARMAN, E.H. 1963. *Pocket Guide to Trees and Shrubs in British Columbia.* British Columbia Provincial Museum, Victoria.

GARTSHORE, M.E., D.A. SUTHERLAND, AND J.D. MCCRACKEN. 1987. *The Natural Areas Inventory of Haldimand-Norfolk.* Vols.1–2. Norfolk Field Naturalists, Simcoe, ON.

GAUDET, J.F., AND W.M. PROFITT. 1958. *Native Trees of Prince Edward Island.* P. E. I. Dep. Agriculture, Charlottetown.

GLEASON, H.A., AND A. CRONQUIST. 1991. *Manual of Vascular Plants of Northeastern United States and Adjacent Canada.* 2nd ed. New York Botanical Garden, New York.

GRANT, W.F., AND B.K. THOMPSON. 1975. Observations on Canadian birches *Betula cordifolia, B. neoalaskana, B. populifolia* and *B. ×caerulea. Canadian Journal of Botany* 53: 1478–1490

HARLOW, W.M. 1959. *Fruit Key and Twig Key to Trees and Shrubs*. Dover Publications, New York.

HARLOW, W.M., E.S. HARRAR, J.W. HARDIN, AND F.M. WHITE 1991. *Textbook of Dendrology: Covering the Important Trees of the United States and Canada*. 7th ed. McGraw-Hill, New York.

HARRIS, R.C., AND G.J. MATTHEWS. 1987. *Historical Atlas of Canada*. Vol. 1. University of Toronto Press, Toronto.

HEREMAN, S., ED. 1868. *Paxton`s Botanical Dictionary.* Bradbury, Evans and Co., London.

HINDS, H.R. 1979. *Annotated Checklist of the Woody Plants of New Brunswick*. Canadian Forestry Service, Information Report M-X-103. Fredericton.

HINDS, H.R. 1986. *Flora of New Brunswick*. Primose Press, Fredericton.

HIRATSUKA, Y. 1987. *Forest Tree Diseases of the Prairie Provinces*. Canadian Forestry Service, Information Report NOR-X-286. Edmonton.

HOSIE, R.C. 1979. *Native Trees of Canada*. 8th ed. Canadian Forestry Service; Fitzhenry and Whiteside, Don Mills.

HOUGH, R.B. 1907. *Handbook of the Trees of the Northern States and Canada*. Hough Publishing, Louisville, NY.

HULTEN, E. 1968. *Flora of Alaska and Neighboring Territories*. Stanford University Press, Stanford, CA.

HUNTLEY, B., AND T. WEBB, EDS. 1988. *Vegetation History*. Kluwer Academic Publishers, Dordrecht and Boston.

IVES, W.G.H., AND H.R. WONG. 1988. *Tree and Shrub Insects of the Prairie Provinces*. Canadian Forestry Service, Information Report NOR-X-292. Edmonton.

JEFFREY, C. 1989. *Biological Nomenclature*. 3rd ed. Edward Arnold, London.

JONCKHEERE, F., ED. 1990. *Rare, Threatened, or Endangered Trees in Haldimand-Norfolk*. Norfolk Field Naturalists, Simcoe, ON.

KARTESZ, J.T., AND R. KARTESZ. 1980. *A Synchronized Checklist of the Vascular Flora of the United States, Canada, and Greenland*. Vol.2. University of North Carolina Press, Chapel Hill.

KELSEY, H.P., AND W. A. DAYTON. 1942. *Standardized Plant Names*. 2nd ed. J. Horace McFarland Co., Harrisburg, PA.

KOZLOWSKI, T.T., P.J. KRAMER, AND S. G. PALLARDY. 1991. *The Physiological Ecology of Woody Plants*. Academic Press, San Diego, CA.

KRAJINA, V.J., K. KLINKA, AND J. WORRALL. 1982. Distribution and ecological characteristics of trees and shrubs of British Columbia. University of British Columbia, Vancouver, BC.

KRÜSSMANN, G. 1983. *Manual of Cultivated Conifers*. 2nd ed. Translated by M. E. Epp. Timber Press, Portland, OR.

KRÜSSMANN, G. 1986. *Manual of Broad-leaved Trees and Shrubs*. 3 vols. Translated by M.E. Epp. Timber Press, Portland, OR.

LAURIAULT, J. 1989. *Identification Guide to the Trees of Canada*. National Museum of Natural Sciences (now Canadian Museum of Nature); Fitzhenry and Whiteside, Markham, ON.

Little, E.L. Jr. 1979. *Checklist of United States Trees (Native and Naturalized)*. USDA Forest Service, Agricultural Handbook 541. Washington, D.C.

Little, E.L. Jr. 1971. *Atlas of United States Trees*. Vol. 1. *Conifers and Important Hardwoods*. 1976. Vol 3. *Minor Western Hardwoods*. 1977. Vol. 4. *Minor Eastern Hardwoods*. 1981. Vol. 6. *Supplement*. USDA Forest Service, Washington, D.C.

Lyons, C.P. 1965. *Trees, Shrubs and Flowers to Know in British Columbia*. Rev. ed. Evergreen Press, Vancouver.

Margulis, L., and K.V. Schwartz. 1988. *Five Kingdoms: An Illustrated Guide to the Phyla of Life on Earth*. 2nd ed. W. H. Freeman and Co., New York.

Marie-Victorin, Fr. 1935. *Flore laurentienne*. Imprimerie de la Salle, Montréal.

Marie-Victorin, Fr., and Rolland-Germain, Fr. 1964. *Flore de L'Anticosti-Minganie*. 2nd ed. Les Presses de l'Université de Montréal.

Mauseth, J.D. 1988. *Plant Anatomy*. The Benjamin/Cummings Publishing Co., Don Mills, ON.

Mauseth, J.D. 1991. *Botany: An Introduction to Plant Biology*. Saunders College Publishing, Philadelphia.

McMinn, H.E., and E. Maino. 1967. *An Illustrated Manual of Pacific Coast Trees*. University of California Press, Berkeley.

Mirov, N.T. 1967. *The Genus* Pinus. Ronald Press, New York.

Mitchell, A.F. 1985. *Conifers*. Forestry Commission Booklet No. 15. Her Majesty's Stationery Office, London.

Moss, E.H. 1959. *Flora of Alberta*. University of Toronto Press, Toronto.

Ouden, D.P., and B.K. Boom. 1965. *Manual of the Cultivated Conifers*. Martinus Nijhoff, The Hague, Netherlands.

Peattie, D.C. 1977. *A Natural History of Trees of Eastern and Central North America*. Houghton Mifflin Co., Boston.

Petrides, G.A. 1958. *A Field Guide to Trees and Shrubs*. Houghton Mifflin Co., Boston.

Phillips, R. 1978. *Trees of North America and Europe*. Random House, New York.

Phipps, J.B., and M. Muniyamma. 1980. A taxonomic revision of *Crataegus* (Rosaceae) in Ontario. *Canadian Journal of Botany* 58:1 621–99.

Preston, R.J. Jr. 1947. *Rocky Mountain Trees*. 2nd ed. Iowa State College Press.

Preston, R.J. Jr. 1966. *North American Trees*. Massachusetts Institute of Technology Press, Cambridge.

Raven, P.H., R.F. Evert, and S. E. Eichorn. 1991. *Biology of Plants*. 5th ed. Worth Publishers, New York.

Rehder, A. 1940. *Manual of Cultivated Trees and Shrubs*. 2nd ed. Reprinted 1990 by Dioscorides Press, Portland, OR.

Richie, J.C. 1987. *Postglacial Vegetation in Canada*. Cambridge University Press, New York.

Roland, A.E. 1944. *The Flora of Nova Scotia*. Proceedings of the Nova Scotia Institute of Science 21 (3): 95–642.

Rosendaul, C.O. 1928. *Betula cordifolia*, a well-marked species in the Lake Superior region. *Journal of Forestry* 26: 878–882.

Rousseau, C. 1974. *Geographie floristique du Québec-Labrador*. Les Presses de l'Université Laval, Québec.

ROWE, J.S. 1972. *Forest Regions of Canada*. Canadian Forestry Service, Ottawa.

RYAN, A.G. 1978. *Native Trees and Shrubs of Newfoundland and Labrador*. Parks Division, Dep. Tourism, St. John's, NF.

SARGENT, C.S. 1922. *Manual of the Trees of North America*. 2nd ed. Dover Publications, New York.

SAUNDERS, G.L. 1989. *Trees of Nova Scotia: A Guide to the Native and Exotic Species*. Rev. ed. Nova Scotia Dep. Lands and Forests, Truro, NS.

SCOGGAN, H.J. 1957. *Flora of Manitoba*. National Museums of Canada, Bulletin 140. Ottawa.

SCOGGAN, H.J. 1978. *The Flora of Canada*. 4 vols. National Museums of Canada, Ottawa.

SECRETARY OF STATE OF CANADA. 1974. *Canadian Flora*. Translation Bureau, Terminology Bulletin 156. Ottawa.

SHERK, L.C. 1971. *Checklist of Ornamental Trees for Canada*. Agriculture Canada, Bulletin 1343. Ottawa.

SHERK L.C., AND A.R. BUCKLEY. 1968. *Ornamental Shrubs for Canada*. Agriculture Canada, Publication 1697. Ottawa.

SHOPMEYER, C.S. 1974. *Seeds of Woody Plants in the United States*. USDA Forest Service, Agriculture Handbook 450. Washington, D.C.

SOPER, J.H., AND M.L. HEIMBURGER. 1985. *Shrubs of Ontario*. 2nd ed. Royal Ontario Museum, Toronto.

STAFLEU, F.A., AND R.S. COWEN, EDS. 1976. *Taxonomic Literature*. 2nd ed. 7 vols. Bohn, Scheltema and Holkema, Utrecht.

STEARN, W. 1972. *A Gardener's Dictionary of Plant Names*. Cassell and Co., London.

STRALEY, G.B. 1992. *Trees of Vancouver*. University of British Columbia Press, Vancouver.

SUDWORTH, G.B. 1908. *Forest Trees of the Pacific Slope*. USDA Forest Service, Washington, D.C.

SUTHERLAND, D.A. 1987. The vascular plants of Haldiman-Norfolk. In *The Natural Areas Inventory of Haldiman-Norfolk*. Volume II: *Annotated Checklist*. Edited by M.E. Gartshore, D.A. Sutherland, and J.D. McCracken. Norfolk Field Naturalists, Simcoe, ON.

SYMONDS, G.W.D. 1958. *The Tree Identification Book*. George J. Mcleod, Toronto.

TRELEASE, W. 1967. *Winter Botany*. 3rd ed. Dover Publications, New York.

VIERECK L.A., AND E.L. LITTLE JR. 1972. *Alaska Trees and Shrubs*. USDA Forest Service, Agriculture Handbook 410. Washington, D.C.

VIERECK, L.A., AND E.L. LITTLE JR. 1975. *Atlas of United States Trees* Vol. 2. *Alaska Trees and Common Shrubs*. USDA Forest Service, Washington, D.C.

VIERTEL, A.T. 1974. *Trees, Shrubs and Vines: A Pictorial Guide to the Ornamental Woody Plants of the Northern United States Exclusive of Conifers*. Syracuse University Press, Syracuse, NY.

WALKER, P.M.B., ed. 1989. *Chambers Biology Dictionary*. New York City.

WATT, M.T. 1986. *Tree Finder*. Nature Study Guild, Berkeley, CA.

WHITE J.H., AND R.C. HOSIE. 1977. *The Forest Trees of Ontario*. Ministry of Natural Resources, Toronto.

ZIMMER, G.F. 1923. *A Popular Dictionary of Botanical Names and Terms*. George Rutledge and Sons, London.

Botanical Authors

When a botanist identifies a new species, it must be classified and named according to the International Code of Botanical Nomenclature. Although the plant may already be known locally and have a common name, its scientific name has to follow the very precise rules outlined in the Code. The name of the person who described or identified the species is part of this naming system. It is cited after the genus and species name. In tracking histories of nomenclature, these attached "authority names" provide the only true means of full identity of the species involved.

The names of authors are often abbreviated. For example, L. following a scientific name indicates that a plant was first classified and named by Linnaeus. To avoid confusing authors with the same or a similar name, it is sometimes necessary to include initials with an abbreviation or to use the name in full.

Ait.
William Aiton (1731–1793), Scottish botanist, director of the Royal Botanic Gardens, Kew, England; works include *Hortus Kewensis* (1789), a catalogue of the plants cultivated at Kew.

Andersson
Nils Johan Andersson (1821–1880), Swedish botanist, professor of botany at the National Museum of Stockholm; specialized in willows and the flora of Lapland.

Arnold
Johann Franz Xavier Arnold (1785– n.d.), Austrian botanist; classified the Austrian pine, commonly planted in Canada.

Ashe
William Willard Ashe (1872–1932), Chair of the U.S. Forest Service Tree Name Committee and specialist in hawthorns; co-author of *Flora of the Southeastern United States* and *Standardized Plant Names*.

Audub.
John James Laforest Audubon (1785–1851), American naturalist and artist; best known for his bird illustrations and the 87-part book, *Birds of America*.

Balf.
John Hutton Balfour (1808–1884), Scottish doctor and botanist, director of the Royal Botanic Gardens, Edinburgh; wrote a number of botany books, including works about the plants of the Bible.

Barratt
Joseph Barratt (1796–1882), American botanist; studied willows, particularly those most suited for crafts and ornamental planting.

Bartr., W.
William Bartram (1739–1823), American botanist, ornithologist, writer and traveller; son of John Bartram (1699–1777), pioneer of American botany.

Beissn.
Ludwig Beissner (1843–1927), inspector at the botanical garden in Bonn, Germany, and teacher at the Academy of Land Research; specialized in conifers.

Benth.
George Bentham (1800–1884), English botanist; commissioned by the British Government in 1857 to produce a series of books on the flora of the British colonies.

Bernh.
Johann Jacob Bernhardi (1774–1850), German botanist and horticulturist; editor of *Allgemeines Deutsches Garten-Magazin*.

Blanch.
William Henry Blanchard (1850–1922), Vermont school-teacher; collected and classified blackberries and the flora of southwestern Maine.

Blume
Carl Ludwig von Blume (1796–1862), German-born Dutch botanist; traveled and worked in Java; director of the National Herbarium in Leiden, Netherlands.

Boiss.
Edmond Pierre Boissier (1810–1885), Swiss naturalist; wrote *Flora Orientalis*; donated his herbarium to the Faculty of Sciences at the University of Geneva.

Boivin
Joseph Robert Bernard Boivin (1916–1985), Canadian research scientist with the federal

Department of Agriculture in Ottawa; wrote the 5-volume *Flora of the Prairie Provinces*, *Enumération des plantes du Canada*, *Flore du Québec méridional et du Canada oriental*, and *Survey of Canadian Herbaria*.

Bong.
August Heinrich von Bongard (1786–1839), German-born botanist; became a member of the Russian privy council in St. Petersburg and compiled a flora of Russia.

Borkh.
Moritz Balthasar Borkhausen (1760–1806), officer at the Forestry College and public official in Darmstadt, Germany; wrote several books on dendrology.

Bosc
Louis Augustin Guillaume Bosc (1759–1828), French botanist; managed the gardens at Versailles and the Jardin des Plantes, Paris; traveled in North America; works include those on oaks.

Brayshaw
Thomas Christopher Brayshaw (1919–), Canadian botanist and curator emeritus at the Royal British Columbia Museum, Victoria; specialist in aquatic plants, and trees and shrubs, especially willows.

Breit.
August Julius Breitung (1913–1987), Canadian plant collector; created a large herbarium of prairie flora; wrote the *Annotated Catalogue of the Vascular Flora of Saskatchewan* (1957), and *A Botanical Survey of the Cypress Hills* (1954).

Britt.
Nathaniel Lord Britton (1859–1934), American botanist; founder and first director of the New York Botanical Garden; wrote many books, including *Illustrated Flora of the Northern United States and Canada* (1896) and *North American Trees* (1908).

BSP
Nathaniel Lord Britton (1859–1934), Emerson Ellick Sterns (1846–1926), Justus Ferdinand Poggenburg (1840–1893), botanists who collaborated on several projects, including *A Preliminary Catalogue of Anthophyta and Pteridophyta Native to the New York City Area* (1888).

Buchh.
John Theodore Buchholz (1888–1951), American botanist and conifer specialist at the University of Illinois.

Buckl.
Samuel Botsford Buckley (1809–1884), state geologist of Texas for nearly 20 years; published reports on regional agriculture and geology.

Bush
Benjamin Franklin Bush (1858–1937), American botanist; co-author of *Manual of the Flora of Jackson County, Missouri* (1902).

Carrière
Élie Abel Carrière (1816–1896), director of the nurseries of the Jardin des Plantes in Paris; specialist in conifers and noted for his *Traité général des conifères*.

Cham.
Ludolf Karl Adalbert von Chamisso (1781–1838), French-born botanist and curator of the natural history museum in Berlin-Dahlem; wrote about his travels around the world.

Cheng
Wan Chun Cheng (1903–1983), Chinese botanist; specialized in conifers; wrote about the flora of the mountains of Sichuan and Xinjiang.

Clausen, J.
Jens Christian Clausen (1891–1969), geneticist at Royal Veterinary and Agricultural College of Copenhagen; later professor of biology at Stanford University.

Cov.
Fredrick Vernon Coville (1867–1937), American taxonomist and phytogeographer; wrote *Botany of the Death Valley Expedition* (1893).

DC.
Augustin Pyramus de Candolle (1778–1841), Swiss botanist; specialized in the classification of plants; works include the 32-part *Plantarum historia succulentarum* (1798–1837).

Dippel
Leopold Dippel (1827–1914), German professor of botany and director of the Darmstadt botanical garden; works include *Handbuch der Laubholzkunde* (1889).

Dode
Louis Albert Dode (1875–1943), French attorney and botanist; studied poplars, walnuts, sycamores, and mountain-ash; secretary of the Société française de dendrologie.

Don, D.
David Don (1799–1841), British librarian and professor of botany at King's College in London, England; wrote a flora of Nepal.

Donn
James Donn (1758–1813), English botanist and curator of the Cambridge Botanic Garden; known for naming *Claytonia perfoliata* and preparing a catalogue of the plants growing at Cambridge.

Dougl.
David Douglas (1798–1834), Scottish botanical explorer of North America and China; took ornamental plants back to Europe; described many North American species of oak and pine in *Journal Kept by David Douglas*, a record of his 1823 travels.

Dum. Cours.
Georges Louis Marie Dumont de Courset (1746–1824), French nobleman and agronomist; works include the 5-volume *Le botaniste cultivateur*, which described the culture and use of native and introduced plants in France and England.

Dunal
Michel Félix Dunal (1789–1856), French botanist; studied under de Candolle and later took over Candolle's chair of botany at Geneva University.

Du Roi
Johann Philipp Du Roi (1741–1785), German botanist; studied trees, shrubs, and bushes, including some of the flora of North America; published a 2-volume work on his findings in 1771.

Eckenw.
James Emory Eckenwalder (1949–), specialist in systematic botany and professor at the University of Toronto; specialist in systematics of Salicaceae and taxonomy of cultivated plants.

Eggl.
Willard Webster Eggleston (1863–1935), American botanist, pharmacist, and taxonomist; wrote *A Botanical and Chemical Study of Dicentra eximia* (1929).

Ehrh.
Jakob Friedrich Ehrhart (1742–1795), German botanist of Swiss origin, pupil of Linnaeus, and overseer at the Herrenhäuser Garden in Hannover.

Endl.
Stephen Friedrich Ladislaus Endlicher (1804–1849), Czechoslovakian-born professor of botany and director of the botanical garden in Vienna, Austria; works include the 18-part *Genera plantarum* (1831–1841).

Engelm.
Georg Engelmann (1809–1884), German-born physician, botanist, and plant collector; worked in St. Louis, MO.

Evans, W.H.
Walter Harrison Evans (1863–1941), American botanist; specialized in trees and shrubs, especially birches.

Fern.
Merritt Lyndon Fernald (1873–1950), director of the Gray Herbarium in Cambridge, MA; an authority on the flora of northeastern America; author of the 8th edition of *Gray's Manual of Botany* (1950) and editor-in chief of the botanical magazine, *Rhodora*.

Forbes, J.
James Forbes (1773–1861), gardener at Woburn Abbey, home of the Duke of Bedford; wrote *Hortus Woburnensis* and *Pinetum Woburnense*.

Franco
João Manual Antonio Pais do Amaral Franco (1921–), professor and agronomical engineer in Lisbon, Portugal; authority on botanical nomenclature; author of *Dendrologia florestal*.

Gaertn.
Joseph Gaertner (1732–1791), German botanist; director of the botanical garden in St. Petersburg, Russia; wrote a standard work on the morphology of fruits and seeds.

Gaudin
Jean François Aimé Gottlieb Philippe Gaudin (1766–1833), Swiss clergyman and botanist; works include *Flora Helvetica*.

Glend.
Robert Glendinning (1805–1862), proprietor of a nursery in Cheswick, England; author of *The Pinetum* (1838), in which he described methods of transplanting large evergreen trees.

Gord.
George Gordon (1806–1879), Irish botanist; superintendent of the Royal Horticultural Society in Cheswick, England, and later at the Royal Botanic Gardens, Kew.

Gray, A.
Asa Gray (1810–1888), American botanist; authority on the flora of the United States; specialist in the classification of plants and author of *Manual of Botany of the Northern United States*.

Greene
Edward Lee Greene (1843–1915), American botanist and clergyman; specialized in plants of California and the southwestern states; works include *Landmarks of Botanical History* (1909) and *Flora Franciscana* (1891).

Guinier
Philibert Guinier (1876–1962), French forester and botanist; director of l'École nationale des eaux et forêts (1924–1941); played a fundamental role in the evolution of forestry education in France.

Hance
Henry Fletcher Hance (1827–1886), British botanist, plant collector, and taxonomist in China while vice-consul in Whampoa and then consul in Canton.

Hand.-Mazz.
Heinrich von Handel-Mazzetti (1862–1940), Austrian botanist and explorer in China and Iraq; professor at University of Vienna and curator at the Museum of Natural History in Vienna.

Hayne
Friedrich Gottlob Hayne (1763–1832), German botanist and pharmacist; professor of botany in Berlin; specialized in dendrology.

Henry, L.
Louis Henry (1853–1913), French botanist; prolific writer for 30 years.

Hill, E.J.
Ellsworth Jerome Hill (1833–1917), American botanist; wrote about the flora of the Midwest; works include *Quercus ellipsoidalis in Iowa* (1899).

Hochst.
Wilhelm Hochstetter (1825–1881), German botanist; author of a guide to the botanical garden at the University of Tübingen and works on conifers.

Hook.
Sir William Jackson Hooker (1785–1865), Regius Professor of Botany at Glasgow University until 1841; later director of the Royal Botanic Gardens, Kew; published *Flora boreali Americana* between 1833 and 1840.

Howell, T.J.
Thomas Jefferson Howell (1842–1912), American botanist; known for his work in the Pacific northwest; wrote *New Species of Pacific Coast Plants* (1895) and *A Flora of Northwest America* (1897).

Hu
Hsen-Hsu Hu (1894–1968), Chinese taxonomist and plant collector, professor of botany in Nanjing; specialist in the woody flora of China.

Huds.
William Hudson (1730–1793), English pharmacist and botanist; wrote *Flora Anglica* (1762).

Jacq.
Baron Nicolaus Joseph von Jacquin (1727–1817); Dutch-born botanist and chemistry professor; worked in Vienna and became director of the botanical garden in Schönbrunn, Austria.

James
Edwin James (1797–1861), American explorer, naturalist, and physician; wrote the 2-volume *Account of an Expedition from Pittsburg to the Rocky Mountains* (1819).

Karst.
Gustav Karl Wilhelm Hermann Karsten (1817–1908), German botanist; professor of botany in Vienna; wrote about Central and South American flora.

Koch, K.
Karl Heinrich Emil Koch (1809–1879), director of the botanical garden in Potsdam, Germany, and professor of botany in Berlin; author of a 3-volume dendrology text.

Kuntze
Carl Ernst Otto Kuntze (1843–1907), German botanist and nomenclature reformer; traveled around the world twice and produced a large collection of botanical illustrations.

Kuzen.
Olga Iakinfovna Kuzeneva (1887–1978), Russian botanist; specialized in conifers.

L.
Carolus Linnaeus [Carl von Linné] (1707–1778), Swedish botanist; author of *Systema naturae*, *Genera platarum*, and *Species platarum*, on which the modern system of binomial nomenclature is founded; through a patent of nobility in 1757 became known as Carl von Linné.

Raup
Hugh Miller Raup (1901–), professor of forestry at Harvard for over 30 years; member of several botanical surveys to the Yukon and Northwest Territories.

Regel
Eduard August von Regel (1815–1892), German botanist; director of the botanical garden in Zurich, Switzerland; attracted to St. Petersburg by the Czar's interest in botany, where he became director of the botanical garden; specialized in birches; author of many dendrological works and founder of *Gartenflora*.

Rehd.
Alfred Rehder (1863–1949), curator of the Arnold Arboretum and Professor of Dendrology at Harvard; best known for his *Manual of Cultivated Trees and Shrubs* (1927).

Rich., A.
Achille Richard (1794–1852), French botanist and physician; works include publications on botanical medicine, Mexican orchids, and botany in Cuba.

Rich., L.
Louis Claude Marie Richard (1754–1821), botanist and explorer in French Guyana, Brazil, and the Antilles; on his return from these travels, became professor of botany at the École de Medicine, Paris; works include a botanical dictionary (1798).

Roem., M.J.
Max Joseph Roemer (1791–1849), German judge and botanist; published handbooks of botany (ca. 1836).

Roth
Albrecht Wilhelm Roth (1757–1834) German physician and botanist; practised medicine in Bremen; works include a 3-volume botany manual.

Rowlee
Willard Winfield Rowlee (1861–1923), American botanist; taught at Cornell University, Ithaca, NY, for 34 years.

Roxb.
William Roxburgh (1751–1815), Scottish botanist and physician; director of the Botanical Gardens in Calcutta; wrote several floras of India.

Rudd
Velva Elaine Rudd (1910–), American botanist; specialized in plant taxonomy and plant geography; curator of the Department of Botany, Smithsonian Institution, Washington, D.C.

Rupr.
Franz Josef Ivanovich Ruprecht (1814–1870), Austro-Bohemian botanist and physician; curator at the St. Petersburg botanical garden; traveled widely in Russia, reporting on regional flora.

Rydb.
Pehr Axel Rydberg (1860–1931), Swedish-born engineer, later botanist in the United States; collected specimens for the New York Botanical Garden; wrote a flora of the Rocky Mountains and a monograph on North American *Potentilla*.

Sab.
Joseph Sabine (1770–1837), British barrister and horticulturist; worked as a tax-inspector; secretary of the Horticulture Society of London, 1816–1830.

Salisb.
Richard Anthony Salisbury (1761–1829), British botanist; published papers on a wide range of botanical subjects, from conifers to peaches.

Sanson
Information about this author is inconclusive, but an M. Sanson is credited with naming *Salix sitchensis* in the 1830s.

Sarg.
Charles Sprague Sargent (1841–1927), professor of horticulture and arboriculture and director of the Arnold Arboretum, Harvard University; wrote *Manual of the Trees of North America* and *The Silva of North America*; specialized in *Crataegus*.

Scheele
Georg Heinrich Adolph Scheele (1808–1864); German botanist and clergyman; wrote *Beiträge zur Flora von Texas* (1848) on plants sent to him by explorers in Texas.

Schneid., C.K.
Camillo Karl Schneider (1876–1951), dendrologist and garden designer in Vienna and Berlin; traveled in East Asia; wrote several botany books.

Schrad.
Heinrich Adolph Schrader (1767–1836), professor of botany at Göttingen University, Germany; director of the Göttingen botanical garden for 34 years; works include *Spicilegium florae Germanicae* (1794).

Scop.
Giovanni Antonio Scopoli (1723–1788), Italian doctor and professor of mineralogy, later a botany professor; wrote on the flora of Austria and Italy.

Seem.
Berthold Carl Seemann (1825–1871), German botanist; trained at the Royal Botanic Gardens, Kew; gathered specimens in the Fiji Islands, Venezuela, and Central America; editor and publisher of *Bonplandia*.

Siebold
Philipp Franz von Siebold (1796–1866), German physician, botanist, and ethnologist; practised medicine in Japan for 7 years and then wrote *Flora Japonica*; founded a nursery in Leiden, Holland.

Simonk.
Lájos von Simonkai (1851–1910), Hungarian botanist; taught natural history in Budapest for 33 years; authority on Hungarian flora.

Sm.
James Edward Smith (1759–1828), British botanist and physician; purchased the Linnaean herbarium and founded the Linnaean Society in 1788, presiding over it for 40 years.

Soul.-Bod.
Étienne Soulange-Bodin (1774–1846), French politician and diplomat; founded a nursery in Fromant, near Paris; president of the Société Linnéenne de Paris.

Spach
Édouard Spach (1801–1879), French botanist at the natural history museum in Paris; wrote a 14-volume treatise on plants.

Spreng.
Kurt Polykarp Joachim Sprengel (1766–1833), professor of botany and physician in Halle-Saale, Germany; works include a history of botany.

Steven
Christian von Steven (1781–1863), Finnish botanist; studied and worked in Russia; established and directed the Nikita botanical garden in the Crimea in 1812; general inspector for Russian agriculture 1841–1850.

Stokes
Jonathan S. Stokes (1755–1831), English botanist and physician; wrote about medicinal plants.

Sweet
Robert S. Sweet (1783–1835), English gardener and botanist; wrote many books on botanical and horticultural subjects, including a hothouse and greenhouse manual that was a standard reference.

Swingle
Walter Tennyson Swingle (1871–1952), American botanist with the U.S. Department of Agriculture in Washington, D.C.; focused on hybridizing *Citrus* and related genera.

Tausch
Ignaz Friedrich Tausch (1793–1848), Bohemian botanist at the university of Prague, Czechoslovakia, and author of *Hortus Canalius.*

Thomas
(Abraham Louis) Emmanuel Thomas (1788–1859), Swiss plant collector, plant dealer, and forester; works include *Catalogue des plantes suisses* (1818).

Thunb.
Carl Pehr Thunberg (1743–1828), Swedish botanist, professor of botany, and pupil of Linnaeus; widely traveled in Asia and Europe; works include *Flora Japonica* and *Flora Capensis.*

Torr.
John Torrey (1796–1873), American botanist and chemist; wrote on the plants of the northern and central states; works include *A Flora of North America*, with Asa Gray.

Trel.
William Trelease (1857–1945), director of the Missouri Botanic Garden; professor of botany at the University of Illinois; specialized in American oaks.

Turcz.
Nicolai Stephanovich Turczaninow (1796–1864), Russian official, later a botanist; prepared a large herbarium from specimens he obtained during botanical explorations in remote regions of Russia.

Turra
Antonio Turra (1730–1796), Italian botanist; best known for his flora of Italy.

Turrill
William Bertram Turrill (1890–1961), curator of the herbarium at the Royal Botanic Gardens, Kew, for 46 years; publisher of *Botanical Magazine*; leading authority on flora of the Balkan Peninsula.

Vahl
Martin Hendriksen Vahl (1749–1804), Danish botanist, pupil of Linnaeus, professor of botany, and inspector of the botanical garden in Copenhagen; wrote 7 parts of *Flora Danica* and the work *Enumeratis plantarum*.

Villars
Dominique Villars (1745–1814), French physician and botanist; a hospital director and botany professor in Grenoble; later professor of botany and medicine at the University of Strasbourg.

Voss
Andreas Voss (1857–1924), German horticulturist and naturalist; published several botanical studies and edited *Salomon's Dictionary of Botany* (1903).

Walt.
Thomas Walter (1740–1789), British-born botanist; emigrated to South Carolina in 1768, where he became a planter, merchant, and politician; wrote *Flora Caroliniana* (1788).

Wangenh.
Friedrich Adam Julius von Wangenheim (1747–1800), German botanist, head forester in Gumbinnen, East Prussia; studied woody plants of North America.

Warder
John Aston Warder (1812–1883), American forester, horticulturist, and physician; practised medicine for 20 years before becoming a farmer and the president of the Ohio Horticultural Society.

Wieg.
Karl McKay Wiegand (1873–1941), American botanist; professor at Cornell University, Ithaca, NY; coauthored *A Key to the Genera of Woody Plants in Winter* (1906).

Willd.
Karl Ludwig von Willdenow (1765–1812), German botanist; wrote over 30 botanical papers and books; best known for *Anleitung zum Selbststudium der Botanik* (1804).

Zucc.
Joseph Gerhard Zuccarini (1797–1848), professor of botany in Munich, Germany; works include papers on the woody flora of Germany and a flora of Japan.

Meanings of Tree Names

a

Abies classical Latin name of the fir; probably derived from *abire* (to arise).

Acer Latin name for maple; derived from the Celtic *ac* (a point), thought to refer to the wood, which was used in making lance heads.

acerifolia with leaves like those of a maple; from the Latin *acer* (maple) and *folium* (leaf).

acuminata sharp-pointed, referring to the leaves; from the Latin *acuminare* (to make sharp).

Aesculus classical Latin name for an oak with edible acorns; from *esca* (food).

Ailanthus from *ailanthos* (tree of heaven), the Indonesian name for *A. moluccana*.

alatus winged, with wings or appendages that appear to be wings.

alaxensis Latin form of "from Alaska".

alba Latin for white.

albicaulis white-stemmed; from Latin *albus* (white) and *caulis* (stem).

albidum whitish.

alleghaniensis Latin form of "from the Allegheny Mountains".

alnifolia with leaves like those of the alder; from the Latin *alnus* (alder) and *folium* (leaf).

Alnus classical Latin name of the alder.

alternifolia with alternate leaves; from the Latin *alternus* (other) and *folium* (leaf).

altissima Latin for very high or the highest.

amabilis Latin for lovely.

Amelanchier from *amélanquier*, the French Provençal name of a species of serviceberry.

americana, americanum Latin form of "American".

amurense from Amur, a region of Siberia.

amygdalus, amygdaloides almond, almond-like, referring to the shape of the leaves; from the Greek *amugdalos* (almond) and *öides* (resembling).

anagyroides bearing curved pods like those of anagyris, a Mediterranean shrub; from the Greek *ana* (backwards), *gyros* (a circle), and *öides* (resembling).

angustifolia with narrow leaves; from the Latin *angustus* (narrow) and *folium* (leaf).

aquifolium holly-leaved, with pointed leaves.

Aralia from *aralie*, the French-Canadian name for these deciduous shrubs.

Araucaria, araucana from Arauco, a district in Chile.

arborea tree-like; from the Latin *arbor* (tree).

arborescens tree-like.

arbusculoides like a little tree.

Arbutus classical Latin name of the strawberry madrone; possibly from the Celtic *arboise*.

argentea silver colored.

aristata having a beard; from Latin *arista* (a point or beard).

Asimina possibly from the Illinois *rassi* (divided) and *mina* (seed).

atlantica Latin for Atlantic.

atropurpureus dark purple, referring to the color of the fruit; from the Latin *ater* (black) and *purpureus* (purple).

aucuparia used by bird hunters, presumably referring to the attractiveness of the tree to birds; from the Latin *aucupare* (to catch birds).

austini after Coe Finch Austin (1831-1880), an American botanist from New York.

avellana drab or dull.

avium of the birds, referring to the tree's attractiveness to birds; from the Latin *avis* (bird).

b

babylonica of Babylon; the weeping willow was thought to be the tree in Babylon under which exiled Jews wept; from the Greek *babulon*.

baccata berried; from the Latin *bacca* (berry).

balsamea pertaining to balsam, a fragrant resin; from the Hebrew *balsam*, the Greek *balsamon*, and the Latin *balsamum*, referring to the resinous pockets in the bark of balsam fir.

balsamifera balsam-bearing, with the odor of balsam; from the Latin *balsamum* (resin) and *ferre* (to bear).

banksiana after Sir Joseph Banks (1743-1823), president of the Royal Society of London.

bartramiana after William Bartram (1739-1823), American botanist who sent mountain serviceberry seeds to Europe.

bebbiana after Michael Schuck Bebb (1833-1895), an American specialist on willows.

Betula classical Latin name of the birch, from *betu*, its Celtic name.

bicolor two-colored, referring to the leaves; from the Latin *bis* (two) and *color* (color).

bignonioides of or like the genus *Bignonia*, trumpet flowers and cross vines.

biloba leaves or anthers divided into two lobes.

brevifolia short-leaved; from the Latin *brevis* (short) and *folium* (leaf).

c

californica from California

callicarpa from the Greek *kallos* (beautiful) and *karpos* (fruit).

calpodendron   Latin for "urn tree", referring to the shape of the fruit, which is longer than it is broad.

canadensis   Latin form of "Canadian".

Caragana   Latinization of the Mongolian name for one of the species of these ornamental shrubs.

carolinia, caroliniana   Latin form of "from the Carolinas".

carnea   flesh-colored.

Carpinus   the classical Latin name of the hornbeams; possibly from the Celtic *car* (wood) and *pin*, an old word meaning "yoke".

Carya   from the Greek *karua* (walnut tree).

Castanea   Latin for the chestnut tree; from the Greek *kastanea* (chestnut tree) which may have been named after Castane, a town in Thessaly.

Catalpa   from the name given to this tree by native North Americans.

cathartica   Latin for purging; from the Greek *katharo* (to cleanse).

Cedrus   named after the river Cedron in Judaea, where cedars grew plentifully.

Celtis   classical Latin name of the African lotus and applied to the hackberries because of their sweet fruit.

Cephalanthus   from the Greek *kephale* (a head) and *anthos* (flower), referring to the round clusters of flowers.

Cercidiphyllum   with leaves like those of the genus *Cercis*; from the Greek *kerkis* (redbud tree) and *phyllum* (leaf).

Cercis   from the Greek *kerkis*, the ancient name for the redbud or Judas-tree (from its legendary biblical association).

Chamaecyparis   from the Greek *chamai* (ground) and *kuparissos* (cypress).

chinensis   Latin form of "Chinese".

chrysocarpa   golden or yellow fruited; from the Greek *chrysos* (golden) and *karpos* (fruit).

cinerea   ashy or ash colored; referring to the bark of the white walnut; from the Greek *konis*, and the Latin *cineres* (ashes).

circinatum   Latin for rounded or circular; referring to leaf shape.

Cladrastis   from the Greek *klados* (a branch) and *thraustos* (fragile); referring to the brittle twigs.

coccinea   Latin for scarlet.

coerulea, cerulea   heavenly blue, sky blue.

coggygria   a false rendering of the Greek *kokkugia* (smoke tree).

columbiana   Latin form of "Columbian"; from the Columbia River east of the Cascade Mountains.

colurna   resembling hazel, from the Latin *Colurnus*.

communis   Latin for common.

commutata   changeable; from the Latin *commutare* (to change).

compacta   Latin for dense or compact.

concolor   of uniform color; from the Latin *con* (together) and *color* (color).

contorta   Latin for contorted or twisted.

copallina   containing copal gum.

cordata, cordiformis   heart-shaped; from the Latin *cordis* (of the heart) and *forma* (shape).

cordifolia   with heart-shaped leaves; from the Latin *cordis* (of the heart) and *folium* (leaf).

Cornus   Latin for horn; referring to the hard wood.

cornuta   horned.

coronaria   from the Latin *corona* (crown or wreath); referring to the attractive flowers of the wild crab apple.

Corylus   from the Greek *korys* (a helmet), referring to the shape of the fruit.

Cotinus   from the Greek *kotinos* (wild-olive).

Crataegus   classical Greek name of the hawthorns; from *kratos* (strength).

crispa   Latin for curled or crimped.

crus-galli   with thorns like a cock's spurs; from the Latin *crus* (shin or leg) and *gallus* (cock).

cuspidata   with a sharp stiff point; from the Latin *cuspis* (point).

Cryptomeria   from the Greek *kryptos* (hidden) and *meris* (a part).

d

daphnoides   resembling daphne.

decidua   deciduous, with leaves shed the same year; from the Latin *decidere* (to fall).

decora   showy, or ornamental; referring to the colorful fruit of the showy mountain-ash; from the Latin *decor* (elegance or beauty).

deltoides   triangular; referring to the shape of the leaves; from the Greek letter delta and *öides* (resembling).

demissa   Latin for low or weak.

dentata   with teeth; from the Latin *dens, dentis* (a tooth), referring to the toothed leaf-margins of the chestnut.

deodara   timber of the gods; from the Sanskrit *deva* (god) and *daru* (wood).

dioicus   two dwellings; referring to the male and female flowers being on separate trees; from the Greek *di* (two) and *oikos* (dwelling).

discolor   with two or more colors; referring to the leaves of the pussy willow; from the Latin prefix *dis* (two) and *color* (color).

distichum   arranged in two rows; from the Greek *distichus* (of two rows).

diversifolia   separated leaves; from the Latin *diversus* (divergent) and *folium* (leaf).

diversiloba   with lobes turned in different directions; from the Latin *diversus* (divergent) and the Greek *lobos* (lobe).

dodgei   after Carroll William Dodge (1895-1988), American botany professor and member of botanical expeditions in North and South America.

douglasii   after David Douglas (1798-1834), famous Scottish plant collector and traveler in North America and China.

e

edule   edible.

Elaeagnus   Greek name originally applied to a willow, from *helodes* (growing in marshes) and *hagnos* (pure), referring to the white fruit.

elata   Latin for tall, lofty.

ellipsoidalis   Greek for "in the shape of an ellipse"; referring to the acorns.

emarginata   with a notched edge; referring to the tip of the petals and sepals of bitter cherry; from the Latin *emarginare* (to deprive of its edge).

engelmannii   after George Engelmann (1809-1884), German-born physician and botanist of St. Louis, who was an authority on conifers.

eriocephala   from the Greek *erion* (wool) and *kephale* (a head), referring to the woolly seed clusters.

Euonymus   "of good name" in Latin, referring ironically to it being poisonous to animals.

europaeus   Latin form of "from Europe".

excelsior   Latin for taller, or higher.

exigua   Latin for small, narrow, insignificant.

f

Fagus   classical Latin name of the beeches; from the Greek *fagein* (to eat); referring to the fruit.

flabellata   Latin for fan-shaped.

flexilis   Latin for reflexed, or bent back; referring to the scales of the open cones of limber pine.

florida   flowering, or abounding in flowers; from the Latin *floris* (blossom).

fontinalis   Latin for fountain-like or fountain-shaped.

fortunei   after Robert Fortune (1812-1860), Scottish horticulturist and collector in China, noted for introducing the tea plant into India from China.

fragilis   Latin for fragile; referring to the easily broken twigs of the crack willow.

frangula   name of the drug obtained from glossy buckthorn; from the Latin *frangere* (to weaken or diminish).

Fraxinus   the classical Latin name of the ash, from *phraxix* (a separation), referring to its use in hedges.

fruticosa   from the Latin *fruticare* (to sprout or become bushy).

fusca   Latin for tawny or brownish-gray.

g

garryana   after Nicholas Garry, director and deputy governor of the Hudson's Bay Company; Fort Garry, now Winnipeg, was named after him in 1822.

giganteum   very large; from the Greek *gigant* (giant).

Ginkgo   from the Japanese *gin* (silver) and *kyo* (apricot).

ginnala   possibly from the Greek *ginnos* (a small mule), referring to the size and hardiness of the tree.

glabra, glabrum   Latin for smooth, without hairs, or bald.

glauca   glaucous, or covered with a bloom; referring to the bluish-green foliage; from the Greek *glaukos* (bluish).

Gleditsia   sometimes spelled *Gleditschia*; after the German botanist Johann Gottlieb Gleditsch (1714-1786), professor and director of the Berlin Botanic Garden.

glutinosa   Latin for sticky or full of glue.

glyptostroboides   resembling the genus *Glyptostrobus*; from the Greek *glypto* (to carve) and *strobilos* (a cone), referring to the pits on the cones.

gmelinii   after Johann Georg Gmelin (1709-1755), a German naturalist and traveller in Siberia.

grandidentata   Latin for large-toothed.

grandifolia   Latin for large-leaved, with large leaves.

grandis   Latin for large or great.

Gymnocladus   naked branch; from the Greek *gymnos* (naked) and *klados* (branch).

h

Hamamelis   from the Greek *hama* (together with) and *melon* (apple), referring to the fact that witch-hazel blooms in autumn, when apples ripen.

heterophylla   Latin for various-leaved, or with leaves of different sizes and shapes; from the Greek *heteros* (different) and *phyllon* (leaf).

hippocastanum   horse-chestnut; from the Greek *hippos* (horse) and *kastanon* (chestnut).

Hippophae   applied to the sea-buckthorn; derived from the Greek name for a different plant, spurge.

holmesiana   after Joseph Austin Holmes (1859-1915), a geologist who lived in North Carolina.

hookeriana   after William Jackson Hooker (1785-1865), noted Scottish botanist.

horizontalis   Latin for growing horizontally or flat on the ground.

humilis   Latin for base or low.

i

Ilex   from Celtic *ac* (a point), because of its prickly leaves.

incana   Latin for pale gray or hoary.

interius   Latin for "of the interior".

Italica   Latin form of "Italian".

j

jackii   after John George Jack (1861-1949), professor of dendrology at the Bussey Institute in Massachusetts.

japonica, japonicum   Latin form of "Japanese".

Juglans   Jupiter's nut; the classical Latin name of the walnut tree; from *Jovis* (Jupiter's) and *glans* (nut).

Juniperus   classical Latin name of the junipers, from the Celtic for rough, referring to the texture of the shrub.

k

kousa   Japanese name for *Cornus*.

kenaica   after the Kenai Peninsula in Alaska.

kaempferi   after Engelbert Kaempfer (1651-1716), German physician who traveled widely in Asia and wrote about the flora of Japan.

l

Laburnum   classical Latin name for these trees.

laciniosa   Latin for torn or shaggy.

laevis   Latin for smooth.

lantana   Latin name for *Viburnum*.

laricina   Latin for larch-like.

Larix   From the Celtic *lar* (fat), referring to the somewhat oily wood.

lasiandra   having woolly stamens; from the Greek *lasios* (shaggy or woolly) and *andron* (male).

lasiocarpa   hairy-fruited; from the Greek *lasios* (shaggy or woolly) and *karpos* (fruit).

latifolia   Latin for "with broad leaves".

lawsoniana   after Charles Lawson (1794-1873), a Scottish agriculturist.

lenta, lentago   flexible or supple; from the Latin *lentus* (pliant).

libani   Latin form of "from the Lebanon mountains".

Liquidambar   from the Latin *liquidus* (liquid) and *ambar* (amber); referring to the resin obtained from the bark.

Liriodendron   lily tree; from the Greek *leirion* (lily) and *dendron* (tree).

lucida   Latin for bright, lustrous, or shining.

lutea   Latin for golden-yellow.

lyallii   after David Lyall (1817-1895), Scottish surgeon and naturalist who accompanied several British expeditions, including the Oregon Boundary Commission (1858).

m

maackii   after Richard Maack (1825-1886), Russian botanist who collected plants in the Amur and Issuri regions.

Maclura   after William Maclure (1763-1840), American geologist.

macrocantha   with large thorns or spines; from the Greek *makros* (large) and *akantha* (thorn).

macrocarpa   with large fruit; from the Greek *makros* (large) and *karpos* (fruit).

macrophyllum   large-leaved; from the Greek *makros* (large) and *phyllon* (leaf).

macrosperma   with large seeds; from the Greek *makros* (large) and *sperma* (seed).

magnifica   Latin for magnificent; referring to the cones of the red fir.

Magnolia   after Pierre Magnol (1638-1715), professor of botany and director of the Botanical Gardens in Montpellier, France.

Malus   the classical Latin name of the apple; from the Greek *melon* (apple or fruit).

mariana   Latin form of "from Maryland"; to botanist Philip

Miller, "Maryland" epitomized North America; hence he named black spruce, *Picea mariana*, although this species does not grow in Maryland.

mas   Latin for male, masculine. Used by early writers in a metaphorical sense to distinguish robust species from more delicate ones.

maximum   Latin for greatest, largest.

melanocarpa   Latin for black-fruited; from the Greek *melas* (black) and *karpos* (fruit).

menziesii   after Archibald Menzies (1754-1842), Scottish physician and naturalist who identified Douglas-fir at Nootka Sound on Vancouver Island.

mertensiana   after Karl Heinrick Mertens (1796-1830), German naturalist and physician who identified mountain hemlock at Sitka, Alaska.

Metasequoia   from Greek *meta* (changed) and sequoia, the tree it is related to.

mollis   Latin for soft.

mollissima   very soft.

monogyna   having a single ovary; from the Greek *monos* (one) and *gyne* (female).

monolifera   necklace-like; strung like beads.

montana   Latin for "from the mountains".

monticola   Latin for inhabiting mountains; from *montis* (of the mountain) and *colere* (to dwell).

Morus   from the Celtic word *mor* (black), referring to the color of the fruit.

mucronatus   Latin for pointed.

muehlenbergii   after Gotthilf Henry Ernest Muhlenberg (1753-1815).

mugo   of unknown origin.

Myrica   From *myrio* (to flow), found on the banks of rivers; or from the Greek *myrike* (tamarisk) which it resembles.

n

negundo from *negundi*, the Sanskrit name of the chaste tree of India, also applied to the Manitoba maple because of the similarity of the leaves.

Nemopanthus from the Greek *nema* (a thread), *pous* (foot), and *anthos* (flower); referring to the long, slender peduncles of this North American shrub.

neoalaskana new Alaskan; referring to *Betula neoalaskana;* the name *Betula alaskana* had already been given to a fossil species of birch.

nigra, nigrum Latin for black.

nootkatensis After Nootka Sound, Vancouver Island, where yellow-cedar was first described by Europeans.

nordmanniana after Alexander von Nordmann (1803-1866), Finnish botanist, professor, and director of the Botanic Garden at Odessa.

nuttallii after Thomas Nuttall (1784-1859), English-American botanist and ornithologist.

Nyssa from Mount Nyssa in Asia Minor, the legendary home of the naiads, or water nymphs, who brought fruitfulness to plants, herbs and mortals.

o

obtusa blunted, obtuse; from the Latin *obtundere* (to dull).

occidentalis Latin for western, of the Western Hemisphere; from *occidere* (to set), as with the sun.

odorata Latin for sweet-smelling, fragrant.

omorika local name of the Serbian spruce.

opaca Latin for dark, shaded, with a dull surface.

opulus Latin name for a kind of maple; referring to the maple-like leaves of *Viburnum opulus*.

oregona Latin form of "from Oregon".

orientalis Latin for eastern, of the Orient.

Ostrya from the Greek *ostrua*, a tree with very hard wood; or *ostruos* (a scale), referring to the scaly catkins.

ovalis, ovata Latin for egg-shaped or oval; referring to the leaves; from *ovum* (egg).

p

palmatum Latin for lobed or divided like a hand.

palustris Latin for "of marshes, swamps or wet places".

papyrifera Latin for paper-bearing, with a papery bark; from the Greek *papyros* (paper reed) and the Latin *ferre* (to bear).

pellita Latin for "clad in skins", referring to the dense, shiny, white pubescence beneath the leaves.

pendula Latin for hanging or drooping; from *pendere* (to hang).

pennsylvanica, pensylvanica, pensylvanicum Latin forms of "from Pennsylvania"; the one "n" is an old spelling of the former colony.

pentandra Latin for five-stamened; from the Greek *penta* (five) and *andron* (male).

petiolaris having a leaf stalk or a particularly long one.

phaenopyrum resembling a pear; from Greek *phaino* (to show) and Latin *pyrum* (pear).

phanerolepis with visible scales, referring to the exposed bracts on the cones of bracted balsam fir; from the Greek *phaneros* (visible) and *lepis* (scale or flake).

Phellodendron from the Greek *phellos* (cork) and *dendron* (tree), referring to the corky bark.

Picea from the Latin *pix* (pitch), referring to a pine; later applied to spruces as the genus name.

pinsapo old Spanish Andalusian word meaning pine-fir, from *pino* (pine) and *sapino* (fir).

Pinus from the Greek *pinos* (pine-tree); possibly from the Celtic *pin* or *pyn* (mountain or rock), referring to the habitat of the pine.

pisifera producing peas or pea-like seeds.

platanoides resembling a plane tree; from the Greek *platus* (broad) and *öides* (resembling).

Platanus plane-tree; from the Greek *platus* (broad), referring to the leaves.

plicata Latin for folded; referring to the flattened twigs, with regularly arranged scale-like leaves, of western redcedar.

pomifera Latin for apple-bearing; referring to the large ball-like fruits of the osage-orange tree.

ponderosa Latin for "of great weight", or imposing; referring to the massive appearance of ponderosa pine.

populifolia with leaves like the poplar; from the Latin *populus* (poplar) and *folium* (leaf).

Populus classical Latin name of the poplars; possibly from *paipallo* (vibrate or shake); or originating in ancient times when the poplar was called *arbor populi* (the tree of the people), because it was used to decorate public places in Rome.

porsildii after Alf Erling Porsild (1901-1977), Danish-Canadian botanist at the National Herbarium of Canada, Ottawa, 1936-1966.

pringlei after Cyrus Guernsey Pringle (1838-1911), American professor, plant breeder, and plant collector.

prinoides with leaves resembling those of the chestnut oak; from the Greek *prinos* (the great scarlet oak) and *öides* (resembling).

prinus the classical Latin name of a European oak.

procera Latin for tall or high.

profunda Latin for thick, dense.

prolixa Latin for long, extended.

pruinosa having a waxy, powdery secretion or bloom on the surface; from the Latin meaning, "full of hoar-frost".

Prunus   the classical Latin name of the plum tree; from the Greek *prunos* (plum).

pseudoacacia   Latin for false-acacia; from the Greek *pseudos* (falsehood) and *akakia*, the Greek name for the Egyptian pod-thorn tree.

pseudoplatanus   Latin for false plane-tree.

Pseudotsuga   false hemlock; from the Greek *pseudos* (falsehood) and the Japanese *tsuga* (hemlock).

Ptelea   Greek for "elm"; used as the genus name of the hoptree because of the similarity of their fruits.

pubens, pubescens   downy, with soft, short hairs; from the Latin *pubes* (covered with soft hair) and *escens* (becoming).

pumila   Latin for dwarf, small or diminutive.

punctata   Latin for covered with dots; referring to the fruit of the whitehaw.

pungens   sharp-pointed; referring to the needles of Colorado spruce; from the Latin *pungere* (to prick).

purpurea   Latin for purple.

purshiana   after Frederick Pursh (1774-1820), German botanist who collected in North America.

pyrifolia   with leaves like pear leaves; from the Latin *pyrum* (pear) and *folium* (leaf).

q

quadrangulata   Latin for four-angled.

Quercus   the classical Latin name of the oaks; probably from the Celtic *quer* (fine) and *cuez* (a tree).

r

racemosa   Latin for full of clusters; referring to fruit or flowers.

radicans   taking root or rooting; referring to the habit of poison-ivy which takes root as it grows over the ground; from the Latin *radix* (root).

regia   from the Latin *regis* (royal, regal, magnificent).

resinifera, resinosa   resin-bearing, resinous; from the Latin *resina* (resin) and *ferre* (to bear).

Rhamnus   from the Greek *rhamnos* (buckthorn) or from the Celtic *rham* (a tuft of branches).

rhamnoides   resembling *Rhamnus*.

Rhododendron   from the Greek *rhodon* (rose) and *dendron* (tree).

Rhus   Latin for sumac; from the classical Greek *rhous*, thought to be from the Celtic *rhudd* (red), referring to the fruit.

rigida   Latin for rigid, or stiff; referring to the cone scales of pitch pine.

Robinia   after Jean Robin (1550-1629) and his son Vespasien Robin (1579-1662), herbalists to Henri IV of France, who first cultivated the locust tree in Europe.

robur   Latin for hard tree or hard wood; applied specifically to the oak.

rubens   Latin for reddish; referring to the reddish-brown cones of red spruce.

rubra, rubrum, rubrus   Latin for red, from the Celtic *rub* (red).

rugosa   Latin for rough, coarsely wrinkled.

s

sabina   Latin for the Sabines, an ancient Italian people, contemporaries of the Latins; *Sabina herba* referred to a kind of juniper, the savin.

saccharinum, saccharum   sweet or sugary; referring to the sap; from the Sanskrit *sarkara* (grit or sugar), through the Greek *sakcharon*, (the sweet liquid obtained from bamboo joints).

Salix   classical Latin name of the willows; probably from the Celtic *sal* (near) and *lis* (water).

Sambucus   classical Latin name of the elders; from the Greek *sambuke*, an ancient musical instrument; referring to the bark of the elder which can

easily be removed in tubes to make a flute or whistle.

sanguinea   Latin for blood red.

sargentii   after Charles Sprague Sargent (1841-1927).

Sassafras   probably adapted by French settlers from a native North American name.

schuettei   after Joachim Heinrich Schuette (1821-1908), a German-American newspaper editor and botanist who willed his large herbarium to the Wisconsin Field Museum.

Sciadopitys   from Greek *skiados* (an umbel) and *pitys* (a fir tree); the leaves appear in whorls like the ribs of an umbrella.

scopulina   shrubby or broom-like; from the Latin *scopula* (little broom) and *ina* (resembling).

scopulorum   Latin for "of rocky cliffs or crags"; referring to the habitat of Rocky Mountain juniper.

scouleriana   after John Scouler (1804-1871), Scottish naturalist and physician who collected plants on the northwest coast of North America.

sempervirens   evergreen, retaining leaves in winter, from *semper* (always) and *virere* (to be green).

Sequoia   after Sequoiah (1770-1843), American of Cherokee and British descent who was noted for producing a written lexicon of the Cherokee language.

Sequoiadendron   from Sequoia (see above) and the Greek *dendron* (tree).

serotina   Latin for late; referring to the late-maturing fruit of black cherry.

serrata   from *serra* (a saw), referring to the toothed leaf-margins.

serrulata   with small, saw-like teeth.

shastensis   Latin form of "from Mount Shasta" of Shasta County, California.

Shepherdia after John Shepherd (1794-1836), curator of the Liverpool Botanic Garden.

shumardii after Benjamin Franklin Shumard (1820 - 1869), state geologist of Texas.

sibirica Latin form of "Siberian".

simonii after Eugène Simon (1871-1967), French botanist who collected in North America.

sinuata wavy margined; referring to the leaves of Sitka alder; from the Latin *sinuare* (to bend or to wave).

sitchensis Latin form of "from Sitka Sound", the area in southeastern Alaska where Sitka spruce was first described.

Sorbus the classical Latin name of the mountain-ash.

soulangiana after Étienne Soulange-Bodin (1774-1846), French politician and horticulturist.

speciosa Latin for beautiful, showy.

spicatum Latin for spiked, spike-like.

strobus the ancient name of an incense-bearing pine; related to the Latin *strobilus* (pine cone), and the Greek *strobos* (whirling around).

styraciflua resinous, containing styrax or gum.

subintegerrima with an almost even margin; from the Latin *sub* (under or less than) and *integer* (complete).

submollis partly covered with soft hairs; from the Latin *sub* (under or less than) and *mollis* (soft).

succulenta succulent, fleshy, filled with juice.

suksdorfii after Wilhelm Nikolaus Suksdorf (1850-1932), German-American botanist and plant collector.

sylvatica of the woods or trees; from the Latin *sylva* (a forest).

sylvestris Latin for "of the forest".

Syringa from the Greek *syrinx* (pipe), referring to the hollow stems.

t

Taxodium from Latin *taxus* (the yew) and Greek *eidos* (resemblance).

Taxus the classical Latin name of the yew, from Greek *taxon* (bow), referring to a former use of yew wood.

tenuifolia thin-leaved; from the Latin *tenuis* (thin) and *folium* (leaf).

ternax from the Latin *terni* (in three), growing in threes.

thomasii after David Thomas (1776-1859), American civil engineer and horticulturist who first classified rock elm.

Thuja, Thuya a form of *thya*, the Latin name for an aromatic tree; from the Greek *thuia* (an aromatic cedar).

Tilia the classical Latin name of the linden or lime tree.

tomentosa Latin for "covered with densely matted hairs".

Toxicodendron from Greek *toxicon* (poison) and *dendron* (tree).

tremuloides having a tendency to tremble; from the Latin *tremulus* (trembling) and the Greek *öides* (resembling).

triacanthos three-thorned; from the Greek *treis* (three) and *akantha* (a spine).

trichocarpa hairy fruited; from the Greek *thrix* (a hair) and *karpos* (fruit).

trifoliata three-leaved; from the Latin *tres* (three) and *folium* (leaf).

triloba, trilobum three-lobed; from the Greek *treis* (three) and *lobos* (lobe).

Tsuga hemlock; from the Japanese name for the native hemlocks of Japan.

tulipifera tulip-bearing; from the Turkish *tulbend* (a turban) and the Latin *ferre* (to bear).

typhina like the cattail (*Typha*); from the Greek *typhe* (reed mace) and the suffix *ina* (resembling).

u

Ulmus the Latin name of the elms, from the Saxon word *ulm*.

v

velutina velvet-like; referring to the young leaves of the black oak; from the Latin *vellus* (a fleece) and *ina* (resembling).

vernix Latin for varnish; so named because poison-sumac was mistakenly thought to be the same tree from which Japanese lacquer is obtained.

verticillata whorled, three or more leaves arranged in a spiral around the same point; from the Latin *verticillus* (whorl of spindle).

Viburnum the classical Latin name of the wayfaring tree of Eurasia; thought to be from *vieo* (to tie), because of the pliability of the branches.

viminalis Latin for "osiers" or "flexible twigs".

virginiana Latin form of Virginian.

viridis green.

vitellina having the color of an egg yolk, from the Latin *vitellus* (egg yolk) and *ina* (resembling).

vulgaris Latin for common, of the common people.

w

watereri after John Waterer Sons, a nursery in Surrey, England.

williamsii after Thomas Albert Williams (1865-1900), American agricultural botanist, professor, and author of *Native Trees and Shrubs* (1895).

x

xanthocarpa with yellow fruit; from the Greek *xanthos* (yellow) and *karpos* (fruit).

z

Zanthoxylum yellow wood; from the Greek *xanthos* (yellow) and *xylon* (wood).

Zelkova from the Caucasian name for these deciduous trees.

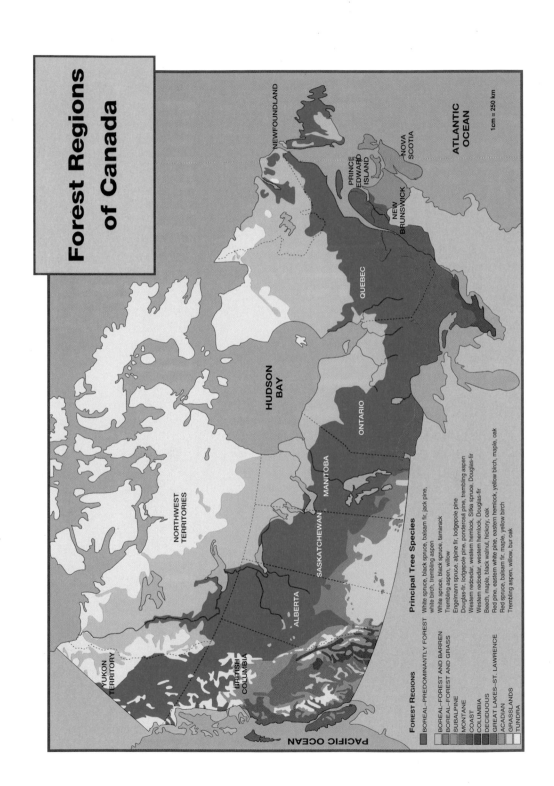

Forest Regions of Canada

NEWFOUNDLAND

PRINCE EDWARD ISLAND

NOVA SCOTIA

NEW BRUNSWICK

ATLANTIC OCEAN

1cm = 250 km

QUEBEC

HUDSON BAY

ONTARIO

MANITOBA

NORTHWEST TERRITORIES

SASKATCHEWAN

ALBERTA

YUKON TERRITORY

BRITISH COLUMBIA

PACIFIC OCEAN

FOREST REGIONS   **Principal Tree Species**

BOREAL—PREDOMINANTLY FOREST   White spruce, black spruce, balsam fir, jack pine, white birch, trembling aspen

BOREAL—FOREST AND BARREN   White spruce, black spruce, tamarack

BOREAL—FOREST AND GRASS   Trembling aspen, willow

SUBALPINE   Engelmann spruce, alpine fir, lodgepole pine

MONTANE   Douglas-fir, lodgepole pine, ponderosa pine, trembling aspen

COAST   Western redcedar, western hemlock, Sitka spruce, Douglas-fir

COLUMBIA   Western redcedar, western hemlock, Douglas-fir

DECIDUOUS   Beech, maple, black walnut, hickory, oak

GREAT LAKES—ST. LAWRENCE   Red pine, eastern white pine, eastern hemlock, yellow birch, maple, oak

ACADIAN   Red spruce, balsam fir, maple, yellow birch

GRASSLANDS   Trembling aspen, willow, bur oak

TUNDRA

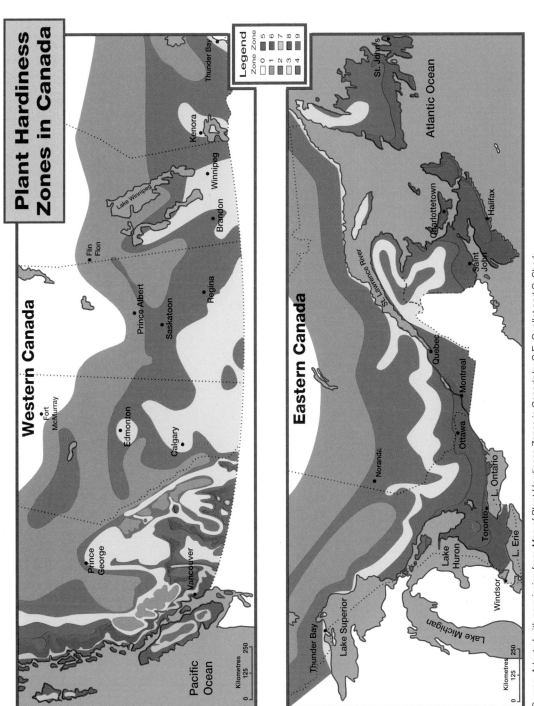

Plant Hardiness Zones in Canada

Legend

Zone Zone
0 5
1 6
2 7
3 8
4 9

Western Canada

Thunder Bay

Kenora

Lake Winnipeg

Winnipeg

Brandon

Flin Flon

Regina

Prince Albert

Saskatoon

Fort McMurray

Edmonton

Calgary

Prince George

Vancouver

Pacific Ocean

Kilometres
0 125 250

Eastern Canada

St. John's

Atlantic Ocean

Charlottetown

Halifax

Saint John

St. Lawrence River

Québec

Montreal

Noranda

Ottawa

L. Ontario

Toronto

L. Erie

Windsor

Lake Huron

Lake Michigan

Lake Superior

Thunder Bay

Kilometres
0 125 250

Source: Adapted with permission from *Map of Plant Hardiness Zones in Canada* by C.E. Ouellet and C. Sherk, Agriculture and Agri-Food Canada, Publication no. 5003, revised 1981.

Index of English and Latin Names

Page numbers in bold face type indicate that the entry is the preferred common or scientific name and refer the reader to the detailed description of the species or genera. Other entries include secondary common names, synonyms of scientific names, and species briefly mentioned in the text.

Index of French Common Names

Group 1.

Scale-like needles or scales; evergreen; closely spaced in opposite pairs or whorls of 3, often overlapping and obscuring the stem; seeds in cones, some cones berry-like

Group 2.

Introduced species hardy in south-western British Columbia; leaves various shapes; seeds in cones

Group 3.

Needles evergreen, in bundles of 2, 3, or 5; seeds in cones

Group 4.

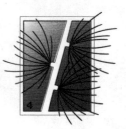

Needles deciduous (or evergreen), in tufts of 10 or more on dwarf shoots, also single on long shoots; seeds in cones

Group 5.

Needles evergreen, single, flat or 4-sided; seeds in cones

Group 6.

Needles evergreen, single, flat; twigs and buds green; single seed in a fleshy cup